LICHENOLOGY IN THE BRITISH ISLES 1568–1975

LICHENOLOGY IN THE BRITISH ISLES
1568–1975

An Historical and Bibliographical Survey

by

D. L. Hawksworth and **M. R. D. Seaward**

The Richmond Publishing Co. Ltd., Richmond, 1977

First published in 1977 by The Richmond Publishing
Co. Ltd., Orchard Road, Richmond, Surrey TW9 4PD

ISBN 0 85546 200 0

Printed in Great Britain by
Kingprint Limited, Richmond, Surrey

TO ARTHUR E. WADE AND ALL FIELD WORKERS PAST AND
PRESENT WHO HAVE CONTRIBUTED TO OUR KNOWLEDGE OF
LICHENS IN THE BRITISH ISLES

But he the man of science and of taste
Sees wealth far richer in the worthless waste
Where bits of lichen and a sprig of moss
Will all the raptures of his mind engross

John Clare (1793–1864), *Shadows of Taste*

PREFACE

Our intention to produce a comprehensive bibliography of titles including references to British lichens cross-indexed on a vice-county basis was conceived almost ten years ago. During this period cards were made out for items encountered which were subsequently scored for vice-counties mentioned. The task proved to be much greater than at first envisaged. We started to uncover more and more works largely or entirely overlooked by previous workers; not only incidental references or records of a few species, but also extensive surveys came to light. Many such items included data pertinent to the study of changes in the British lichen flora or proved to be unfamiliar to colleagues currently working on county or regional lists; we thus became increasingly convinced of the need for such a study. The compilation of an adequate bibliography is essential in the production of detailed regional surveys and in determining past and present distribution patterns.

As neither of us has been able to devote the major part of his time to lichenology in recent years, the checking of citations, abstracting of journals and examination of floras, natural histories and other books proceeded at a very slow rate; by 1972 it was clear that in order to make the work as comprehensive as was necessary full-time assistance would be required. Through a contract between the Nature Conservancy Council and the Commonwealth Mycological Institute, Mr David G. Reid was employed as a Research Assistant on this project from 1 December 1975 to 31 March 1976. As a result of this the bulk of the major searches required was completed and many elusive references were tracked down. At the conclusion of the contract period, *A Preliminary Bibliography of British Lichens*, listing some 2347 items, was compiled and copies sent to colleagues likely to be able to draw our attention to omissions. This stage in our study would never have been attained without David Reid's assistance and we would like to record our gratitude both to him for all the painstaking work he so willingly and conscientiously undertook on our behalf and to the Nature Conservancy Council for making his employment possible.

In the course of our own studies on lichens in the British Isles, we located a number of lichen herbaria, the whereabouts of which had not been generally known. As it is often necessary to verify the identification of specimens on which literature reports are based, we considered that it would be valuable to include a list of the known locations of the specimens of at least the major collectors here. In order to supplement data already known to us, Mr Peter M. Earland-Bennett solicited further information through the Museums Association's *Bulletin* in 1975; we are very grateful to him for placing the information he received at our disposal.

As no account of the development of the study of lichen systematics and distribution in the British Isles had appeared since an important but little known and brief summary by Miss Annie Lorrain Smith issued in 1922, we thought it would be desirable to complement our bibliographic and herbarium surveys with an historical account.

In view of the appearance in 1971 of Dr Michael E. Mitchell's bibliography of Irish lichenology, which lists some 422 titles and indexes them by vice-county, we decided to limit our treatment of Ireland to brief mentions in our section *History* and noting some omissions and recent publications in our section *Bibliography*.

In preparing a book of this type, it is necessary to call upon the assistance of numerous colleagues and librarians. They are too many to mention individually but we would particularly like to thank Mr R. H. Bailey, Mr G. D. R. Bridson, Mr B. J. Coppins, Mr A. Henderson, Miss V. A. Hinton, Mr P. W. Lambley, Dr M. E. Mitchell and Mr M. Walpole. The vice-county cross-index was kindly prepared by Mr M. Dowell and Mr D. Buckland of the Computer Centre of Bradford University. For permission to photograph and reproduce portraits or other items in their care, we are also indebted to the British Museum (Natural History) (Figs. 3, 8 and 14 *left*), Linnean Society of London (Figs. 6 *left* and 9 *right*), Royal Botanic Gardens, Kew (Figs. 1, 2, 6 *right*, 7, 11, and 14 *right*), and Dr E. Neal and Taunton School (Fig. 15 *right*).

We would also like to acknowledge the encouragement received from Mr P. W. James and Mr J. R. Laundon throughout the preparation of this book, and Mr D. R. Clark and Mr F. S. Dobson of the Richmond Publishing Co. Ltd. for their assistance in its publication.

D. L. Hawksworth
M. R. D. Seaward

Kew and Bradford
January 1977

CONTENTS

HISTORY

Pre-1650: The Herbalists

References to lichen-covered trees in eastern England feature in some of the early Anglo-Saxon charters, for example the account of the boundary of Hurstbourne Priors, Hampshire, dated 901 A.D. (see Rackham, 1976, p. 49), but not until after the invention of printing and the publication of the first printed book (Gutenberg's *Bible* of 1455) were attempts at their description made.

Botanical science itself gradually evolved from the practice of medicine; prior to the sixteenth century the study of plants was almost entirely the province of physicians and herbalists. Lichens were little used in the medicinal practices of the day,

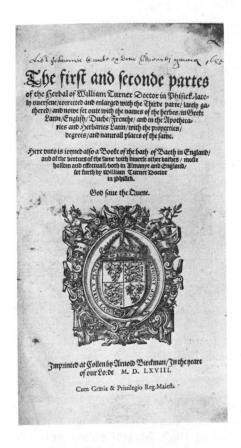

Fig. 1. Title-page and p. 56 of the *thirde parte* of the *Herbal* of William Turner (*c.* 1508–1568), referring to the occurrence of Lungwort (*Lobaria pulmonaria*) in England, published in 1568[2393].

1

and therefore received scant attention in the printed, and often illustrated, lists of reputedly medicinally important plants, called *Herbals*, which originated on the Continent in the late fifteenth century. The systematic treatment of plants in such works was non-existent, but it is possible to identify a few lichen species from the often crude and stylized woodcut illustrations.

Although foreign herbals were certainly to be found in the major centres of learning in Britain by the early decades of the sixteenth century, the first specifically British *Herbal* did not appear until late in that century; it was written by William Turner (*c*. 1508–1568), a Protestant zealot born in Morpeth who travelled in self-exile on the Continent during the reign of Mary Queen of Scots (1553–1558), and completed in 1568[2393]*, the year of his death. Turner is often recognized as the father of British botany, and he appears to have been the first author to mention a British lichen (*Lobaria pulmonaria*, Fig. 1) in print. However, the best known British herbal is probably the *Herball or Generall Historie of Plantes*, published in 1597[764], by John Gerard[e] (1545–1612), a surgeon who either owned or supervised gardens in Holborn and the Strand. This herbal included only one more lichen than Turner's work: a member of the *Cladonia pyxidata* group. The first localized lichen records are attributable to Thomas Johnson (1595/7–1644), an apothecary with a 'physic' (medicinal herb) garden in Snow Hill, London, who extended his 'herbarizings' from the immediate precincts of London to include Kent and even Wales. His publications and manuscripts for the period 1629–38[806, 1161–1163, 1165] included lichen records and, unlike his predecessors, he started to produce lists of plants in specific areas—the foundation of the British local flora tradition still in evidence today. Johnson issued an enlarged and amended edition of Gerard's *Herball* in 1633[765] (Fig. 2) which included some further lichen records, but unfortunately the identity of many of his lichens remains uncertain in the absence of illustrations. John Parkinson's (1567–1650) *Herball* or *Theatrum botanicum*, published in 1640[1745], included eight pages with lichen records and his fuller descriptions enable most to be identified; however, it is difficult to ascertain whether many were based on British material, one at least being from Italy.

The publication of William How's (1620–1656) *Phytologia Britannica* in 1650[1082] continued Johnson's innovation, namely the recording of plants for their intrinsic interest and not merely for their medicinal uses; he listed five lichens as British. Although the paths of the herbalists and botanists were now starting to diverge, this is not to say that the tradition of the herbalists ceased. The 1652 *English Physitian* [*sic*][540] of Dr Nicholas Culpeper (1616–1654) included *Lobaria pulmonaria* which continues to feature in herbals even to this day, as, for example, in F. and V. Mitton's *Practical Modern Herbal* first published in 1976.

1651–1750: The influence of Ray and Dillenius

Christopher Merret[t] (1614–1695) published his *Pinax rerum naturalium Britannicarum* in 1666[1592]; the first printing was largely destroyed during the Great Fire of London and a new impression appeared in 1667. Merret endeavoured to list his plants alphabetically and included five lichens amongst them. Robert

* Superscript numerals refer to the number of entries in the *Bibliography* section of this book (pp. 46–174); other references cited are included on pp. 201–203.

Morison (1620–1683), a Scotsman who fought in the Civil War for the king, became Botanist Royal at Oxford after a period abroad. In his monumental *Plantarum historiae universalis Oxoniensis*[1627], published posthumously in 1699 through the efforts of Jacob Bobart the Younger (1641–1719), Morison attempted to devise a new system of plant classification; about fifty lichens were recognized here as a

Fig. 2. Title-page of the 1633 edition of Gerard's *Herball*[765] enlarged by Thomas Johnson (*c*. 1596–1644).

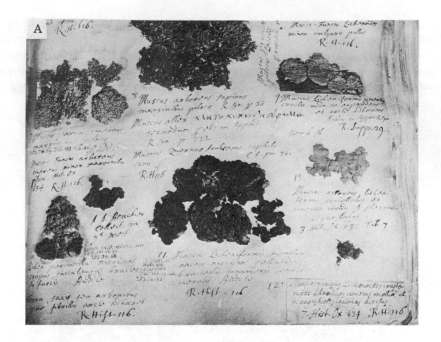

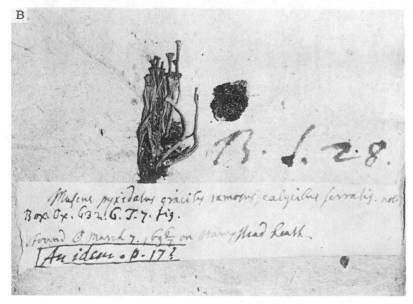

Fig. 3. Some of the earliest extant British lichen herbarium specimens now in the Sloane Herbarium of the British Museum (Natural History). A, Portion of the Rev. Adam Buddle's (*c.* 1660–1715) '*Hortus siccus*' (BM-Sloane **115** sheet 6). B, Specimen of *Cladonia gracilis* found on Hampstead Heath on 7 March 1696/7 labelled in James Petiver's (1664?–1718) hand but perhaps collected by Buddle (BM-Sloane **285** sheet 32).

single group, *Musco-fungus*, ordered into five sections. At least some of Morison's specimens are still preserved in Oxford (OXF) and these, together with the lichens of the Rev. Adam Buddle (*c.* 1660–1715) now in the Sloane herbarium (BM; Fig. 3) and which date from around 1700, and those of Thomas Willisel (d. ?1675) who had collected for Merret in the Sherardian herbarium (OXF), are amongst some of the oldest lichen herbarium material extant. Other important collections made during this period include those of James Petiver (1664?–1718) who collected omnivorously and published catalogues of his holdings; some of his specimens were destroyed by fire but others are now in the Sloane herbarium. Buddle, Petiver,

Fig. 4. Illustration of *Cladonia coccifera* in *The Natural History of Staffordshire* by Robert Plot (1640–1696) published in 1686[1836] (Plate 14 fig. 1): '. . . Scarlet-headed Cup or Chalice-Moss . . . grows thick upon mole hills . . . and is so certainly an undescribed plant that I can find nothing like it in any of the books . . . I rather chuse to reckon it among the *fungus's*' (p. 199).

and some other botanists interested in cryptogams, such as William Sherard (1659–1728), Samuel Doody (1656–1705), Leonard Plukenet (1642–1706), and Hans Sloane (1660–1753), were members of the Temple Coffee House Botanic Club which started in about 1689; Doody's lichen specimens are now in the Sloane Herbarium.

John Ray (1627–1705), an eminent naturalist whose numerous and varied publications show him to have been one of the most outstanding scholars of the period, included lichens in many of his botanical works. In 1660, the year the Royal Society was founded, Ray published a list of plants growing around Cambridge[1919] but in 1670 he expanded this into a *Catalogue* embracing the whole of the British Isles[1920]. These texts gave way to Ray's *Synopsis methodica stirpium Britannicarum*, the first edition of which appeared in 1690[1923], and this soon became the standard British flora of the period. Ray himself, however, perhaps had rather little time for lichens as in the *Synopsis* most of the information on cryptogams was contributed by Doody[604-605]. Similarly, the first two volumes of Ray's *Historiae plantarum*[1922, 1924] contained very few lichen records and in the third (1704) volume[1926], where the lichen section is increased considerably, almost all were attributed to Doody, Petiver or Sherard.

Other local lichen records were also being published in this period, for example those by Sir Robert Sibbald (1641–1722), founder of the Edinburgh Botanic Garden, in his *Scotia illustrata* of 1684[2178], in Robert Plot's (1640–1696) *Natural History of Staffordshire* of 1686[1836] (which included a drawing of *Cladonia coccifera;* Fig. 4), and in George C. Deering's (*c.* 1695–1749) catalogue of plants around Nottingham of 1738[576].

In Florence in 1729 Micheli (1679–1737) published his *Nova plantarum genera*[1598] which included an entirely revised system of lichen classification and descriptions of numerous hitherto undescribed species. William Sherard, who visited Micheli in Florence in 1717, was undoubtedly impressed by this work, then being prepared, and brought some of his herbarium material to Oxford. Following Ray's death, he was instrumental in bringing John James [Jacob] Dillenius (1684–1747) from Giessen in Germany to Oxford in 1721. Although Dillenius appears to have been poorly paid he devoted a great deal of his spare time to the cryptogams, particularly mosses and lichens; one of his sources of income appears to have been the preparation of plates for James Sherard's (1666–1737) never completed *Hortus Elthamensis*. The preparation of a third edition of Ray's *Synopsis*, published in 1724[1927], was Dillenius' first major contribution to British botany. Dillenius made numerous changes to the text and was responsible for all 24 plates included but did not allow his name to appear on the work as, to judge from a letter from Sherard to Richardson of 26 December 1723, he thought a British flora by a German might be treated with some scorn by his contemporaries. Following William Sherard's death, Dillenius became the first Sherardian Professor of Botany at Oxford in 1728. He then appears to have had more time to devote to cryptogams, for in March 1742 his monumental *Historia muscorum*[591] was published. This was an outstanding piece of work influenced in no small measure by the ideas of Micheli whose polynomials are included amongst his synonyms and some of whose illustrations were used. About 190 lichens, mainly British, were treated, most being illustrated (Fig. 5); the original drawings, some coloured, from which the plates for this work were prepared are now in BM[591] and specimens used to prepare each of these are preserved in OXF. His specimens and comments still prove very valuable in the typification and characterization of names used by Linnaeus and later workers.

Dillenius evidently carried out a considerable amount of fieldwork himself both in southern England and Wales[590] and also received specimens and comments from many other important botanists of the period, including the Yorkshiremen Richard Richardson (1663–1741) and Samuel Brewer (1670–1743). The wide range of counties from which Dillenius cites lichen material (27 vice-counties in all) is in marked contrast to the two earlier editions of Ray's *Synopsis* which noted few taxa outside the immediate environs of London.

Ireland remained relatively little known botanically up to this period and the single specific Irish flora then available, the 1727 *Synopsis stirpium Hibernicarum*

Fig. 5. Plate 17 from the *Historia muscorum* of John J. Dillenius (1684–1747) published in March 1742[591] and including illustrations of *Cetraria islandica* and species of *Ramalina, Roccella, Sphaerophorus* and *Stereocaulon*.

of Caleb Threlkeld (1676–1728) listed only five lichens. In the 1801 edition of Withering's *Arrangement*[2671] this number had risen only to nine. In contrast, the considerable attention Leicestershire was to receive in the succeeding centuries was initiated by Richard Pulteney (1730–1801) who first produced manuscripts of plants (including some lichens) growing around Loughborough, Leicestershire[1874, 1876–1877], at the age of 17, and much more substantial works later that century. An undated manuscript check-list of the British flora by Pulteney[1871] included some 142 lichens and his 1790 two volume history of the rise of interest in botany in England[1875] is a remarkable source of information on this period.

1751–1785: The impact of Linnaeus
The recognition of several genera-like groupings within the lichens by Micheli and Dillenius was also attempted by Sir John Hill (1716?–1775) in his *History of Plants* of 1751[1021] but this trend and the recognition of increasing numbers of lichen species suffered a setback with the publication of Carl Linnaeus' *Species plantarum* in 1753[1435]. While the introduction of binomial nomenclature by Linnaeus simplified the communication of plant names at a time when polynomials were becoming increasingly cumbersome, his use of the single generic name *Lichen* (apart from a few in *Byssus* and *Mucor*) and acceptance of only about eighty species in it held back, in some respects, the study of the group for several decades. In his treatment of the mosses and lichens Linnaeus drew heavily on the work of Dillenius, whom he had visited in Oxford in 1736 and with whom he subsequently corresponded. The reduction in the number of lichen species recognized from some 298 in Micheli's book to the mere 80 in *Species plantarum* was, perhaps not surprisingly, welcomed by many botanists, for example the eminent physician Sir William Watson (1715–1787) who forcefully stated his views in 1759[2530].

Hill adopted the Linnaean system in his *Flora Britannica* of 1760[1022] but his work was soon followed in 1762[1086] (with later editions in 1778[1087] and 1798[1088]) by the particularly important *Flora Anglica* of William Hudson (1734–1793), son of a Kendal innkeeper who once worked at the Chelsea Physic Garden. Hudson's flora included localized records of eighty-five lichens in the single genus *Lichen*; many of his specimens were destroyed in a fire at his apothecary's business in the Haymarket in 1783. The provision of localities for the species listed was also a feature of William Withering's (1741–1799) *Botanical Arrangement*, the first edition of which appeared in 1776[2668]; this evidently proved very popular as it ran to seven editions[2669–2674] and appears to have been published in considerable numbers. The 1777 *British Flora* of Stephen Robson (1741–1779), which used 'archil' as the colloquial name for lichens, accepted some 87 species with the conservative species concepts seen in Hudson's *Flora*.

Several important local lichen lists continued to appear; that in Watson's *The History and Antiquities of the Parish of Halifax* of 1775[2477] is now generally attributed to James Bolton (fl. *c.* 1758–d.1799), better known as a mycologist, and is particularly extensive. Some botanists interested in cryptogams were becoming restless with the inadequacies of Linnaeus' system by this time and when the Rev. John Lightfoot (1735–1788) of Uxbridge published his *Flora Scotica* in 1777[1401] after a summer's tour through Scotland with Pennant in 1772, he felt disposed to recognize 103 lichens in all, several of which were newly described; others, for example the Rev. Richard Relhan (1754–1823) in the first edition of his *Flora Cantabrigiensis* of 1785[1938], remained conservative.

1786–1840: J. E. Smith, Dickson, Borrer and Turner; the influence of Acharius

Following the death of Linnaeus in 1778, Sir James Edward Smith (1759–1828; Fig. 6), apparently partly at the instigation of Sir Joseph Banks (1743–1820), secured Linnaeus' collections and library and brought them to England, founding the Linnean Society of London in 1788. Lichenology, like most branches of natural history in Britain at this time, was pursued with renewed vigour. In the case of botany, the large library and herbarium Banks was amassing undoubtedly acted as a major stimulus as he made these available for consultation by his colleagues; the Banks collections and library were housed at his home in Soho Square during his life-time, but subsequently provided the nucleus of the Department of Botany of the British Museum (Natural History). Banks, who was President of the Royal Society from 1778 until 1820 and who had circumnavigated the world with Capt. James Cook in 1768–71, had little personal interest in lichens although he did collect them on his foreign voyages.

On the continent Georg H. Weber (1752–1828), Christiaan H. Persoon (1761–1836), Georg Franz Hoffmann (1760–1826) and Friederich Ehrhart (1742–1795)

Fig. 6. James Dickson (173?8–1822, *left*) and Sir James Edward Smith (1759–1828, *right*).

were starting to recognize increasing numbers of lichens in the 1780s and 1790s and this trend was soon taken up in Britain. The Scottish nurseryman James Dickson (1738–1822; Fig. 6), for example, issued two exsiccatae[586–587] (Fig. 8B), and described many new lichens in his *Plantarum cryptogamicarum Britanniae* between 1785 and 1801[585]. Access to the Linnean collections enabled Dickson to match his specimens with them and his first exsiccata was subtitled 'named on the authority of the Linnean herbarium'[586]; most of the lichen epithets he coined are still in use today (e.g. *Buellia canescens* (Dicks.) de Not., *Dimerella lutea* (Dicks.) Trevis.). The Rev. Hugh Davies (1739–1821) paid particular attention to the lichens of Anglesey at this time, describing several species new to science in 1794[564] and later providing a detailed treatment of the lichen flora of that island[566]. The Forster brothers, Benjamin M. (1764–1829), Edward (1765–1849) and Thomas F. (1761–1825), of Walthamstow, made botanical contributions to the 1789 edition of Camden's *Britannia*[585] (edited by R. Gough); this massive work included lichens and listed the more noteworthy species on a county basis.

James Sowerby (1757–1822), who had illustrated Dickson's book[585], joined Sir J. E. Smith in producing *English Botany*[2220], one of the most outstanding botanical works ever written; the first edition of this comprised thirty-six volumes and was issued in the years 1790–1814. A coloured illustration and detailed text were provided for each species and several lichens were described as new to science. Lichen specimens for this work were sent to Smith and Sowerby by very many collectors including Dickson, John Harriman (1760–1831), James Dalton (1764–1843), Charles Lyell (1767–1849), the Rev. John Burgess (1725–1795; also his son James B. MacGarroch, 1765–1782) and William Brunton (1775–1806). The Swedish lichenologist Erik Acharius (1757–1819) was rapidly becoming recognized as the foremost worker on the group at this time. Acharius described many new genera and species and corresponded with numerous lichenologists around the world, including several in Britain. Smith was quick to realise the importance of Acharius' work and his ideas were rapidly introduced into *English Botany* (apart from his generic concepts which were largely not taken up in this work until its *Supplement* (see below)); Smith also arranged with Acharius to send a set of his specimens to the Linnean Society (of which he was made an honorary member) in exchange for a set of the Society's *Transactions*. This set of specimens was dispatched to London in 1807 as Acharius' *Lichenographia Universalis*[6] was being sent to the printers (although that work was not issued until 1810). After lengthy difficulties with the customs officials, the collection arrived at the Linnean Society in the spring of 1809. William Borrer (1781–1862; Fig. 7) of Henfield, Sussex, then 28 years old and already proving to be an exceptional field botanist, had sent specimens to Acharius and was quick to examine the 894 Acharian species sent and the extensive notes he made on them (now preserved with the specimens in the British Museum (Natural History)) were completed by May of that year. Borrer had already contributed to Dawson Turner (1775–1858; Fig. 7) of Yarmouth and Lewis W. Dillwyn's (1778–1855) *Botanist's Guide* of 1805[2391] which listed the most noteworthy plants to be found in all English and Welsh counties at that time; this *Guide* is an extremely valuable compilation as it included much hitherto unpublished data, partly obtained by sending out a printed questionnaire to botanists throughout England and

Wales, and was thus not merely gleanings from earlier works. Some counties received more attention than others and, for example, while Sussex had many lichen records contributed by Borrer, Kent had none. Turner, referred to by J. E. Smith as 'that exquisite cryptogamist' in 1813, worked closely with Borrer on lichens and described many new species from Borrer's material in the period 1804–08[2386], [2388–2389]. Together Turner and Borrer planned to produce a comprehensive account of British lichens and a check-list by Turner was compiled in 1800–06[2383] but never published. By 1812 the proofs of some 208 pages of their proposed monograph, *Lichenographia Britannica*, had been widely circulated, not only in England but also abroad, but, due to the death of the printer and other difficulties the bulk remained unwritten. In January 1839 Turner, then 64, had such manuscripts as were written printed and issued in book form[2390] together with the printed sheets prepared some 26 years earlier so as '. . . to rescue [them] from dust and vermin . . .', and also to '. . . remain a monument of your [i.e. Borrer's] industry, your ability, and your profound knowledge of the Family of Lichens'.

Fig. 7. William Borrer (1781–1862, *left*) and Dawson Turner (1755–1858, *right*).

The extant literature scarcely bears due testimony to the role that Turner and Borrer played in the development of British lichenology. Their influence can, however, be seen in the works of many other authors of the time and both Turner and Borrer had voluminous correspondence[260, 2385] with an extremely wide range of naturalists not only in England but also abroad; in 1859, after Turner's death, his manuscript library was sold by auction—this was described in the 308 page auction catalogue as including a 'matchless collection of upwards of forty thousand autograph letters' and realized £6,558 8s. 0d. Borrer's contribution to lichenology was clearly recognized by Acharius who named a genus, *Borrera* Ach., after him in 1809, '. . . dedi in honorem Cel. D. Borrer. Angli, Lichenologi eximii, . . .'[6]. Borrer is perhaps best known for his contributions to the *Supplement* to *English Botany* edited by Sir William J. Hooker (1785–1865, who married one of Turner's daughters) and issued in 1831–66[2220]. The taxonomic insight shown in this, and his contributions to earlier parts of this work and the *Lichenographia Britannica*, were well in advance of their time; indeed many of his comments can still be read with profit today.

Another botanist who appears to have been a particularly able lichenologist in the early nineteenth century and who also published little was Richard Deakin (1809–1873). He evidently planned a major work on European lichens, the notes and beautifully executed coloured drawings for which are still preserved[574]. Deakin spent a considerable time in Italy, but while in Sheffield where he practised as a doctor he was living close to John Bohler (1797–1872) of South Wingfield, Derbyshire, and contributed extensively to Bohler's exsiccata *Lichenes Britannici* issued in 1835–37[255]. Bohler's exsiccata included a considerable text as well as specimens and some drawings.

Dawson Turner's brother James (1786–1820) was also reputed to be a keen lichenologist as was Robert Leyland (1784–1847) of Halifax.

Although too much reliance cannot be placed on the nature poets of the late eighteenth and early nineteenth centuries in terms of firm records, nevertheless one should perhaps not entirely ignore the natural history observations to be found in their verses. It is quite obvious, for example, that the use of the term 'moss' by Robert Southey (1744–1843) in at least two of his poems undoubtedly refers to lichens: verse five of *Written on the First of December* (prepared in Bath in 1793) in volume two of his *Poetical Works of Robert Southey, collected by himself* (1837, London: Longman, pp. 48–49) reads:

> And see the spangled branches shine;
> And mark the moss of many a hue
> That varies the old tree's brown bark,
> Or o'er the grey stone spreads.

and his *Sonnet no. 15* (written in Westbury in 1799) in the same volume (p. 97) contains the following lines:

> A wrinkled, crabbed man they picture thee,
> Old winter, with a rugged beard as grey
> As the long moss upon the apple-tree.

Numerous other references to lichens in poetry, prose and more scientific works up to this period are no doubt obscured by non-specific terminology, especially in the misapplication of the word 'moss', but some others certainly referring to lichens are reproduced in the books of Lindsay[1415] and Plues [1837–1838]. The general enthusiasm for natural history observation engendered through poetry and prose was considerable, and the cryptogamic botanist of today can but reiterate the feelings of John Clare (1793–1864) expressed in his poem *Shadows of Taste* (in E. Robinson and G. Summerfield, eds, 1966, *Clare: Selected Poems and Prose*, London: Oxford University Press, pp. 148–153; reproduced on p. v above).

Lichens received increasing attention in compendia of British plants in this period. In the posthumous 1819 edition of John Galpine's (1769 ?–1806) *Synoptical Compend of British Botany*[741] 37 lichen genera were accepted, following Acharius. Increasing numbers of species were included in later lists, for example 394 in Samuel F. Gray's (1766–1828) *Natural Arrangement of British Plants* of 1821[826], and 439 in the second (1844) edition of Sir William J. Hooker's *English Flora*[1060]. The tradition of local 'floras' embracing all plant groups gained popularity in the early part of the nineteenth century and from a trickle had increased to a flood by the 1860s, which D. E. Allen (1976) described as the golden age for detailed floras. To itemise all here would be an impossible task but the following merit particular mention for their extensive lichen coverage.

One of the foremost flora writers of the period was Nathaniel John Winch (1768–1838) who produced major works on Durham and Northumberland in the years 1807–37[2659–2661, 2663] and on Cumberland in 1833[2662]. His observations on lichens and their distributions in the regions he studied were remarkably accurate and his floras are still useful field-guides 150 years later[2052]; Winch's voluminous correspondence and his own annotated copies of his books are preserved in LINN [2657–2659]. Borrer and Turner's local observations came to be included in floras by other authors; for example, the 110 Sussex lichens in Thomas H. Cooper's *The Botany of the County of Sussex*[427] which were certainly supplied by Borrer, and Turner's lichen records in J. M. Druery's *Historical and Topographical notices of Great Yarmouth*[614]. Some of the earliest English local lichen lists were, however, those of John Sibthorp (1758–96) for Oxfordshire in 1794[2179], Charles Abbot (1761–1817) for Bedfordshire in 1798[4], Erasmus Darwin (1731–1802; grandfather of Charles) for Derbyshire in 1789[1835], Robert Teesdale (*c*. 1740–1804) for Yorkshire in 1794[2340] (emphasizing the Castle Howard area), and the Rev. Richard Polwhele (1760–1838) for Devonshire in 1797[1848]. Accounts of some other collections made in this were period were not, however, published at the time: for example, those of William Robertson (d. 1846/7) of Newcastle upon Tyne who was said to have been a 'very accurate investigator of lichens', Sir Thomas Gage (1781–1820) of whom J. E. Smith wrote in 1814 'Few botanists are more deeply versed in this difficult tribe of vegetables' and who collected in Ireland and Suffolk, Jonathan Salt (1759–1810) whose lichens from the Sheffield area date from 1795 to 1807[916], and Edward Forster (see p. 10), who made notes[720] and collections in Epping Forest[970] and around his home in Walthamstow[1309], north London. Later English local floras of particular note appearing prior to 1840 include those of Thomas F. Forster for Tonbridge in 1816[721], Thomas Purton (1768–1833) for the east Midlands in 1817[1878]

14

and 1822[1879], William Baxter (1787–1871) for Oxfordshire in an exsiccata of 1825–28[213], Godfrey Howitt (1800–1875) for Nottinghamshire in 1839[1083] (the lichens in this were probably contributed by J. Bohler), and Charles C. Babington (1808–1895) for the Channel Islands in the same year[126] (the lichens in this were attributed largely to F. C. Lukis).

In the Scottish border country George Johnston (1797–1855), a doctor in Berwick, who corresponded with Winch, produced a particularly detailed flora of Berwickshire, the cryptogamic volume of which was issued in 1831[1202]; Johnston

A

100. Lichen Oreſtœus.

On fir-trees near Forfar; and on fir-trees at Caroline Park, near Edinburgh, from whence my ſpecimens were collected. I ſent this lichen to Dr Smith, who is of opinion that it is L. Oreſtœus of Acharius growing on wood, it being uſually found, in Sweden, growing on rocks. I have alſo obſerved it on rocks in Ravelſton Wood near Edinburgh; and Dr Smith's opinion is therefore probably correct. Having never had an opportunity of ſeeing the work of Acharius on Lichens, I cannot give any reference to a figure of this plant.

B

23. Lichen læte virens *Lightf. Flo. Scot.*

herbaceus *Hudſ. Flo. An.*

Derbyſhire,

Fig. 8. Examples of labels of two early British exsiccatae including lichens. A, *Herbarium Britannicum*, Fasc. 4 no. 100 (1805[602]) by George Don (1764–1814). B, *A Collection of Dried Plants*, Fasc. 1 no. 23 (1789[586]) by James Dickson (173?8–1822).

For examples of labels of other exsiccatae including British lichens see pp. 201–223.

was aware of shortcomings in his treatment of the lichens and in a letter to J. Hardy (see p. 20) of 29 February 1840 regretted that he had so little time for these 'pets of Flora'. Johnston was one of the most important figures of the early

Berwickshire Naturalists' Club, the first of its kind, which he helped to found in 1831. Such societies and field clubs were to play an increasingly important role in the development of British natural history later in the century.

Hooker produced a general account of the Scottish lichens in his *Flora Scotica* of 1821[1058] but few local floras for that country appeared in this period. Notable exceptions were Robert Kaye Greville's (1794–1866) *Flora Edinensis* of 1824[832] and George Don's (1764–1814) contributions of an important exsiccata in 1804–12[602] (Fig. 8A) and a list of the lichens of Angus in 1813[603]. Greville, although primarily remembered as a cryptogamist, had little time for lichens, to judge from the almost negligible mention they received in his major work, the six volume *Scottish Cryptogamic Flora* of 1822–28[833]. Don is often regarded as the pioneer of Scottish upland botany; he reached many generally inaccessible sites by sleeping in the open and feeding on bread, cheese and oatmeal for up to a week at a time.

Wales received scant attention in this period apart from the studies of the Rev. H. Davies in Anglesey[566] already referred to. In Ireland Thomas Taylor (*c.* 1787–1848) appears to have been the only major worker although he published little on British and Irish lichens with the notable exception of his contribution to J. T. Mackay's *Flora Hibernica* of 1836 which must rank as the first major listing of the Irish lichen flora.

1841–1865: Lindsay, Mudd and Salwey; the influence of de Notaris, Körber and Massalongo

While the tenor of lichenology in the early decades of the nineteenth century was set by Acharius who had had little time for the microscope, the realization of the value of microscopic and particularly spore characters for the delimitation of lichen taxa started to open up new vistas in the 1840s and 1850s. The major European lichen text at the start of this period was Elias Magnus Fries' (1794–1878) *Lichenographia Europaea reformata*[732] of 1831. Fries, best known for his mycological compendia, had largely followed the concepts of Acharius (although including later work) and it was left to the Italian Giuseppe de Notaris (1805–1877) to recognize such new genera as *Bacidia* and *Buellia* in publications over the period 1846–61. The value of spore characters was quickly appreciated by other workers, particularly Abramo B. Massalongo (1824–1860), also an Italian, and the German Gustav Wilhelm Körber (1817–1885) whose *Systema lichenum Germaniae* (1855) was the first major national flora employing spore characters.

William Lauder Lindsay (1829–1880; Fig. 9), born in Edinburgh, worked as a physician at Murray's Royal Institution for the Insane (now the Murray Royal Psychiatric Hospital) in Perth; he was a proponent of 'non-restraint' in the treatment of mental illness, regarded the study of natural history as therapeutic, was concerned with higher education reform (on which he published in 1878), and edited *Excelsior*, the literary gazette of his Institution. Lindsay will be remembered by lichenologists for a variety of reasons as he made important contributions to diverse aspects of the subject. He was very much aware of the possible economic value of lichens, particularly as sources of dyes (then a declining industry in Scotland), and published many papers on this topic in the years 1853–68[1409–1414, 1424–1426]. His *Popular History of British Lichens* of 1856[1415] (Fig. 9), with numerous

16

A

POPULAR HISTORY

OF

BRITISH LICHENS,

COMPRISING

AN ACCOUNT OF THEIR STRUCTURE, REPRODUCTION,
USES, DISTRIBUTION, AND CLASSIFICATION.

BY

W. LAUDER LINDSAY, M.D.,

FELLOW OF THE BOTANICAL AND ROYAL PHYSICAL SOCIETIES OF EDINBURGH, ETC.

LONDON:
LOVELL REEVE, 5, HENRIETTA STREET, COVENT GARDEN.
1856.

Fig. 9. William Lauder Lindsay (1829–1880) and title-page of his *Popular History of British Lichens* published in 1856[1415].

coloured drawings, is still a most readable and fascinating account of lichens and certainly must have played an important role in popularising lichenology in this period; it was the first book of its type ever to have been produced in Britain. Fired with enthusiasm for the microscope, he produced extremely carefully illustrated and documented accounts of various lichenicolous fungi[1418, 1430, 1431] and lichen pycnidia and spermatia[1419, 1434] over the years 1857–72 which are also of much value today. Lindsay collected lichens extensively around Perth for use in his anatomical investigations (his very rich collections are now in E) but contributed little to local floristic studies[1417] and described rather few taxa (and most of these as a result of a visit to New Zealand; he also wrote a history of Otago). Although his work overlapped in time with the British lichenologists most strongly influenced by Nylander, he strongly opposed Nylander's views on some topics (see p. 23).

The popularisation of lichenology in Britain during this period was also aided by the eminent mycologist the Rev. Miles J. Berkeley (1803–1889) in his *Introduction to Cryptogamic Botany*[221] (1857), the Rev. Hugh Macmillan (1833–1903) in his *Footnotes from the Page of Nature*[1489] (1861) and other works, and Margaret Plues (*c.* 1840–*c.* 1903), later a nun, in her most readable *Rambles in Search of*

Flowerless Plants[1837] (1864) which included coloured illustrations of many lichens; the latter proved so popular that it had to be re-issued in 1865[1838].

It was, however, William Mudd (1830–1879) who produced the major systematic treatment of British lichens of the period. Born at Bedale, Yorkshire, he became a gardener at Great Ayton, Cleveland, and published an account of the lichens of Cleveland in 1854[1641]. His major work, *A Manual of British Lichens*[1643] (Fig. 10), appeared in 1861 and included descriptions of all 497 species then known in the British Isles. Particular emphasis was placed on spores in this work, many of them illustrated with coloured plates; unfortunately many of his spore measurements appear to be inaccurate. Mudd collected extensively in Cleveland and his flora has a strong regional bias. He issued an exsiccata[1642] of some 300 specimens to accompany his *Manual*; critical notes on the exsiccata were, however, published by Nylander[1675] (see p. 21) shortly after they appeared. In 1865 Mudd was Curator of the Botanic Garden in Cambridge and it was from there he issued his *Monograph of British Cladoniae*[1644] which comprised both a printed text and representative specimens of the taxa treated. Mudd was a particularly enthusiastic fieldworker, and some measure of this can be obtained from the following account by William Foggitt (1835–1917) which appears in W. Herbert Smith's *Walks in Weardale*[2230a] (1885, pp. 66–67):

> The road from the High Force in Teesdale over Swinhope Fell comes out at Westgate. An interesting instance of enthusiasm in the cause of science is associated with this road; some years ago, four botanists were exploring Teesdale, they were J. G. Baker, F.R.S., of Kew, W. Mudd, Esq., late Curator of Cambridge University Gardens, W. Foggitt, Esq., of Thirsk, and J. Watson, Esq., of Ayton; Mr. Mudd was an eminent lichenologist and all day long was busy chipping off fragments of lichen-covered rock, which were duly deposited in bags slung round his person; when the other gentlemen retired for the night, they left Mr. Mudd still chipping and dressing his specimens, and in the morning as soon as they awoke, they heard the chip of Mr. Mudd's hammer already at work. After breakfast, they walked over Swinhope Fell and caught the morning coach at Westgate for Frosterley, which was then the terminus of the line; on arriving at the station, Mr. Mudd's bags were overhauled by one of the porters, who said Mr. Mudd would have to pay for "excess luggage"; the party protested against this, and said it was impossible that the bags could be heavy, as Mr. Mudd had carried them from the High Force Inn over Swinhope to Westgate; however, the load was placed on the scales, and it weighed over 8 stones or upwards of 1 cwt.!

Isaac Carroll (1828–1880), better known for his work on Irish lichens, published three lists of species noted in Britain since Mudd's *Manual* in 1865–67[371–373].

Mention must also be made here of the Rev. William Allport Leighton (1805–1889; Fig. 11) of Shrewsbury, an hotel keeper's son who had been at school with Charles Darwin and whose lichenological publications extend from 1837 to 1879; he became an ardent follower of Nylander and is thus treated further below (p. 21). In the present period he was, like Mudd, starting to make use of spore characters and his *British Species of Angiocarpous Lichens* of 1851[1343] was subtitled 'elucidated by their sporidia'; the taxa dealt with in this book were discussed in considerable detail and coloured drawings of ascocarp sections and spores were included. Such attention to spores enabled Leighton to produce an important account of the British lirellate lichens in 1854[1345] and of the arthonioid species in

A MANUAL

OF

BRITISH LICHENS,

CONTAINING

DESCRIPTIONS OF ALL THE SPECIES AND

VARIETIES,

AND

FIVE PLATES, WITH FIGURES OF THE SPORES OF ONE

HUNDRED AND THIRTY SPECIES, ILLUSTRATIVE

OF THE GENERA.

BY WILLIAM MUDD.

PRINTED FOR THE AUTHOR BY HARRISON PENNEY, PREBEND ROW,
DARLINGTON.
1861.

Fig. 10. Title-page of *A Manual of British Lichens* by William Mudd
(1830–1879) published in 1861[1643].

1856[1347]. Leighton also issued an exsiccata over the years 1851–67[1344] and corresponded widely with collectors in many parts of the British Isles. His exsiccata was revised in detail by Arnold[113–114], a leading continental lichenologist, who considered many specimens incorrectly named.

Also living in Shropshire at this time was the Rev. Thomas Salwey (1791–1877), Rector of Oswestry in 1833–72, who collected very extensively and sent material to Borrer, Mudd and Leighton; Borrer collected with him in Merioneth on at least one occasion (July 1837). Salwey made important contributions to our knowledge of the lichens of north Wales[2077, 2079, 2087] and Shropshire[2078, 2080, 2083] and published more briefly on species encountered on visits to Cornwall[2082], Guernsey[2081] and the Isle of Wight[2084]. Salwey appears to have been a very keen fieldworker and carefully localized the records in his generally very comprehensive lists; his 1863 account of the lichens of Barmouth and its surrounding areas[2087], for example,

Fig. 11. The Rev. William A. Leighton (1805–1889, *left*) and Edward M. Holmes (1843–1930, *right*).

included 169 species. He also prepared one[2085], or possibly two[2076], exsiccatae but these appear to have had a very limited distribution. The publications of Salwey were by no means restricted to lichenology; he also issued the *Duty of a Christian Magistrate* in 1835 and *Gospel Hymns* in 1847. The combination of natural history with Christian and social fervour was evidently not uncommon in this period (see Allen, 1976), the pursuit of natural history being regarded almost as a part of one's Christian commitment.

Other important compilers of regional lichen floras in this period included

John G. Baker (1834–1920, who worked at Kew 1866–99 and was the father of L. Porter, see p. 33) in Yorkshire[172–175] (although his major work on Yorkshire botany did not appear until 1907[176]), Robert Garner (1809–1890) in Staffordshire[758], Edwin Lees (1800–1887) in the Malvern Hills in particular[1325–1326,1328–1331], William Curnow (18?09–1887) in Cornwall[542], Philip B. Ayres (1813–1863) in Oxfordshire[123], the Rev. John S. Henslow (1796–1861) and Edmund Skepper (1825–1867) in Suffolk[981], both Benjamin Carrington (1827–1893) and Louis C. Miall (1842–1921) in the West Riding of Yorkshire[1597], and Frederick P. Marratt (1820–1904) around Liverpool[1542]. In an account of the flora of the area around Manchester in 1859[839], Leopold H. Grindon (1818–1904) drew attention to the diminution of the lichen flora in southern Lancashire which he attributed to air pollution.

An important list of Channel Island lichens was contributed by Charles du Bois Larbalestier (1838–1911) to Anstead and Latham's 1862[103] account of those islands and Larbalestier also issued an exsiccata of lichens from that area in 1867–72[1290]. In the Scottish borders, James Hardy (1815–1898) continued the traditions of the Berwickshire Naturalist's Club[864–885], started by George Johnston (see p. 14) to whom he sent specimens. Meanwhile, in Scotland itself, due in no small measure perhaps to Lindsay's influence, floristic studies also increasingly included lichens; amongst these the work of the following merits particular mention: William Gardiner (1808–1852), an umbrella-maker in Dundee, on the Angus flora[755–756], Professor George Dickie (1812–1882) in Aberdeen[583], Thomas Edmondston (1825–1846; killed by natives in Ecuador when 'naturalist' on H.M.S. Herald) in Shetland[657], Alexander Croall (1809–1885) in Braemar[450], John Sadler (1837–1882) the curator of the Edinburgh Botanic Garden around Edinburgh[187], and Professor J. H. Balfour (1808–1884) mainly in western Scotland[189] and the Hebrides[195]. In Wales, apart from the studies of Salwey already referred to above, little progress was made in this period apart from John W. G. Gutch's (1809–1862) investigations in the Swansea area[850–851].

Lichens were beginning to feature increasingly not only in general county floras and reports of field meetings in the 1840s–60s but also in popular guide books to holiday resorts or sites of historic importance. The evidence suggests that lichenology was now becoming very acceptable in botany and that the popularity then attained and continued in our next period was not to be approached again for almost a century.

Every period appears to have had workers whose publications scarcely do them justice. The Rev. Andrew Bloxam (1801–1878), Perpetual Curate of Twycross, Leicestershire, from 1839 to 1871 and Rector of Harborough Magna, Warwickshire, from then until his death, was such a person. He made important collections of almost all plant groups but seems to have had particular interests in lichens and microfungi; Bloxam corresponded with most leading botanists of the time and sent some material from which many new species were described. He contributed lichen records to accounts of the floras of the Isle of Wight[252] and the area around Burton-on-Trent[336] but was mainly concerned with Leicestershire, many of his lichen records from the Charnwood Forest being published in a general account of that region in 1842[1853]. Bloxam was also almost certainly responsible for the long list of Leicestershire lichens published by the Rev. William H. Coleman (181?6–1863)[415] in 1863, but an extremely detailed manuscript of Leicestershire cryptogams prepared in about 1854[251] and attributed to him remains unpublished. This manuscript was also contributed to by the Rev. Churchill Babington (1821–1889) of whom little is known; Babington must, however, have been an exceptionally able lichenologist as Borrer, in a letter dated 11 November 1839, spoke of him at 18 as one who '. . . has gone more deeply into the subject of Lichens than I have, & with the aid of modern authors with some of whom I am unaquainted'. Some other manuscripts of Bloxam's, perhaps formerly at Calke Abbey, Derbyshire, and still known to exist in 1903, now appear to be lost. A study of Bloxam's collections and manuscript showed that he discovered at least 106 lichens in Gopsall Park, near Twycross[2246]; this in itself testifies to his skill in the field.

1866–1899: Crombie, Leighton, and Stirton; the influence of Nylander

1866 was a remarkable year for lichenology; H. A. de Bary (1831–1888) suggested that lichens were either fruiting forms of algae or the form algae take when attacked by parasitic fungi (S. Schwendener's (1829–1919) 'dual hypothesis' followed in 1867); William Nylander (1822–1899), who discovered that iodine was a useful reagent in lichenized and non-lichenized ascomycetes in 1865, reported on the still routinely used bleaching powder and potassium hydroxide reagents; and, in the same year, Nylander also suggested that lichens might be used as indicators of air quality. These three discoveries were to have profound effects on lichenology.

Leighton, whose earlier work has already been mentioned (p. 17), was quick to realize the importance of Nylander's work which started to flow from that recluse's home in Paris as numerous books, articles and pamphlets over the period 1852–99. He translated Nylander's papers on lichen chemistry and air pollution into English shortly after they appeared, and adopted chemical characters as well as many of Nylander's taxonomic ideas in the first edition of his *Lichen Flora of Great Britain, Ireland, and the Channel Islands* in 1871[1366]. This impressive flora included some 781 species and soon became the standard British text, the last edition appearing in 1879[1383]. Leighton, and more particularly Crombie (see below) and Charles du Bois Larbalestier (see p. 20), sent many specimens to Nylander who described some of them as new; Leighton published English translations of many of Nylander's papers including British taxa[1351,1352,1354,1357,1361, etc.] as well as continuing his own important studies on both British and foreign lichens until 1884.

The Rev. James Morrison Crombie (1831–1906), who was a lecturer at St. Mary's Hospital in London in 1879–91, became the major figure in British lichenology in the 1870s and 1880s publishing some 50 British works during the forty years 1861–1901; he also published extensively on foreign lichens sent to the British Museum for determination. He appears to have had little contact with Leighton, although the reasons for this are obscure. Most of Crombie's publications were taxonomically rather than floristically orientated but his accounts of Middlesex[456] and Epping Forest[511] lichens merit particular attention; he also monographed various genera in Britain[466, 482], issued an important exsiccata[483], checklist[512–514], and two books on lichens: *Lichenes Britannici* in 1870[461], and *A Monograph of Lichens found in Britain* in 1894[515] (Fig. 12). The latter work included

A MONOGRAPH

OF

LICHENS

FOUND IN BRITAIN:

BEING

A DESCRIPTIVE CATALOGUE

OF THE SPECIES IN THE

HERBARIUM OF THE BRITISH MUSEUM.

BY THE

REV. JAMES M. CROMBIE, M.A.,
F.L.S., F.G.S., &c.

PART I.

LONDON:
PRINTED BY ORDER OF THE TRUSTEES.

SOLD BY

LONGMANS & Co., 39 PATERNOSTER ROW;

B. QUARITCH, 15 PICCADILLY; DULAU & Co., 37 SOHO SQUARE, W.;

KEGAN PAUL, TRENCH, TRÜBNER, & Co., 57 LUDGATE HILL;

AND AT THE

BRITISH MUSEUM (NATURAL HISTORY), CROMWELL ROAD, S.W.

1894.

Fig. 12. Title-page of *A Monograph of Lichens found in Britain* by James M. Crombie (1831–1906) published in 1894[515].

details of all British material in BM together with descriptions and illustrations; this was a monumental undertaking and Crombie saw only the first volume issued in his lifetime, the second being completed for publication by Miss A. L. Smith (see below) in 1911[2204].

While Leighton and Crombie had been ready to follow Nylander on most matters, including a refusal to entertain the idea that lichens might be dual organisms (on which topic Crombie was particularly outspoken), Lindsay (see p. 15) was more cautious. Original investigations had been the hallmark of Lindsay's work and so he tried to ascertain the value of Nylander's chemical tests by his own experiments, details of which were issued in 1869[1429]; the results led him to be sceptical[1429,1432] and he also urged others to put such ideas to the test before deciding whether to accept them or not.

James Stirton (1833–1917) of Glasgow, the other leading lichen systematist in the British Isles at this time (publishing in the years 1873–99), while utilising many of Nylander's characters, particularly chemical ones, appears to have had little personal contact with Nylander and worked largely on his own; he did, however, receive considerable amounts of material from James M'Andrew (1836–1917) working on the Dumfriesshire and Kirkcudbrightshire lichen floras[1461, etc.]. Stirton collected extensively in western and central Scotland and, in addition to describing many species as new to science, prepared some important floristic lists, of which that of 1876 dealing with western Scotland[2275] is of particular note.

Some collecting was carried out in most regions of the British Isles during this period and numerous local lists were published. One of the most astute collectors of the last quarter of the nineteenth century was Edward Morell Holmes (1843–1930; Fig. 11); born in Wendover, Buckinghamshire, the son of a chemist, he moved to Plymouth and later London where he was curator of the Materia Medica Museum of the Pharmaceutical Society for the fifty years 1872–1922. Holmes had a remarkably keen eye and made major contributions not only to British lichenology, but also to our knowledge of bryophytes and especially marine algae, in addition to publishing about 350 articles on drugs in the *Pharmaceutical Journal*. George C. Druce (1850–1932) referred to him as '. . . a rapid and indefatigable walker, his quick sight detecting plants as he was racing along at five miles an hour' and he collected extensively until the loss of a foot in a coaching accident in 1922 limited his fieldwork; after his retirement and faced with heavy doctor's fees he advertised himself as a 'consulting botanist and pharmacognosist' from his home in Sevenoaks, Kent, '. . . open to give expert botanical assistance, on business terms, to merchants and others'. He was an ardent collector and in addition to his main lichen herbarium in Nottingham (in which the specimens are mounted in individual glass-topped boxes) material of his is to be found in many other lichen herbaria in Britain. Holmes prepared detailed accounts of the lichens of Devon and Cornwall in 1872[1034], Kent in 1878[1037], and Hertfordshire in 1887[1868], but will perhaps be best remembered for his outstanding contributions to the *Victoria History of the Counties of England*; for this he provided synopses and check-lists of the lichens of several counties: Bedfordshire[1046], Cornwall[1047], Devonshire[1048], Dorset (manuscript only[1049]), Kent[1051], Somerset[1050] and Surrey[1045] over the years 1902–07. Holmes also assisted in the preparation of a general introduction to collecting cryptogams in 1886[1053] and worked in co-operation with William Joshua (1828–1898) of Cirencester, Gloucestershire, in producing a set of 48 microscope slides of lichens which sold at £3 in 1879[1217–1218].

In addition to Holmes' work in south-west England, John Ralfs (1807–1890) and Richard V. Tellam (1826–1908) published accounts of the lichens of West Cornwall[1899] and East Cornwall[2345], respectively. Edward Parfitt (1820–1893), Librarian at the Devon and Exeter Institution for 1861–93, published over 200 papers dealing with extremely diverse aspects of Devonshire's natural history and dealt with the lichens in 1883[1743]. Harvey B. Holl (1820–1886) corresponded with Holmes and made important collections in south-west England, Wales and Worcestershire but published little apart from a list of additions to the Malvern Hills lichen flora in 1870[1032]; however, his herbarium material, now in the British Museum (Natural History), testifies to his not inconsiderable knowledge of lichens.

The north of England had several lichenologists working very actively in the late nineteenth century. Joseph A. Martindale (1837–1914), who lived in the Lake District and whose main collection is now in Kendal Museum, published some 17 articles, including one booklet[1560] on the lichens of that area in 1872–1906. From Yorkshire the Rev. William Johnson (1844–1919), a Methodist minister, produced important papers dealing not only with Yorkshire but also Cumberland[1168], Durham[1852] and Northumberland[1195] and also issued an important exsiccata of 520 numbers[1196]. William West (1848–1914), a chemist, studied Yorkshire cryptogams, including the lichens, and published a major contribution to Shetland lichenology[2585]. Frederick A. Lees (1847–1921) worked on lichens in both Lincolnshire[1338] and West Yorkshire[1337] and included them in his floras of these counties. The Craven area of Yorkshire was also treated in some detail by John Windsor (1787–1868) in a posthumous publication of 1873[2665]. An important list of Leicestershire lichens was compiled by the Leicester Literary and Philosophical Society in 1886[1634].

Excursion reports by Hardy continued to appear in the journal of the Berwickshire Naturalists' Club. The number of local scientific societies in Britain reached its peak during this period and in 1873 there were at least 169 most of which produced journals, proceedings, annual reports, or transactions; 104 of these were nominally field clubs (Allen, 1976). Brief notes of particular lichen records or species detected on society excursions soon found their way into print; Henry F. Parsons (1846–1913), William D. Roebuck (1851–1919), and Albert Shackleton (1830–1916), as well as West and Lees (already referred to), for example, were all publishing lichen records in *The Naturalist* in this period. Freeman C. S. Roper (1819–1896) published regular addenda to his 1875 *Flora of Eastbourne*[1997] in the local journal and this was a common pattern of the period.

In Wales John E. Griffith (1843–1933) included a listing of Anglesey and Caernarvonshire lichens in his flora of 1895[836]. Daniell A. Jones (1861–1936) prepared a manuscript on the botany of Merionethshire in 1898 including many lichen records, but it remained unpublished[1206]; William H. Wilkinson (d. 1918) of Birmingham dealt with that same county in 1900[2633]. Wilkinson also made major contributions to our knowledge of lichens in the Isle of Man[2632], Hampshire and the Isle of Wight[2634], and Radnorshire[2635]. Guernsey was investigated in some detail by Ernest D. Marquand (1848–1918), whose major publications appeared in 1893[1536] and 1901[1537]; Marquand also revised an early lichen collection made in Essex by Ezekiel G. Varenne (1811–1887)[1535]. Work in Ireland continued less intensely, the only major worker being Larbalestier (see p. 20); important contributions on the lichens of the Mourne Mountains by Henry W. Lett (1838–1920) in 1890, and of Dublin and Wicklow by Greenwood Pim (1851–1906) in 1878 did, however, appear (see Mitchell, 1971).

Although this period was one of the more active lichenologically in the British Isles, the overall impression is one of scattered workers having limited contact with one another. The following sentiments expressed by Martindale in 1889[1560] would probably have been echoed by many of his contemporaries:

> To speak from my own experience, during the twenty years that I have been engaged in the examination of lichens, I have only twice had the pleasure of meeting and speaking with anyone that had taken up the study with heartiness. Of course, scattered up and down the country—alas, at too wide intervals—are several persons more or less enthusiastic in the matter; and with some of these I have enjoyed pleasurable and instructive correspondence. But the fact remains, that comparatively few English botanists, of late years at all events, ever give more than a cursory glance at lichens.

By the 1880s and 1890s a swing away from taxonomy was starting to occur in the British universities. Whereas Chairs of Botany had previously been held almost exclusively by systematists, they now began to be occupied by physiologists. This trend was to have profound effects on the development of natural history in Britain, and subjects such as lichen taxonomy became increasingly the province of the amateur and museum botanist rather than the don. That this is still true in the 1970s, almost a century later, testifies to the impact of the 'discovery' of physiology.

1900–1935: A. L. Smith, Horwood, Paulson, Wheldon, Travis and Wilson; the decline sets in

The impetus botanical studies, including lichenology, had attained in the 1870s and 1880s in the British Isles was slow to wane and in fact revived slightly in about 1905–1915 before reaching its nadir. About the turn of the century the *Victoria History of the Counties of England* was starting to be issued and volumes of it are still being produced in the 1970s although natural history is no longer included. Not all counties had accounts of their natural history prepared and of those which did, lichens were not featured in all, this being a matter for the county compilers. Holmes prepared accounts of the lichens of seven counties for this project (see p. 23) and assisted with this section for others (e.g. Buckinghamshire[610]). Other counties whose lichens were treated (all prior to 1912) were Berkshire (C. G. Druce[611]), Cheshire (O. V. Darbishire, in manuscript only[556]), Durham (M. C. Potter[1852]), Essex (J. C. Shenstone[2168]), Hampshire (W. H. Wilkinson[2634]), Hertfordshire (J. Hopkinson *et al.*[1065]), Lancashire (H. Fisher[707]), Leicestershire (A. R. Horwood[1074]), Norfolk (J. M. Crombie[516]), Northamptonshire (H. N. Dixon[597]). Nottinghamshire (J. W. Carr[368]), Shropshire (W. Phillips[1826]), Staffordshire (J. E. Bagnall[136]), Suffolk (E. N. Bloomfield[247]), Sussex (F. H. Arnold[117]), Warwickshire (J. E. Bagnall[135]), Westmorland (J. A. Martindale, in manuscript only[1565]), Worcestershire (J. E. Bagnall[134]), and Yorkshire (J. G. Baker[176]). In the first decade of the century major studies on some other counties also appeared (see also p. 28): examples are Edwin N. Bloomfield's (1827–1914) account of Norfolk and Suffolk lichens[244], John Amphlett (1845–1918) and Carlton Rea's (1861–1946) Worcestershire flora[28], in which the Rev. J. H. Thompson and [?W.] Holl were thanked for help with the lichen section, and Arthur Reginald Horwood's (1879–1937) studies on Leicestershire[1070, 1075, etc.]. Bloomfield also made important contributions to Sussex lichenology over the years 1898–1914.

Horwood, subcurator at Leicester City Museum for 1902–22, suggested the formation of a *Lichen Exchange Club* in 1907[1071] and this was to become the first society devoted exclusively to the study of lichens. By the end of 1907 16 members had enrolled but although this number had increased to 27 by September 1908, its membership never exceeded 29. Reports were published annually (Fig. 13) and specimens distributed between members; the packets sent out bear a mauve impression of an oval rubber stamp familiar to all dealing with British herbaria today. The Rev. Henry P. Reader (1850–1929) lived at Holy Cross Priory in Leicester and was one of the mainstays of the Club of which he was the first 'Distributor', Horwood being the Secretary. Reader published little apart from compilations in the Club's *Report* but made many collections in Leicestershire and others in Staffordshire to which he moved after the demise of the Club. Notes on specimens submitted were provided in later issues of the *Report* by the Rev. Philip G. M. Rhodes (1885–1934) who also published accounts of the lichens of the Channel Islands in 1910[1945], Cambridgeshire in 1911[677] and Hartlebury Common, Worcestershire, in 1931[1947], and by J. A. Wheldon (see below). The Club ceased to function in 1914, perhaps partly due to controversy amongst its members following the publication by Miss Smith of the second volume of Crombie's *Monograph* in 1911 which received a not too favourable review in the Club's *Report*. One of the final tasks of the Club was the publication of a check-list of British lichens accepting about 1259 species compiled by Horwood[1080].

Annie Lorrain Smith (1854–1937; Fig. 14), born in Dumfriesshire, was originally a governess but became an 'unofficial worker' at the British Museum (Natural History) for the 46 years 1888–1934 (women were then not officially admitted into the Civil Service). She was particularly interested in microfungi and published extensively on them in the first two decades of the century. The 1911 second volume to Crombie's *Monograph*, although appearing under her name[2204], is now known to have been almost entirely the work of Crombie and not of Smith (although many of her contemporaries, such as E. M. Holmes, were evidently unaware of this). Miss Smith carried out limited lichenological field work, to judge from both material in the British Museum (Natural History) and her publications, but did provide some lichen lists from British Mycological Society forays she attended[2206] and was involved in the preparation of an account of the lichens of Clare Island in Co. Mayo in 1911. In 1918[2205] and 1926[2210] she prepared a second edition of Crombie's *Monograph* which is still an important reference work. The *Monograph* suffered from a lack of workable keys and to remedy this she published her own set of keys to the British species in 1921[2207]; these were to remain the only keys available to all known British lichens until Dahl's keys of 1952 which, however, dealt only with macrolichens. Miss Smith will, however, be remembered also for

The
Lichen Exchange Club
of the
British Isles.

REPORT FOR 1908.

Parcels for 1909 should be sent before March 31st, 1909, to the Distributor, the Rev. H. P. READER, M.A., Holy Cross Priory, Leicester.

The Subscription, 5/- per annum, should be paid on the 1st January to the Secretary and Treasurer, A. R. HORWOOD, Leicester Corporation Museum.

PRICE SIXPENCE.

LEICESTER :
T. H. JEAYS & SONS, PRINTERS, 7 ST. MARTIN'S.

Fig. 13. Title-page of the first, 1908, report of *The Lichen Exchange Club of the British Isles.*

Fig. 14. Annie Lorrain Smith (1854–1937, *left*) and James A. Wheldon (1862–1924, *right*).

her historical studies; her text on *Lichens*, largely prepared during the First World War but not published until 1921[2208] (re-issued 1975) remains the standard English work on the development of the subject. She also published a separate account of the development of lichenology in the British Isles[2209], and in a series of papers summarized literature appearing after 1921 until 1933.

Meanwhile, in the period 1904–30, an active school of lichenology had been developing in the Liverpool area. This included James A. Wheldon (1862–1924; Fig. 14), Albert Wilson (1862–1949), John W. Hartley (1866–1939) and William G. Travis (1877–1958), all of whom published major floristic studies, often in conjunction with one another. Their studies are too numerous to cite individually here, but the following may be mentioned as representative: Isle of Man (Hartley and Wheldon, 1927[905]), Isle of Wight (Wheldon, 1909[2593]), Isle of Arran (Wheldon and Travis, 1913[2609]), South Lancashire (Wheldon and Travis, 1915[2611]), West Lancashire (Wheldon and Wilson, 1904[2614], 1907[2615]) and Perthshire (Wheldon and Wilson, 1915[2618]). These authors were also responsible for numerous short notes in local journals.

The integration of mycology and lichenology was strongly advocated by Miss Smith who, in 1918, was able to report that at a recent meeting of members of the British Mycological Society it had been decided that '. . . there was good reason for enriching the scope of our work by associating the study of lichens with that of mycology.' One result of this decision was that lichens started to be collected on Mycological Society forays and the lichen lists published in reports of these in 1919–35 are an extremely valuable record of the lichen flora at that time. The major contributors of lichen data to these foray reports were the school teachers Robert Paulson (1857–1935) who lived in Pinner, north-west London, and Henry H. Knight (1862–1944); Paulson published an important account of the Hertfordshire lichen flora in 1919[1767] but will perhaps be best remembered for his investigations with Percy G. Thompson (1866–1953) on lichens in Epping Forest, Essex, in 1911–20[1774–1776]; Knight published little apart from foray reports with the exception of an account of the Isle of Wight species in 1933[1251] but W. Watson (see below) collated a major manuscript on Gloucestershire lichens which Knight was working on and this was published in 1950[1255].

Other floristic studies appearing in 1910–35 included those by the Rev. David Lillie on Caithness lichens in 1912[1403], the Rev. Arthur Mayfield (186?8–1956) on Suffolk in 1917[1583] and 1930[1584], D. A. Jones on Ingleton in 1925[1207], and Thomas Hebden (1849–1931) on Harden Beck, Yorkshire, in 1916[977]. Hebden built up a particularly extensive lichen herbarium[2129], corresponded with several leading foreign lichenologists (including Nylander), and was evidently much respected by his contemporaries whom he assisted in their identifications[2497]; his influence is belied by his meagre publications.

Some workers active in the latter decades of the nineteenth century continued to publish into the early part of the twentieth, for example M'Andrew (to 1913) and W. Johnson (to 1918). Furthermore, the tradition of reports of species noted on field excursions in local journals continued, although the number of such journals had declined considerably by this time; of particular note are the reports of W. P. Hiern (1839–1925) for Devonshire[1008–1015], Cecil P. Hurst (d. 1956) for Wiltshire[1094–1109], and William E. L. Wattam (1872–1953) for Yorkshire[2540–2565].

A new development in lichenology seen in this period was a growing awareness of lichens as part of the ecosystem and several major ecologically rather than floristically orientated studies were made. The most notable of these were by Walter Watson (1872–1960), for most of his life a teacher at Taunton School, Somerset, who was to become one of the mainstays of British lichenology in later decades (see below). Watson sent many of his early collections to Hebden for critical study, but his lichenology was perhaps largely self-taught. His series of ecological papers, running from 1918–36[2481–2483, 2489, 2499, 2506–2507], took particular habitats in turn and described the lichens and bryophytes to be found in them. Other ecological studies in the period to include lichens were those of R. C. McLean on Blakeney Point[1482], and E. Price Evans (1882–1959) on Cader Idris[679]. Watson soon came upon taxonomic problems and began to publish a series of papers including taxonomic and distributional data which spanned the years 1917–42; in 1922 he also issued a field key to the more common lichens[2485]. Watson was very interested in the lichen flora of Somerset, where he settled, and published a detailed account of the lichens of that county in 1927–29[2491, 2493, 2495] (reissued in a single volume in 1930[2497]).

In Ireland John Adams (1872–1948) prepared a synopsis of lichen distribution

30

in 1909, which was to be succeeded in 1929 by the publication of the most detailed account to date of the Irish lichen flora: *The Lichens of Ireland* by Matilda C. Knowles (1864–1933), who from 1907 until her death worked at the National Museum of Ireland in Dublin as a 'Temporary Assistant' and followed Miss Smith's views.

1936–1957: Sowter, Wade and Watson; the leanest years

Lichenology in the British Isles over this period was held together by the above three workers. Watson was by this time acting as the mentor of British lichenology and his journal publications continued to flow after he retired from teaching in 1939; the most important of these were perhaps his re-examination of British material preserved in the Edinburgh herbarium of 1935–45[2505, 2510, 2517, 2519], his annotated check-list of lichen parasites of 1948[2523], and his compilation of the data on Yorkshire lichens of 1946[2521]. He also published Knight's account of the lichens of Gloucestershire in 1950[1255] and revised Dunston and Dunston's paper on the lichens of part of Wiltshire prior to publication in 1940[636]. Various Devonshire

Order 4. PELTIGERALES

Family 21. STICTACEAE

72 STICTINA Nyl.
  979 **fuliginosa** (Dicks.) Nyl.   C, S, 1–6, 10, 15, 36, 37, 40, 45, 46, 48, 49, 52, 62, 64, 66, 69, 70, 72, 73, 75, 77, 81, 83, 87–90, 92, 97–101, 104, 105, 108, 110. Ir. 1–3, 6, 16, 20, 21, 27–29, 33, 39, 40.
  980 **limbata** (Sm.) Nyl.   C, 1–5, 10, 11, 13–15, 23, 29, 36, 37, 45, 46, 48–50, 52, 62, 64–70, 72, 73, 77, 80, 83, 85–90, 92, 97–99, 101, 104, 110. Ir. 1–6, 20, 27, 39.
  981 **sylvatica** (Huds.) Nyl.   1–6, 10, 21, 34, 36, 37, 40, 46, 48–50, 52, [55], 64–70, 72, 73, 77, 85–88, 90, 92, 96–100, 108. Ir. 1–3, 7, 16, 20, 21, 39, 40.
  982 **dufourei** (Del.) Nyl.   1–4, 48, 73, 98, 101. Ir. 1–3, 8, 39.
  983 **intricata** (Del.) Nyl.   var. *thouarsii* (Del.) Nyl.   3, 48, 73, 87, 88, 97, 98, 101, 104. Ir. 1, 2, 13, 16, 39.
  984 **crocata** (L.) Nyl.   1, 3, 4, 83, 87, 88, 90, 96–98, 101. Ir. 1, 20, 39.

73 STICTA Schreb.
  985 **damaecornis** Ach. (*sinuosa* Pers.)   Ir. 1–3.
  986 **aurata** Ach.   C, S, 1–4, 9–11.   Ir. 2, 3.

74 LOBARINA Nyl.
  987 **scrobiculata** (Scop.) Nyl.   C, S, 1–5, 10–12, 14–16, 36–38, 40, 45, 47–50, 52, [55], 62–65, 67–73, 81, 83, 85, 87, 88, 90, 92, 96–101, 108, 109.   Ir. 2, 3, 6, 16, 39.

Fig. 15. Walter Watson (1872–1960) and an example (part of p. 59) of the arrangement of entries and vice-county distributions in his *Census Catalogue of British Lichens* published in 1953[2528].

lichen records by Watson were included in the 'Botany Reports' of the Devonshire Association's journal and he completed a manuscript on the lichens of that county in about 1954 which was destined never to appear[2529]. The name of Watson is perhaps most familiar to lichenologists today for his *Census Catalogue of British Lichens*[2528] (Fig. 15), published under the auspices of the British Mycological

Society in 1953 when he was 81. This listed each of the 1466 species he accepted and also gave its known distribution with reference to the vice-county system (see p. 175). Data for this work came from literature and herbarium sources, and from field observations of not only Watson himself but also his contemporaries. Watson's papers extended over 45 years (1909–54) but today he is generally regarded as not being sufficiently critical. However, this is mainly true only for his later works and he himself recognized his memory was failing after the publication of his *Census Catalogue*. To disagree with some of his taxonomic conclusions is not to belittle the role he played in keeping lichenology going in Britain at this time; he is undeniably one of the key figures in the history of British lichenology.

Frederick Archibald Sowter (1899–1972) and Arthur E. Wade, in contrast to Watson, were first introduced to lichens as boys in Leicester when they were taken out on excursions by A. R. Horwood (see p. 26) to various sites in the county. Sowter was first concerned mainly with flowering plants and later bryophytes and started to turn his attention to lichens in the early 1940s; he collected extensively and sent many records to Watson, of whom he was a close colleague, but published very little on British lichens apart from his detailed account of the Leicestershire and Rutland species which appeared in 1950[2243]. Wade, in contrast, collected lichens from about 1914 and his first publication with a lichen record is dated 1919[2426]; he was a friend of both Watson and Sowter, regularly visiting the latter (until Sowter's death in 1972) on visits to Leicester to see his sister. Although Wade contributed records to Sowter's Leicestershire lichen flora (especially from Rutland), he was employed in the Botany Department of the National Museum of Wales in Cardiff from 1920 until his retirement in 1961, and so was able to make important contributions to the botany of South Wales. His first major lichenological publication was with Watson in 1936 on the Glamorganshire lichen flora[2450]; he went on to prepare accounts of the species in Pembrokeshire in 1954[2431] and Carmarthenshire in 1958[2437] in addition to publishing many shorter notes and working on a supplement to Watson's *Census Catalogue* which was never published [2434].

Elke Mackenzie (formerly I. Mackenzie Lamb) worked on lichens at the British Museum (Natural History) during the years 1935–43 prior to taking up various posts abroad. In this short period Mackenzie, who is one of the world's leading lichenologists today, published an important series of taxonomic notes on British species in 1936–41[1269–1270, 1272–1274], and produced a most fascinating account of a visit to western Scotland in 1942[1275] which hinted at the richness of the lichen vegetation in an area that would not be surveyed thoroughly until the 1970s. In East Anglia Samuel A. Manning and Edward A. Ellis started to note lichens in 1938[1526] and 1934[669], respectively, as they continue to do today. Wilson (see p. 28) issued his *Flora of Westmorland* in 1938[2650] which included a considerable amount of information on the lichens of that county. The available data on Hertfordshire lichens was compiled by Peter H. Gregory in his study of the fungi in that county in 1951[829], and F. C. Rimington investigated the Scarborough district in 1953[1964]. Christopher A. Cheetham (1875–1954) contributed lichen records to the Yorkshire *Naturalist* in 1941–47.

Sir Arthur G. Tansley (1871–1955), founder of the British Ecological Society,

LICHEN STUDY GROUP

In order to promote the study of lichens in Britain a 'Lichen Study Group' is proposed. The object of the group will be to assist its members in all branches of lichenology and especially to help members to see as much lichen material as possible.

To this end I shall ask all interested botanists to send to me two named lichens with all taxonomic data such as chemical reactions, spore size, internal structure of the apothecium etc. together with all ecological data possible. This information should be given on a sheet of paper attached to the packet containing the specimen. Any other information that will help the student in the identification should be added and any relevant literature cited. Members will be encouraged to comment on the specimens but it is obvious that they cannot be allowed to dissect any specimens without the consent of the owner.

On receipt of members' specimens I shall pack them in a strong box and post them to the first name on the list of members who after examination will forward them to the next on the list. When all members have seen them they will be returned to their owners and a further two specimens requested. The length of time that members will be allowed to retain the specimens for study will depend on the number of members in the group. At first we will confine our material to British species (not necessarily British material as it would be interesting to see British species from non British localities) but later we could extend our studies to non British species. Dr. W. Watson (Taunton) has kindly consented to help the group in any way possible and if you have any suggestions I shall be pleased to receive them for consideration and action. It might be possible at some future date to arrange a meeting when we could all get together for discussion and field work.

I am sending this circular to all those botanists who I know are interested in the lichens but if you know of any people who might be pleased to join us I shall be glad to hear from you.

The question of exchange or 'distribution' of lichens has been considered but in the opinion of the writer this tends to 'over-collecting' which is not desirable. A 'Lichen News Sheet' will go out with each box of specimens and members will be invited to add to the sheet any lichen 'news' which will be of interest to other members in the group. I should like to hear from you as soon as possible so that we can get started with the scheme if you think the proposals are acceptable. Postage for the return of specimens will be appreciated.

F.A. Sowter,
9, North Avenue,
LEICESTER.

Fig. 16. Circular distributed by Frederick A. Sowter (1899–1972) in January 1953 launching the *Lichen Study Group*.

summarized the available information on the lichens of different habitats in his monumental *The British Islands and their Vegetation* of 1939[2335], the lists of Watson and other workers for particular sites often being reproduced in full. A particularly noteworthy ecological paper of this period dealing primarily with lichens was that of Eustace W. Jones' 1952[1208] study concerning species on tree trunks in the Midlands which showed clearly how the lichen flora on these decreased as areas were approached which were affected by air pollution; Watson sent Jones some of his own observations from Taunton for inclusion in this paper.

In 1948, Lilian E. Porter (née Baker, b. 1885, the daughter of the well-known botanist J. G. Baker, 1834–1920, see p. 20), whose lichenological papers extend over the period 1917–55, published a comprehensive supplement to Knowles' 1929 account of the Irish lichen flora. There were few other botanists interested in lichens in Ireland in this period apart from Alsager F.-G. Fenton who made contacts with the eminent Swedish lichenologists A. H. Magnusson (1885–1964) and G. Degelius; Fenton's Ph.D. thesis of 1953 was an account of the lichens of Northern Ireland, but much of these data remained unpublished until 1969.

In 1952 Professor Eilif Dahl, the well-known Norwegian ecologist and lichenologist, spent some time collecting lichens in many parts of the British Isles and issued a set of duplicated keys[545] for the identification of British lichens from the Botany School at Cambridge where he had been based. Critical notes on several species he discovered for the first time in Britain together with other important records were treated in a paper he published in 1953[546]; with the exception of Watson's *Usnea* notes of 1951[2527], this was the first major taxonomic (as opposed to floristic) work on British lichens to appear since 1942.

The tide was now beginning to turn and in January 1953 Sowter initiated a *Lichen Study Group* (Fig. 16) which circulated named specimens; it soon had 18 members including many later to rise to lichenological prominence such as U. K. Duncan, P. W. James, J. R. Laundon, G. Salisbury and G. D. Scott. Jack R. Laundon joined the staff of the British Museum (Natural History) in 1952; although he was not concerned principally with lichens there until 1961, he carried out studies on the lichen flora and vegetation of both his native county, Northamptonshire, and Bookham Commons in Surrey in 1953–58. Peter W. James, born in Sutton Coldfield, who met Sowter many times during a period of national service in Leicestershire, was appointed as lichenologist at the British Museum (Natural History) in 1955.

Post-1957: The British Lichen Society; popularity regained
By the end of 1957 the number of people interested in British lichens had risen from about five immediately after the end of World War II to almost 30. One person who had recently become interested was Dr T. D. V. Swinscow who, as he was walking through Borrowdale, Cumberland, on 10 October 1957, began to consider the possibilities of establishing a British Lichen Society[2320]. At his instigation a meeting was held at the British Museum (Natural History) on 1 February 1958; 24 people were present and it was agreed to establish a British Lichen Society. A. E. Wade was elected Secretary, P. W. James as Editor and Recorder, J. H. G. Peterken (1893–1973) as Treasurer, D. C. Smith as Librarian, and T. D. V. Swinscow

VOLUME I   1958   PART I

THE

LICHENOLOGIST

PUBLISHED

BY

THE BRITISH LICHEN SOCIETY

Copyright

Fig. 17. Title-page of the first issue of *The Lichenologist*, 1958.

as Curator and Assistant Editor. The first issues of a *Bulletin* and a journal, *The Lichenologist* (Fig. 17), appeared that same year and continue to thrive. The Officers, Council members and Referees of the Society from 1958 to 1976 are shown in Fig. 18.

From the beginning, the Society was conceived as one active in the field and not only in the herbarium; spring, summer and autumn meetings in various parts of the British Isles and Ireland were arranged, and reports of some of these have been published in *The Lichenologist*. Meetings have also been held abroad, in Brittany, Norway and Spain. A major difficulty facing the Society in its early days was the totally inadequate literature in English on lichen identification. In order to rectify this situation, the Society started to distribute Dahl's keys to macrolichens (see p. 33) and to issue keys reprinted from *The Lichenologist* to particular genera, for example *Alectoria*[2440], *Cladonia*[2329], *Collema*[2435], *Physcia* and *Anaptychia*[2442], *Ramalina*[2445], *Stereocaulon*[1229], *Umbilicaria*[1230] and *Usnea*[2330]. Notes on species new to the British flora and other taxonomic notes also started to appear in *The Lichenologist*. The lack of a comprehensive identification manual remained a problem and Ursula K. Duncan published a *Guide to the Study of Lichens* to help fill this gap in 1959[622]; a volume of illustrations to accompany this was issued in 1963[627]. The Field Studies Council also started to hold field courses on lichens in 1955 (the first led by A. E. Wade at Malham Tarn Field Centre), and this played an important role in improving accuracy in identification and making the subject more widely known; many of those who attended Kenneth L. Alvin's early courses at the Slapton Ley Field Centre, Devonshire, for example, are still very active in the subject. Alvin, in conjunction with Kenneth A. Kershaw, published a well-illustrated elementary text for identification in 1963[27] which could serve as a base from which to proceed to Duncan's more specialized book. Broadly-based reviews of the subject by Frank H. Brightman[283] and Frederick N. Haynes[975] also appeared, and parcels of specimens and journals were circulating amongst the Society's Study Group and Reading Circle, respectively. The Society's Referees (see Fig. 18), and the identification service of the British Museum (Natural History), also played, as they still do today, a vital role in maintaining standards of lichen identification in this country; several thousand specimens pass through their hands each year.

By 1964 lichenology in the British Isles had started to take on a new and re-vitalized complexion and confidence was growing. In that year Peter James completed his check-list of British lichens, which was issued in 1965[1133] and accepted some 1347 species (fewer than in Watson's *Census Catalogue* of 1953, see p. 30, as a result of a great deal of critical study). With an up-to-date check-list in the course of preparation and stimulated by progress in the 10×10 km grid square mapping of vascular plants and bryophytes in Britain, in September 1964 the Society decided to undertake a Society Distribution Maps Scheme under the direction of Mark R. D. Seaward. Some members of the Society were cautious about the value of such a scheme at the start but it has proved to be more successful than even the most optimistic members of Council in 1964 could possibly have forseen. The Scheme nevertheless had a slow start but in 1968, when the proofs of a new identification manual by Duncan (assisted by James) entitled *Introduction to British Lichens* became available, a new cross-off mapping record card was designed.

36

Alvin, K. L. Council 1959–60, 1962, 1965–66, 1968–69

Bailey, R. H. Council 1971–72, 1976
Bowen, H. J. M. Council 1972–73
Brightman, F. H. Council 1959–60, 1962–64, 1973–74; Curator 1964–70; Referee 1966–76; Vice-president 1976
Brown, D. H. Treasurer 1966–70; Librarian & Reading Circle Secretary 1971–76; Vice-president 1972–73; President 1974–75
Burnet, A. M. Council 1974–75

Catcheside, D. G. President 1959–61; Council 1963–66
Coppins, B. J. Council 1971–72, 1976; Referee 1971–76

Davey, S. R. Council 1976
Dobbs, C. G. Council 1961–62, 1964–65; Vice-president 1966–67; President 1968–69
Dobson, F. S. Council 1975–76
Duncan, U. K. Referee 1958–67

Earland-Bennett, P. M. Council 1973–74
Embrey, J. Council 1970–71

Farrar, J. F. Council 1971–72
Ferry, B. W. Council 1967–69
Fletcher, A. Council 1973–74

Gilbert, J. L. Council 1961–62
Gilbert, O. L. Council 1972–73; Vice-president 1974–75; President 1976; Referee 1975–76
Guiterman, J. D. Council 1966–67; Curator 1971–76

Hawksworth, D. L. Assistant Editor 1970–76; Referee 1970–76
Haynes, B. D. Council 1959–61
Haynes, F. N. Council 1959–60, 1962–64; Conservation Officer 1964–74; Chairman of Conservation Committee 1975–76
Hill, D. J. Council 1965–66, 1970; Referee 1966–76

James, P. W. Editor 1958–76; Recorder 1958–72; Vice-president 1968–69; President 1970–71; Referee 1958–76

Kershaw, K. A. Council 1961–62, 1967–69; Referee 1960–63, 1966–76

Lambley, P. W. Council 1974–75; Assistant Treasurer 1975–76
Laundon, J. R. Council 1960–61, 1963; Secretary 1964–76; Referee 1958–76
Lindsay, D. C. Council 1975

Manning, S. A. Council 1960–61; Referee 1960–64; Treasurer 1971–74
Milne, A. R. Council 1970
Morgan-Jones, G. Assistant Editor 1966–69; Referee 1966–68

Peterken, J. H. G. Treasurer 1958–64; President 1966–67

Ranwell, D. S. Council 1967–69
Richardson, D. H. S. Council 1966–67; Regional Treasurer (North America) 1971–74
Rose, F. Council 1970–71, 1975–76; Referee 1976

Salisbury, G. Referee 1975–76
Seaward, M. R. D. Mapping Recorder 1964–76
Sheard, J. W. Council 1963, 1965; Referee 1966–76
Side, A. G. Council 1968–69, 1972–73
Smith, D. C. Librarian 1958–70; Reading Circle Secretary 1966–70; Vice-president 1970–71; President 1972–73
Sowter, F. A. Council 1958–60; Referee 1958–64
Swinscow, T. D. V. Curator 1958–63; Assistant Editor 1958–66; Treasurer 1965; President 1962–63; Council 1970, 1974–75; Referee 1962–76

Tallis, J. H. Council 1959; Acting Librarian 1960
Tallowin, S. N. K. Treasurer 1975–76
Topham, P. B. Council 1970–71, 1976
Townsend, C. C. Council 1962–64

Wade, A. E. Secretary 1958–63; Referee 1958–76; President 1964–65
Wallace, E. C. Council 1958–59, 1966–69
Wallace, N. Council 1965–66; Referee 1966–71
Wilson, S. Council 1958–59

Fig. 18. Officers, Council members and Referees of the
British Lichen Society 1958–1976.

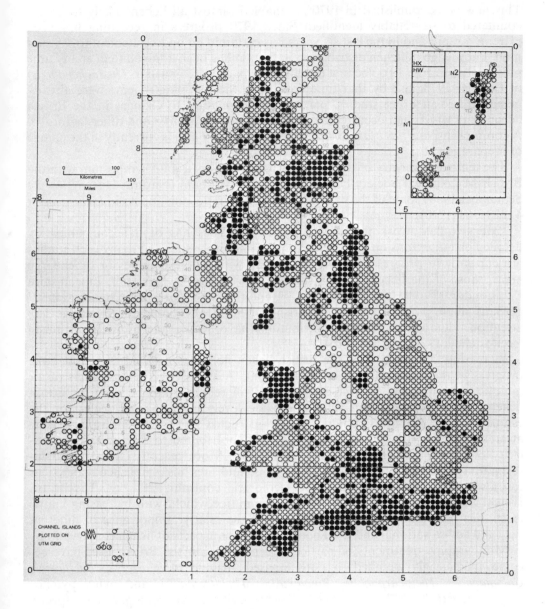

Fig. 19. British Lichen Society Distribution Maps Scheme, summary of numbers of species on record cards received by November 1976.

Cards have been received from 2740 10 × 10 km squares, 712 have over 100 records, 809 have 50–100 records and 1219 have under 50 records. A comparison of these data with the situation in April 1975 (*Lichenologist* **7**: 180, 1975) shows there to have been more than 23% and 36% increases in the areas covered and records accumulated respectively, with an average up-grading of records per square of over 10%. Solid circles = 100 species or more; open circles = less than 100 species.

This new book, published in 1970[630], enabled almost all lichens likely to be encountered to be reliably identified. Since 1970 progress in recording has been extremely rapid and in the last six years it is certainly no understatement to say that more knowledge of lichen distribution in Britain has been attained than at any time previously (see Fig. 19). The first distribution maps were issued in *The Lichenologist* in October 1973, but by that time data from the mapping scheme were already being utilized in other studies, particularly in relation to changes in the British lichen flora which had occurred as a result of air pollution[970] and other factors[963]. A comprehensive *Atlas of the British and Irish Lichen Flora* is currently in the course of preparation.

The tenth anniversary of the Society was marked by a joint symposium with the British Mycological Society held at the British Museum (Natural History) on 27 September 1968; some of the papers read at this symposium appear in *Lichenologist* **4**: 323–368 (1970).

Extensive field work throughout the British Isles has been accompanied by detailed investigations on particular counties and other sites and many such studies are currently in progress. Of surveys dealing with whole counties or other large areas, those of the lichens of Berkshire[265], Derbyshire[920], Dorset[269], Lincolnshire[2127], London[1309, 1311], the New Forest[2053], and the West Yorkshire Conurbation[2144] perhaps merit particular mention. Irish lichens have continued to receive less attention, although important contributions have been made by M. E. Mitchell (see Mitchell, 1971) and Seaward[2145,2148].

Increasing familiarity with the British lichen flora has resulted in considerable progress in the interpretation of distributions. A knowledge of the effects of air pollution in particular and the establishment of relationships between the extant lichen vegetation and mean sulphur dioxide levels have enabled scales for the estimation of levels of this pollutant to be constructed[781, 968–969, etc.]; an increasing awareness of this use of lichens has certainly contributed to the increased popularity of lichenology in the British Isles today. More recently, the detailed information now available from very many woodland sites has established that lichens can also act as valuable indicators of ecological continuity[2038, 2049]. The British Lichen Society also has a Conservation Committee which, with increasing knowledge, has been able to make assessments of the relative importance of different sites and advise national and local bodies on diverse aspects of lichen conservation.

Monographic treatments of particular genera within the British Isles have become increasingly detailed; groups revised include *Alectoria*[931], *Buellia*[2159], *Caloplaca*[2448], *Thelocarpon*[2071], *Rinodina*[2162], *Thelotrema*[2072], and many pyrenocarpous genera[2304, 2307, 2313, 2318, 2321–2322, etc.]. Critical studies on particular species and accounts of species additional to the British flora continue to be produced[1138, 1306, 1310, etc.] and, as with regional lists, further papers are currently being prepared. Keys for the identification of sterile crustaceous species on soil[1303] and bark and wood[1304] have also proved exceptionally useful, as have keys dealing with species in particular habitats, for example the marine and maritime[711–712] and shaded acid rock underhangs[1137].

The Society, in conjunction with the Systematics Association, organized a major international symposium at the University of Bristol in 1974 which was attended

Fig. 20. Officers and Council of the British Lichen Society, 1976.

Front row (seated, *left to right*): J. R. Laundon (Secretary), P. B. Topham, D. L. Hawksworth (Assistant Editor), S. N. K. Tallowin (Treasurer) and F. N. Haynes (Chairman, Conservation Committee). Back row (standing, *left to right*): M. R. D. Seaward (Mapping Recorder), F. H. Brightman (Vice-president), F. Rose, P. W. James (Editor), D. H. Brown (Librarian and Reading Circle Secretary), P. W. Lambley (Assistant Treasurer), R. H. Bailey and S. R. Davey. Insets (*top to bottom*): O. L. Gilbert (President), J. D. Guiterman (Curator), B. J. Coppins and F. S. Dobson.

by some 70 persons from 15 nations; this resulted in the publication in 1976 of a volume reviewing in depth many aspects of modern lichenology (*Lichenology: Progress and Problems*, D. H. Brown, D. L. Hawksworth and R. H. Bailey, eds.). In 1971 the Society had also played an important role in the organization of the First International Mycological Congress at Exeter, one meeting of which gave rise to the publication of a book, *Air Pollution and Lichens*, in 1973 (B. W. Ferry, M. S. Baddeley and D. L. Hawksworth, eds). That such lichenological gatherings could be organized in Britain was almost unthinkable a few years ago.

At the end of 1976, the British Lichen Society had 492 members (excluding subscribers), its greatest ever number. The Society is now more active than it ever has been in terms of studies in progress and in press; the latter include a major multi-author survey entitled *Lichen Ecology* (M. R. D. Seaward, ed.) scheduled for publication in late 1977. An analysis of the titles in this *Bibliography* reveals that 25% of them appeared in the ten years 1966–75 (Fig. 21). It is interesting to note a

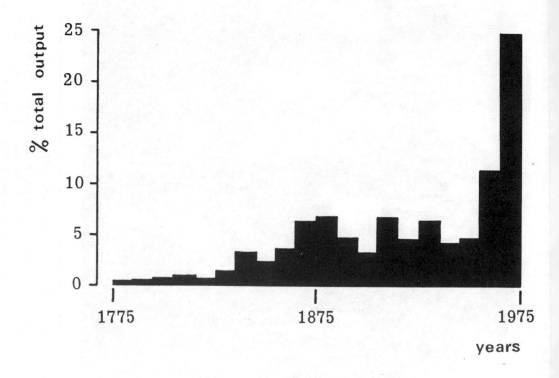

Fig. 21. Decadal output of publications up to the end of 1975 (excluding manuscripts and exsiccatae) relating to the British lichen flora.

Analysis made of 2483 entries in this *Bibliography* (pp. 46–174); those published prior to 1775 (i.e. 51) have been omitted from the chronological presentation but not the percentage calculation.

similar, but less dramatic, increase in output of lichenological publications during the period up to 1885, mainly due to the efforts of Crombie, Leighton, and their colleagues (see pp. 21–25).

As persons currently active in the Society, it is not for us to attempt to evaluate its role or the contributions of ourselves and our colleagues to the growth of lichenology in the British Isles; that task must fall to a future historian. We do, however, feel proud to have been involved in the growth of the Society, to have worked in such stimulating circumstances, and to have known some of those who kept lichenology in the British Isles alive in the lean years prior to the Society's formation.

BIBLIOGRAPHY

Perfection or near-perfection is impossible to achieve in this work: the words simply do not apply.

F. A. Stafleu, *Taxonomic Literature* (1967, p. x).

Scope

The aim of this *Bibliography* is to provide a comprehensive list of titles including references to British lichens which appeared before the end of 1975; also included are such titles published during 1976 as became available before this book went to press. For the purposes of this *Bibliography*, books, articles in journals, exsiccatae, theses and manuscripts are all interpreted as constituting 'titles'. Theses and manuscripts have, however, generally only been listed if copies are preserved in museum, university or other institutional libraries, or in the library of the British Lichen Society. Works relating solely to Ireland have been omitted in view of M. E. Mitchell's *A Bibliography of Books, Pamphlets and Articles relating to Irish Lichenology, 1727–1970* which was published in 1971, except for more recent papers and some omissions from Mitchell's book which have been indexed separately (p. 194).

Titles listed here have been located by a variety of methods. We have endeavoured to examine all titles thought to contain possible references to British lichens mentioned in the works of Culberson (1954 onwards), Grummann (1974), Mitchell (1971) and Watson (1953) as well as those in all other publications examined in the course of this study. These data have been supplemented by the following searches: (a) British floras in BM and K, (b) general natural history books and guidebooks in BM, (c) county and local histories in BM and the British Library Reference Division Reading Room, (d) all books and reprints in the lichen sectional libraries in BM and K (including bound opusculae), (e) reprints and books of the British Lichen Society, (f) our personal collections of reprints and books, (g) the card-index to manuscripts in LINN (in so far as it has been completed), (h) items in the Archives Room at K, (i) the index to manuscripts in BM (Sawyer, 1971), (j) entries under selected authors' names in the thesauri of Ciferri (1957–60) and Lindau and Sydow (1908–18), and (k) complete runs or very substantial parts of 152 journals (see pp. 44–46). Suggestions were also received from various colleagues to whom we sent a copy of our *Preliminary Bibliography* in April 1976.

Our original criterion for the inclusion of a title here was that at least one named lichen localized down to county level was mentioned. It soon became apparent that this would exclude many items containing pertinent information, and so in the final analysis titles including specifically British taxonomic or ecological observations were included even if no locality down to county level was provided; these have been enumerated in the Index under the heading 'General' (p. 194), but it should be noted that this category also includes some works which contain both localized and unlocalized records (such works also being indexed under the appropriate vice-counties). We also considered that it was desirable to list a few items largely, or only, of historical interest referred to in the preceding section of this book. Chemical, anatomical or physiological papers have consequently been cited only where the material used in them was localized.

Every item listed in this Bibliography has been checked in the original by ourselves, by David Reid, or by one of our colleagues, unless otherwise indicated in parentheses after the title.

All titles included were also scored for vice-counties covered with the exception of those listed as 'not abstracted' (p. 195), these being mainly collections of letters, or exsiccatae intact sets of which have not been studied. Important exceptions are Nos. 515, 1366, 1370, 1383, 2204, 2205 and 2210 which should be searched by **all** engaged in the preparation of regional lichen floras in Britain.

Lastly, while we have done our utmost in the time available to make this work as comprehensive as possible, we are conscious that there will enevitably be omissions and that in some cases references to particular vice-counties may have been overlooked. We would very much like to receive details of any omissions and corrections which users may find; these will be useful for any future revisions. Titles most likely to have been missed are (a) manuscripts in provincial libraries and museums, (b) university theses, and (c) local nineteenth century guide-books designed for tourist use (in the latter case we have in some instances been able to examine only a few of the numerous editions of some to be issued).

List of Journals scanned

The number of local British natural history society journals ideally requiring examination in their entirety is enormous. N. D. Simpson, in the introduction to his *Bibliographical Index of the British Flora* (1960), listed 616 such journals of which he had examined 204 thoroughly. To check many of the earlier journals is an extremely time-consuming and often unproductive task not only because they are almost invariably not (or not adequately) indexed, but also because locating complete sets of them can be an almost impossible task in itself. In order to obviate workers on regional studies having to re-examine journals which we have already checked either in their entirety or studied substantial parts of, we list these here. The years for which issues have been examined are indicated but for those still being published (indicated by an arrow) it should be noted that some of the 1976 issues of these had not been checked at the time of going to press.

Ann. Anderson. Nat. Soc. 1893–1914.
Ann. Mag. nat. Hist. 1841–90.
Ann. nat. Hist. 1838–40.
Ann. Scot. nat. Hist. 1892–1911.

Beds. Nat. 1946 →.
Biol. Conserv. 1968 →.
Biol. J. Linn. Soc. 1969 →.
Blthca Lichenol., Lehre 1974 →.
Bot. J. Linn. Soc. 1969 →.
Bradford Sci. J. 1904–12.
Bryologist 1898 →.
Bull. Br. Lichen Soc. 1958 →.

Bull. Br. mycol. Soc. 1967 →.
Bull. Kent Fld Club 1956 →.

Devonia 1895–96.

Edinb. J. nat. geogr. Sci. 1829–31.
Edinb. New phil. J. 1826–64.
Edinb. phil. J. 1819–26.
Envir. Poll. 1970 →.
Essex Nat. 1887 →.

Falcon 1887–91.
Fld Nat. 1954–69.
Fld Stud. 1959 →.

Gdnrs' Chron. 1841–77.
Grevillea 1872–94.

Halifax Nat. 1896–1904.
Hastings E. Suss. Nat. 1906 →.
Herzogia 1968 →.
Hist. Berwicksh. Nat. Club 1831 →.

Ir. nat. J. 1970 →.

J. anim. Ecol. 1932 →.
J. appl. Ecol. 1964 →.
J. Biogeogr. 1974 →.
J. Bot., Lond. 1834–42, 1863–1942.
J. Bryol. 1972 →.
J. Cairngorm Club 1896–1953.
J. Camborne-Redruth nat. Hist. Soc. 1963–74.
J. Cheltenham Distr. Nat. Soc. 1948–56.
J. Derbysh. archaeol. nat. Hist. Soc. 1879–1960.
J. Ecol. 1913 →.
J. Gloucs. Nat. Soc. 1974 →.
J. Ipswich Distr. Fld Club 1908–21.
J. Ipswich Distr. nat. Hist. Soc. 1925–35.
J. Linn. Soc. (Bot.) 1867–1968.
J. Proc. Linn. Soc. 1855–62.
J. Ruislip Distr. nat. Hist. Soc. 1953–73.
J. Scott. Mountaineering Club 1891–1959.
J. Scunthorpe Mus. Soc. 1964 →.
J. Torquay nat. Hist. Soc. 1909–22.
Jl N. Gloucs. Nat. Soc. 1957–73.

Lancs. Chesh. Nat. 1914–25.
Lancs. Nat. 1907–14.
Lichen 1972 →.
Lichenologist 1958 →.
Lond. Nat. 1921 →.

Midland Nat. 1878–93.
Miscnea bryol. lichen., Nichinan 1955 →.
Mycol. Pap. 1925 →.
Mycologia 1909 →.
Myxotaxon 1974 →.

Nat. Hist. Book Rev. 1976 →.
Nat. Sci. in Schools 1963 →.
Naturalist, Hull 1864 →.
Naturalist [F. O. Morris] 1851–58.
Nature Cambs. 1958 →.
Nature Lancs. 1970–74.
New Phytol. 1902 →.
Newsl. Int. Ass. Lichenol. 1967 →.
Norw. J. Bot. 1971 →.
Nov. Sist. Nizsh. Rast. 1964 →.
Nova Hedwigia 1959 →.
Nova Hedwigia, Beih. 1962 →.
NWest. Nat. 1926–55.

Persoonia 1959 →.
Phil. Trans. R. Soc. 1665–1807.
Phytologist 1841–63.
Proc. bot. Soc. Br. Isl. 1954–69.
Proc. bot. Soc. Edinb. 1836–44.
Proc. Cleveland Nat. Fld Club 1889–1909.
Proc. Cotteswold Nat. Fld Club 1847 →.
Proc. Coventry Distr. nat. Hist. scient. Soc. 1948 →.

Proc. Coventry nat. Hist. scient. Soc. 1915–17, 1930–40.
Proc. Isle Wight nat. Hist. archaeol. Soc. 1920 →.
Proc. Leeds phil. lit. Soc., sci. sect. 1925 →.
Proc. Linn. Soc. Lond. 1838–1968.
Proc. lit. phil. Soc. Manchr 1857–77.
Proc. Llandudno Distr. Fld Club 1909–50.
Proc. Manchr lit. & phil. Soc. 1877–87.
Proc. R. Ir. Acad. 1970 →.
Proc. Somerset. archaeol. nat. Hist. Soc. 1849 →.
Proc. Trans. Beds. nat. Hist Soc. Fld Club 1877–85.
Publs Hartley bot. Labs Lpool Univ. 1924–39.

Rabenh. Krypt.-Fl. **8–9** 1930–60.
Rep. Alvecote Pools Nat. Res. 1960 →.
Rep. Bardsey Bird Fld Obs. 1953–68.
Rep. botl Loc. Rec. Club 1875–87.
Rep. botl Soc. Exch. Club Br. Isl. 1928–47.
Rep. Eaton Coll. nat. Hist. Soc. 1930–63.
Rep. Lichen Exch. Club 1909–13.
Rep. Lundy Fld Soc. 1947 →.
Rep. Marlboro. Coll. nat. Hist. Soc. 1865–1965.
Rep. nat. Sci. archaeol. Soc. Littlehampton 1924–38.
Rep. Soc. guernes. 1922 →.
Rep. Trans. Bgham nat. Hist. & microsc. Soc. 1869–87.
Rep. Trans. Devon. Ass. Advmt Sci. 1862 →.
Rep. Trans. Glasgow Soc. Fld Nat. 1872–78.
Rep. Trans. Guernsey Soc. nat. Sci. 1882–1921.
Rep. Trans. N. Staffs. Fld Club 1887–1915.
Revue bryol. 1874–1931.
Revue bryol. lichén. 1932 →.

Sch. Nat. Study 1906–24, 1954–62.
Sci. Gossip 1865–1902.
Scot. bot. Rev. 1912.
Scott. J. Sci. 1965 →.
Scott. Nat. 1871–91, 1912–64.

Taxon 1951 →.
Trans. a. Rep. N. Staffs. Fld Club 1915–60.
Trans. Aberd. wkg Men's nat. Hist. scient. Soc. 1901–16.
Trans. bot. Soc. Edinb. 1839–1890.
Trans. Br. bryol. Soc. 1946–71.
Trans. Br. mycol. Soc. 1896 →.
Trans. Burton-on-Trent nat. Hist. archaeol. Soc. 1889–1933.
Trans. Carlisle nat. Hist. Soc. 1909 →.
Trans. Cumberland Ass. Advanc. Sci. 1875–93.
Trans. Edinb. Fld Nat. microsc. Soc. 1886–1915.
Trans. Edinb. nat. Fld Club 1881–86.
Trans. Epping Forest Essex Nat. Fld Cl. 1880–82.
Trans. Essex Fld Club 1883–87.
Trans. Herts. nat. Hist. Soc. Fld Club 1879 →.
Trans. Hull sci. Fld nat. Club 1898–1919.
Trans. J. Proc. Dumfries. Galloway nat. hist. antiq. Soc. 1862 →.
Trans. Leeds Nat. Cl. scient. Ass. 1886–92.
Trans. Leicester lit. phil. Soc. 1835 →.
Trans. Lincs. Nat. Un. 1893 →.
Trans. Linn. Soc. Lond. 1791–1940.
Trans. Malvern Nat. Fld Club 1855–79.

Trans. nat. Hist. antiq. Soc. Penzance 1845–65.
Trans. nat. Hist. Soc. Northumb. 1868–1913.
Trans. Norfolk Norwich Nat. Soc. 1869 →.
Trans. Penzance nat. Hist. antiq. Soc. 1880–98.
Trans. Proc. bot. Soc. Edinb. 1890 →.
Trans. Proc. Perthsh. Soc. nat. Sci. 1886 →.
Trans. (Proc.) Torquay nat. Hist. Soc. 1922 →.
Trans. Suffolk Nat. Soc. 1929 →.
Trans. Yorks. Nat. Un. 1877–1946.

W. Nat. 1972–74.
Watsonia 1949 →.
Wesley Nat. 1887–89.
Wilts. archaeol. nat. Hist. Mag. 1854–91.
Wimbledon & Merton Annual 1903–10.

Yorks. Nat. Rec. 1872–73.
Young Nat. 1879–90.

Bibliography

All titles are listed alphabetically by author, and all co-authors, editors, etc. are cross-referenced. Dates, authors' names, or other items placed in squared parentheses ([ ]) do not appear, at least in that form, on the work itself. Abbreviations for the location of manuscripts and annotated copies of books follow those employed for herbaria in this book (pp. 199–200) with the addition of BL (British Library Reference Division, Great Russell Street, London WC1 3DG). In the case of books with extremely long titles, only the main titles have been given except in a few instances where the subtitles will prove of value in making their contents clearer; full citations for such works are generally readily available in other bibliographic compilations (see pp. 201–203). Journal titles have generally been abbreviated according to the *World List of Scientific Periodicals* (1964) and British Museum (Natural History) (1975). Entries have been numbered to facilitate cross-referencing and indexing; entry numbers followed by 'a' or 'b' were inserted at a late stage of preparation of this book after the index had been produced.

1 **Abbayes, H. des** (1939) Revision monographique des *Cladonia* du sous-genre *Cladina* (Lichens). *Bull. Soc. scient. Bretagne* **16**(2): 1–156.

2 **Abbot, C.** [179–] *Catalogus plantarum in comitatu Bedfordiae sponte crescentium*. Manuscript in LINN.

3 —— (1795) *Plantae Bedfordiensis*. Manuscript in LINN.

4 —— (1798) *Flora Bedfordiensis*. Bedford: W. Smith. [Lichens pp. 256–270; a copy annotated by Abbot is now in Luton Museum.]
—— —See also No. 843.

5 **Acharius, E.** (1803) *Methodus qua omnes detectos Lichenes*. Stockholm.

6 —— (1810) *Lichenographia Universalis*. Göttingen.

7 —— (1814) Monographie der Lichen-Gattung *Pyrenula* mit Abbildungen aller bisher bekannten Arten. *Ges. nat. freunde Mag. Berlin* **6** (1812): 4–28. [Original coloured drawings in BM.]

8 —— (1814) *Synopsis methodica Lichenum*. Lund.

9 **Acton, E.** (1909) *Botrydina vulgaris*, Brébisson, a primitive lichen. *Ann. Bot.* **23**: 579–585.

10 **Adam, P., Birks, H. J. B., Huntley, B. and Prentice, I. C.** (1975) Phytosociological studies at Malham Tarn moss and fen, Yorkshire, England. *Vegetatio* **30**: 117–132.

11 **Adams, G. M.** (1974) *Lichen zonation at Portishead*. B.Sc. thesis, University of Bristol.

11a **Agnew, A. D. Q., Bywater, J., Kershaw, K. A. and Smithson, S.** (1956) A survey of the vegetation of Bardsey Island. *Rep. Bardsey Bird Fld Obs.* **4**: 29–46.

Ahmadjian, V.—See Nos. 788, 1219 and 1890.

12 **Ahti, T.** (1961) Taxonomic studies on reindeer lichens (*Cladonia*, subgenus *Cladina*). *Suomal. eläin-ja Kasvit.Seur.van.Kasvit.Julk.* **32**(1): 1–160.

13 —— (1965) Some notes on British *Cladoniae*. *Lichenologist* **3**: 84–88.

14 —— (1965) Notes on the distribution of *Lecanora conizaeoides*. *Lichenologist* **3**: 91–92.

15 —— (1966) *Parmelia olivacea* and the allied non-isidiate and non-sorediate corticolous lichens in the Northern Hemisphere. *Acta bot. fenn.* **70**: 1–68.

16 —— (1966) Correlation of the chemical and morphological characters in *Cladonia chlorophaea* and allied lichens. *Ann. bot. fenn.* **3**: 380–390.

17 —— (1973) Taxonomic notes on some species of *Cladonia*, subsect. *Unciales*. *Ann. bot. fenn.* **10**: 163–184.

18 —— **and Henssen, A.** (1965) New localities for *Cavernularia hultenii* in Eastern and Western North America. *Bryologist* **68**: 85–89.

19 **Ainsworth, W.** (1827) Sketch of the physical geography of the Malvern Hills. *Edinb. New phil. J.* **4**: 91–100.

20 **Allan, J.** (1877) 25th July, 1876. Excursion. Ben Lawers. *Trans. Glasgow Soc. Fld Nat.* **5**: 184.

21 **Allott, C.** (1971) Report on the lowland vegetation of Foula. August 1971. *Brathay Fld Stud. Rept* **11**: 48–57.

22 **Almborn, O.** (1948) Distribution and ecology of some south Scandinavian lichens. *Bot. Notiser, Suppl.*, **1**(2): 1–252.

23 **Almquist, S.** (1880) Monographia Arthoniarum Scandinaviae. *K. svenska VetenskAkad Handl.* **17**(6): 1–69.

24 **Alvin, K. L.** (1959) Lichens of Hatfield Forest. *Essex Nat.* **30**: 166–169.

25 —— (1960) Observations on the lichen ecology of South Haven Peninsula, Studland Heath, Dorset. *J. Ecol.* **48**: 331–339.

26 —— (1961) Skippers Island papers (3). Lichens of Skippers Island. *Essex Nat.* **30**: 330–335.

27 —— **and Kershaw, K. A.** (1963) *The Observer's Book of Lichens*. London and New York: F. Warne.

28 **Amphlett, J. and Rea, C.** (1909) *The Botany of Worcestershire*. Birmingham: Cornish Brothers. [Lichens pp. 488–501.]

29 **Anderson, G. and Anderson, P.** (1834) *Guide to the Highlands and Islands of Scotland, including Orkney and Zetland*. London: J. Murray.

Anderson, P.—See No. 29.

30 **Anon.** (1824) Botanical excursions to the Scottish mountains in June and July 1824. *Edinb. phil. J.* **11**: 413–415.

31 —— (1828) *The New Harrogate Guide*. Ed. 5. Harrogate: Langdale's Library. [Lichen p. 150.]

32 —— (1830) Botanical tours. *Edinb. J. nat. geogr. Sci.* **3**: 56–57.

33 —— (1830) Dyeing. In *The Edinburgh Encyclopedia* (D. Brewster, ed.) **8**:

207–283. Edinburgh: W. Blackwood.

34 —— (1833) *The Tourist's Companion; being a concise Description and History of Ripon*. Ed. 6. Ripon: E. Langdale. [Lichens pp. 148–152.]

35 —— [1836] Contributions to the flora of Berwickshire. *Hist. Berwicksh. Nat. Club* **1**: 72–73.

36 —— (1841) Analytical notice of no. 54 of the 'Supplement to the English Botany' of Sir J. E. Smith and Mr Sowerby. *Phytologist* **1**: 27–29.

37 —— (1841) *A Guide to Ilfracombe and the neighbouring towns*. New edition. Ilfracombe: J. Banfield.

38 —— (1844) *A series of Select Views in Perthshire*. London, Edinburgh & Dublin: A. Fullarton. [Lichens p. lxxii.]

39 —— [c. 1850] *A Guide to North Devon*. Ilfracombe.

40 —— (1851) Names of plants. *Gdnrs' Fmrs' J., n.s.* **6**: 455.

41 —— (1853) *The Hand Book of Western Cornwall. Penzance, Falmouth, and Neighbourhoods*. Exeter: H. Besley and Son. [Extracted from the *Route Book of Cornwall*.] [Lichens pp. xxiv-xxvii.]

41a —— (1855) Todmorden Botanical Society. *Hebden Bridge Record* **1855** (December): 2.

41b —— (1856) Todmorden Botanical Society. *Hebden Bridge Chronicle* **1856** (May): 2.

42 —— (1856) Proceedings of the Botanical Society for February, 1856. *Scott. Gdnr* **5**: 99–100.

43 —— (1858) *A Handbook for Travellers in Kent and Sussex*. London: John Murray. [Lichens p. 255.]

44 —— (1860) Lichenes. In *The North Devon Hand-Book*. Ed. 2. (G. Tugwell, ed.): 257–258. Ilfracombe: John Banfield.

45 —— (1862) Malvern botany. *Phytologist, n.s.* **6**: 182–187.

46 —— (1864) Correspondence. *Bot. Chron.* **8**: 64.

47 —— (1865) Lichens in Ireland. *Bot. Chron.* **15**: 115.

48 —— (1874) 13th May, 1873. Specimens exhibited. *Trans. Glasgow Soc. Fld Nat.* **2**: 28. [Lichens by J. Stirton.]

49 —— (1874) 27th May, 1873. Excursion. *Trans. Glasgow Soc. Fld Nat.* **2**: 29.

50 —— (1875) 2nd June, 1874. Specimens exhibited. *Trans. Glasgow Soc. Fld Nat.* **3**: 63–64. [Lichens by J. Stirton.]

51 —— (1875) 6th October, 1874. Specimens exhibited. *Trans. Glasgow Soc. Fld Nat.* **3**: 72–73. [Lichens by J. Stirton.]

52 —— (1878) Severn Valley Naturalists' Field Club. *Midland Nat.* **1**: 258–259.

53 —— (1890) Field meeting. 26th May, 1888. St. Peter's, St. Albans. *Trans. Herts. nat. Hist. Soc. Fld Club* **5**: xxi–xxii.

54 —— (1890) Field meeting, 16th June, 1888. Broxbourne and Hoddesdon. *Trans. Herts. nat. Hist. Soc. Fld. Club* **5**: xxv.

55 —— (1890) Field meeting, 19th October, 1889. Bricket Wood. *Trans. Herts. nat. Hist. Soc. Fld Club* **5**: xlvii–xlviii.

56 —— (1892) Field meeting, 8th October, 1890. Hatfield Park. *Trans. Herts. nat. Hist. Soc. Fld Club* **6**: xxxviii–xxxix. [Lichens by G. Massee.]

57 —— (1892) Birmingham Natural History and Microscopical Society. *Midl.*

Nat. **15**: 164–165. [Lichen by T. Clarke.]

58 —— (1895) Transactions of the Society. Monthly meeting held on January 10th, 1894. *Rep. Trans. Guernsey Soc. nat. Sci.* **2**: 315.

59 —— [1905] Reports of the meetings of the Berwickshire Naturalists' Club for 1904. Kielder, Northumberland. *Hist. Berwicksh. Nat. Club* **19**: 117–122.

60 —— (1910) Additional records of Scottish cryptogamic plants, 1909. *A. Conf. crypt. Soc. Scotl.* **34–35**: 12–16.

61 —— (1910) Liverpool Botanical Society. *Lancs. Nat.* **3**: 155.

62 —— (1910) Liverpool Botanical Society. *Lancs. Nat.* **3**: 245–246.

63 —— (1910) Liverpool Botanical Society. *Lancs. Nat.* **3**: 291–292.

64 —— (1911) Liverpool Botanical Society. *Lancs. Nat.* **4**: 199–200.

65 —— (1913) Hale Point in May. *Lancs. Nat.* **6**: 73–74.

66 —— (1914) Liverpool Botanical Society. *Lancs. Chesh. Nat.* **7**: 57–59.

67 —— (1914) Liverpool Botanical Society. *Lancs. Chesh. Nat.* **7**: 98.

68 —— (1915) Liverpool Botanical Society. *Lancs. Chesh. Nat.* **8**: 31.

69 —— (1916) Liverpool Botanical Society. *Lancs. Chesh. Nat.* **9**: 56.

70 —— (1916) Liverpool Botanical Society. *Lancs. Chesh. Nat.* **9**: 153.

71 —— (1918) Liverpool Botanical Society. *Lancs. Chesh. Nat.* **11**: 59–60.

72 —— (1918) Botany—Sectional Report. *Trans. a. Rep. N. Staffs. Fld Club* **52**: 121–126. [Lichens by H. P. Reader.]

73 —— (1919) Liverpool Botanical Society. *Lancs. Chesh. Nat.* **12**: 27–29.

74 —— (1920) Liverpool Botanical Society—Field Meetings. *Lancs. Chesh. Nat.* **13**: 38–39.

75 —— (1922) The botanical section. *Proc. Somerset archaeol. nat. Hist. Soc.* **68** (1): lxxvii–lxxxii.

76 —— (1923) Micro-fungi on lichens. *Lancs. Chesh. Nat.* **15**: 215.

77 —— (1924) Liverpool Botanical Society. *Lancs. Chesh. Nat.* **16**: 237–239.

78 —— (1924) Liverpool Botanical Society. *Lancs. Chesh. Nat.* **17**: 42–43.

79 —— (1924) The botanical section. *Proc. Somerset archaeol. nat. Hist. Soc.* **70** (1): lxi–lxix.

80 —— (1926) The botanical section. *Proc. Somerset archaeol. nat. Hist. Soc.* **72** (1): lxxii–lxxvii.

81 —— (1927) [Report]. *A. Conf. crypt. Soc. Scotl.* **48**: 1–9.

82 —— (1929) The botanical section. *Proc. Somerset archaeol. nat. Hist. Soc.* **75** (1): lvi–lxii.

83 —— (1930) The botanical section. *Proc. Somerset archaeol. nat. Hist. Soc.* **76** (1): lxx–lxxv.

84 —— (1931) List of species gathered during the foray. *A. Conf. crypt. Soc. Scotl.* **50**: 5–10.

85 —— (1931) The botanical section. *Proc. Somerset archaeol. nat. Hist. Soc.* **77** (1): lxxxix–xcviii.

86 —— (1932) The botanical section. *Proc. Somerset archaeol. nat. Hist. Soc.* **78** (1): lxv–lxx.

87 —— (1933) The botanical section. *Proc. Somerset archaeol. nat. Hist. Soc.* **79** (1): lxxxv–xcii.

50

88 —— (1934) The botanical section. *Proc. Somerset archaeol. nat. Hist. Soc.* **80** (1): lxvi–lxxiv.

89 —— (1935) The botanical section. *Proc. Somerset archaeol. nat. Hist. Soc.* **81**: 250–255.

90 —— (1936) The botanical section. *Proc. Somerset archaeol. nat. Hist. Soc.* **82**: 230–235.

91 —— (1938) The botanical section. *Proc. Somerset archaeol. nat. Hist. Soc.* **84**: 162–168.

92 —— (1939) The botanical section. *Proc. Somerset archaeol. nat. Hist. Soc.* **85**: 231–237.

93 —— (1948) Land flora. *Rep. Lundy Fld Soc.* **2**: 34–38. [Lichens by W. A. Gliddon.]

94 —— (1949) Miscellaneous notes—terrestrial and marine. *Rep. Lundy Fld Soc.* **3**: 29–31. [Lichens by W. A. Gliddon.]

94a —— (1958) Liverworts, mosses and lichens. *Fld Nat.* **3**: 42–43.

95 —— (1960) Reports of field meetings held during 1959. May 3rd. To Dungeness. *Bull. Kent Fld Club* **5**: 8.

96 —— (1960) Reports of field meetings held during 1959. October 11th—Cranbrook area. *Bull. Kent Fld Club* **5**: 14–16.

97 —— (1965) Lichens killed by artificial fertiliser. *Bull. Br. Lichen Soc.* **16**: 5.

98 —— (1967) Field meetings, 1966. 2nd July, Ebbor Gorge. *Proc. Somerset archaeol. nat. Hist. Soc.* **111**: 76–77.

99 —— (1969) No lichen secrets here. *Dartington Hall News* **42** (2543): 5.

100 —— (1972) Elm epiphytes in danger. *Bull. Br. Lichen Soc.* **31**: 10.

101 —— (1973) Spring field meeting at Bristol, 1974. *Bull. Br. Lichen Soc.* **33**: 4–5.

102 —— (1976) Caister lichen 'find'. *Eastern Daily Press* no. 32695 (15 March 1976): 3.

103 **Anstead, D. T. and Latham, R. G.** (1862) *The Channel Islands.* London: W. H. Allen. [Lichens pp. 187–190 by C. Larbalestier.]

104 —— —— (1893) *The Channel Islands.* Ed. 3 [Revised E. T. Nicolle.]. London: W. H. Allen. [Lichens pp. 151–152.]

Anthony, J.—See No. 1227.

105 **Armitage, E.** (1947) Notes on plants mentioned in "Collections towards the History and Antiquities of the County of Hereford" by John Duncumb, A.M., (Vol. 1. Hereford, 1804). *Trans. Woolhope Nat. Fld Club* **1942–45**: 271–275. [Lichens partly by H. H. Knight.]

106 **Armstrong, P.** (1975) *The Changing Landscape.* Lavenham: T. Dalton.

107 **Armstrong, R. A.** (1973) Seasonal growth and growth rate-colony size relationships in six species of saxicolous lichens. *New Phytol.* **72**: 1023–1030.

108 —— (1974) *The structure and dynamics of saxicolous lichen communities.* D. Phil. thesis, University of Oxford.

109 —— (1974) The descriptive ecology of saxicolous lichens in an area of south Merionethshire, Wales. *J. Ecol.* **62**: 33–45.

110 —— (1974) Growth phases in the life of a lichen thallus. *New Phytol.* **73**: 913–918.

111 —— (1976) Studies on the growth rate of lichens. In *Lichenology: Progress and Problems* (D. H. Brown, D. L. Hawksworth and R. H. Bailey, eds.): 309–322. [Systematics Association Special Volume No. 8.] London, New York and San Francisco: Academic Press.

111a —— (1976) The influence of the frequency of wetting and drying on the radial growth of three saxicolous lichens in the field. *New Phytol.* **77**: 719–724.

Armstrong, W.—See No. 253.

112 **Arnold, F. C. G.** (1859–1900) *Lichenes exsiccati*. Nos. 1–1816. Eichstadt; Munich. [Complete set not available to us for indexing; includes a few British specimens e.g. 1204, 1205.]

113 —— (1861) Lichenes britannici exsiccati. Herausgegeben von Rev. W. A. Leighton, nach Massalongo's System zusammengestellt. *Flora, Jena* **44**: 435–443, 465–472, 497–507, 534–539, 657–671, 673–679, 697–704, 721–723, 755–756.

114 —— (1863) Leighton Lichenes britannici exsiccati. *Flora, Jena* **46**: 325–330.

115 —— (1867) Lichenologische Fragmente. *Flora, Jena* **50**: 129–143.

116 **Arnold, [F. H.]** (1882) Ordinary meeting, August 9th, 1881. *Rep. Chichester nat. Hist. microsc. Soc.* [Meetings individually paged.]

117 —— (1905) Botany. In *The Victoria History of the Counties of England, Sussex* (W. Page, ed.) **1**: 41–69. London: Constable. [Lichens pp. 64–66.]

118 **Arnold, G. A. and Arnold, M. A.** (1965) *Alvecote Pools, Warwickshire. West Midlands Trust Nature Reserve. Sixth Annual Report for the year ending 31st December* 1964. Birmingham: West Midlands Trust. [Lichens pp. 9–10.]

119 —— —— (1973) *Alvecote Pools Nature Reserve. Fourteenth Annual Report for the year ending 31st December* 1972. Warwick: Warwickshire Nature Conservation Trust. [Lichens pp. 9–12.]

120 —— —— (1976) *Alvecote Pools Nature Reserve. Seventeenth Annual Report for the year ending 31st December* 1975. Warwick: Warwickshire Nature Conservation Trust.

Arnold, M. A.—See Nos. 118, 119 and 120.

121 **Asprey, G. F.** (1976) *Lepraria chlorina* in Wales. *Lichenologist* **8**: 97.

Auden, G. A.—See No. 2637.

122 **Aveling, J. H.** (1870) *The History of Roche Abbey, from its Foundation to its Dissolution.* Worksop: R. White. [Lichens p. 185 by J. Bohler.]

123 **Ayres, P. B.** (1843) List of the cryptogamic plants of Oxfordshire. *Phytologist* **1**: 661–664, 702–704.

124 **[Babington, C. ?]** (18—) *Remarks on British lichens and fungi.* Manuscript in LINN.

125 **Babington, C.** (1839) Remarks on British lichens and fungi, principally on species or varieties new to our flora. *Proc. Linn. Soc. Lond.* **1**: 32.

—— —See also No. 1853.

126 **Babington, C. C.** (1839) *Primitiae Florae Sarnicae, or an Outline of the Flora*

of the Channel Islands. London: Longman. [Lichens pp. 120–124 by F. C. Lukis.]

—— —See also Nos. 194 and 195.

127 **Baddeley, M. S., Ferry, B. W. and Finegan, E. J.** (1971) A new method of measuring lichen respiration: response of selected species to temperature, pH and sulphur dioxide. *Lichenologist* **5**: 18–25.

128 —— —— —— (1972) The effects of sulphur dioxide on lichen respiration. *Lichenologist* **5**: 283–291.

129 —— —— —— (1973) Sulphur dioxide and respiration in lichens. In *Air Pollution and Lichens* (B. W. Ferry, M. S. Baddeley and D. L. Hawksworth, eds.): 299–313. London: Athlone Press of the University of London.

—— —See also Nos. 432, 703, 787, 941, 970, 1139, 1315, 1622, 1673 and 2037.

130 **Bagnall, J. E.** (1879) The cryptogamic flora of Warwickshire. *Midl. Nat.* **2**: 220–224.

131 —— (1886) A half-day's ramble in the Arrow District. *Midl. Nat.* **9**: 117–119.

132 —— (1886) The mosses, hepatics and lichens. In *Handbook of Birmingham* (W. Mathews, ed.): 331–334. Birmingham: Hall & English.

133 —— (1891) *The Flora of Warwickshire.* London: Gurney and Jackson. [Lichens pp. 385–386.]

134 —— (1901) Lichenes (Lichens). In *The Victoria History of the Counties of England, Worcestershire* (J. W. Willis-Bund and H. A. Doubleday, eds.) **1**: 67–68. Westminster: Constable.

135 —— (1904) Botany. In *The Victoria History of the Counties of England, Warwickshire* (H. A. Doubleday and W. Page, eds.) **1**: 33–66. London: Constable. [Lichens pp. 57–58.]

136 —— (1908) Botany. In *The Victoria History of the Counties of England, Staffordshire* (W. Page, ed.) **1**: 41–76. London: Constable. [Lichens pp. 65–68.]

137 **Bailey, R. H.** (1963) The lichens of the Ruislip Local Nature Reserve. *J. Ruislip Distr. nat. Hist. Soc.* **12**: 18–21.

138 —— (1964) *Studies on the dispersal of lichen propagules.* M.Sc. thesis, University of London.

139 —— (1966) Studies on the dispersal of lichen soredia. *J. Linn. Soc. (Bot.)* **59**: 479–490.

140 —— (1967) Notes on Gloucestershire lichens—1. *Jl N. Gloucs. Nat. Soc.* **18**: 154–156.

141 —— (1967) Notes on Gloucestershire lichens—2. *Jl N. Gloucs. Nat. Soc.* **18**: 178–180.

142 —— (1968) Lichens of the Brassey Nature Reserve—a preliminary report. In *Brassey Nature Reserve*: sine pagin. Stonehouse: Gloucestershire Trust for Nature Conservation.

143 —— (1968) Notes on Gloucestershire lichens—3. *Jl N. Gloucs. Nat. Soc.* **19**: 284–286.

144 —— (1969) Notes on Gloucestershire lichens—4. *Jl N. Gloucs. Nat. Soc.* **20**: 19–20.

145 —— (1969) Notes on Gloucestershire lichens—5. The lichens of the Badgeworth Nature Reserve. *Jl N. Gloucs. Nat. Soc.* **20**: 127–129.

146 —— (1970) Notes on Gloucestershire lichens—6. *Jl N. Gloucs. Nat. Soc.* **21**: 174–175.

147 —— (1970) Notes on Gloucestershire lichens—7. *Jl N. Gloucs. Nat. Soc.* **21**: 223–225.

148 —— (1970) Notes on Gloucestershire lichens—8. *Jl N. Gloucs. Nat. Soc.* **21**: 246–247.

149 —— (1970) Newtown and Wigpool, F. of D., 8th August. *Jl N. Gloucs. Nat. Soc.* **21**: 251–252.

150 —— (1970) Animals and the dispersal of soredia from *Lecanora conizaeoides* Nyl. ex Cromb. *Lichenologist* **4**: 256.

151 —— (1971) Notes on Gloucestershire lichens—9. Maritime lichens in Gloucestershire. *Jl N. Gloucs. Nat. Soc.* **22**: 426–428.

152 —— (1971) Some lichens from northern Spain. *Revue bryol. lichén.* **37**: 983–986.

153 —— (1971) The British Lichen Society meeting at Cheltenham. *Rep. N. Gloucester. nat. Soc.* **7**: 31–34.

154 —— (1971) Some lichens from the Olchon Valley, Herefordshire. *Trans. Woolhope Nat. Fld Club* **39**: 463–466.

155 —— (1973) Some Irish lichen records. *Ir. Nat. J.* **17**: 392–394.

156 —— (1974) Notes on Gloucestershire lichens—10. Lichens on glass. *J. Gloucs. Nat. Soc.* **25**: 374–375.

157 —— (1974) Distribution maps of lichens in Britain. Map 8. *Cyphelium inquinans* (Sm.) Trev. *Lichenologist* **6**: 169–171.

158 —— (1975) Lichens become mobile in the motor age. *Bull. Br. Lichen Soc.* **36**: 10.

159 —— (1975) Lichens from the Skellig rocks. *Ir. Nat. J.* **18**: 163–164.

160 —— (1975) *Report on a survey of the lichen vegetation likely to be affected by proposed extension to gravel workings at Dungeness, Kent.* Manuscript. Report to the Nature Conservancy; copy with BLS.

161 —— (1976) Ecological aspects of dispersal and establishment in lichens. In *Lichenology: Progress and Problems* (D. H. Brown, D. L. Hawksworth and R. H. Bailey, eds.): 215–247. [Systematics Association Special Volume No. 8.] London, New York and San Francisco: Academic Press.

162 —— (1976) *Report on the lichen flora of Cleevely Wood, Gloucestershire.* Manuscript. Report to the Nature Conservancy; copy in Gloucester Museum.

163 —— (1976) *Report on the lichen flora of Dowdeswell Reservoir Nature Reserve, Cheltenham, Gloucestershire.* Manuscript. Report to the Gloucestershire Trust for Nature Conservation; copy in Gloucester Museum.

164 —— (1976) *Report on the lichen flora of Five Acre Grove Nature Reserve, Gloucestershire.* Manuscript. Report to the Gloucestershire Trust for Nature Conservation; copy in Gloucester Museum.

165 —— (1976) *Report on the lichen flora of Lassington Wood Nature Reserve, Gloucestershire*. Manuscript. Report to the Gloucestershire Trust for Nature Conservation; copy in Gloucester Museum.

166 —— (1976) *Notes on some lichens at Poor's Allotment Nature Reserve, Gloucestershire*. Manuscript. Report to the Gloucestershire Trust for Nature Conservation; copy in Gloucester Museum.

167 —— **and Stott, P. A.** (1973) A contribution to the lichen flora of the Wirral Peninsula, Cheshire. *Naturalist, Hull* **1973**: 101–105.

168 —— **and Taylor, A. M.** (1972) The lichen *Umbilicaria pustulata* (L.) Hoffm. in Gloucestershire. *Jl N. Gloucs. nat. Soc.* **23**: 127–128.

—— —See also Nos. 111, 161, 331, 433, 703, 984, 1144, 1673, 2049, 2051 and 2146.

169 **Bainbridge, F.** (1843) Note on a new British lichen. *Phytologist* **1**: 616.

170 **Baird, A.** [1834] Address read at the second anniversary meeting of the Berwickshire Naturalists' Club, held at Dunse, September 18, 1833. *Hist. Berwicksh. Nat. Club* **1**: 12–18.

171 **Baker, J. G.** [18—] [*Correspondence*]. Letters in K (1 vol.). [Not abstracted.]

172 —— (1854) Contributions to British lichenology. *Phytologist* **5**: 191–194.

173 —— (1855) Two days in Wensleydale. *Naturalist* (F. O. Morris) **5**: 121–127.

174 —— (1857) Descriptions of new British lichens. *Phytologist, n.s.* **2**: 106–109.

175 —— (1863) *North Yorkshire; studies of its botany, geology, climate and physical geography*. London: Longman.

176 —— (1907) Botany. In *The Victoria History of the Counties of England, Yorkshire* (W. Page, ed.) **1**: 111–172. London: Constable. [Lichens pp. 171–172.]

Balderston, B. M.—See No. 177.

177 **Balderston, R. R. and Balderston, B. M.** (1888) *Ingleton: Bygone and Present*. London: Simpkin & Marshall. [Lichens pp. 193–196.]

178 **Balfour, I. B.** (1900) Eighteenth century records of Scottish plants. *Ann. Scot. nat. Hist.* **1900**: 169–174.

179 —— (1900) Eighteenth century records of Scottish plants. *Ann. Scot. nat. Hist.* **1900**: 237–243.

180 —— (1901) Eighteenth century records of Scottish plants. *Ann. Scot. nat. Hist.* **1901**: 37–48.

181 **Balfour, J. H.** (1845) Account of a botanical excursion to Ailsa Crag [*sic*], in July 1844. *Phytologist* **2**: 257–263.

182 —— (1848) Botany of the Bass. In *The Bass Rock* (T. M'Crie, ed.): 411–431. Edinburgh: W. P. Kennedy.

183 —— (1855) Account of a botanical excursion to the Braemar mountains in August, 1854. *Proc. bot. Soc. Edinb.* **1855**: 3–7.

184 —— (1855) Notes on the flora of the Bass Rock. *Proc. bot. Soc. Edinb.* **1855**: 30–32.

185 —— (1858) Short account of a botanical trip in the Island of Arran, with pupils, in 1857. *Trans. bot. Soc. Edinb.* **6**: 3–6.

186 [——] (1863) Notice of a botanical excursion to Kielder and Deadwater Fell, Northumberland, on 4th July 1863. *Trans. bot. Soc. Edinb.* **7**: 581–582.

187 —— [assisted by J. Sadler] (1863) *Flora of Edinburgh*. Edinburgh: Adam and Charles Black. [Lichens by Sadler pp. 165–169; lichens unchanged in Ed. 2 (1871); in E there is an interleaved copy of Ed. 1 annotated by by Sadler and labelled "third edition" in his hand.]

188 —— (1868) Notice of plants collected during recent excursions with pupils. *Trans. bot. Soc. Edinb.* **9**: 480–481.

189 —— (1870) Account of botanical excursions in the Island of Arran during August and September 1869. *Trans. bot. Soc. Edinb.* **10**: 355–365.

190 —— (1873) Botanical excursions in July and August 1870, with pupils. *Trans. bot. Soc. Edinb.* **11**: 68–75.

191 —— (1873) Notes of a botanical excursion to the Breadalbane Mountains, in July 1871. *Trans. bot. Soc. Edinb.* **11**: 353–356.

192 —— (1873) Notice of botanical excursions made in 1872 and 1873 (No. 1). *Trans. bot. Soc. Edinb.* **11**: 511–516.

193 [——] (1876) Notice of botanical excursions made to different parts of Scotland in 1875. *Trans. bot. Soc. Edinb.* **12**: 448–450.

194 —— **and Babington, C. C.** (1844) Account of a botanical excursion to Skye and the Outer Hebrides; during the month of August 1841. *Trans. bot. Soc. Edinb.* **1**: 133–144.

195 —— —— (1844) A catalogue of the plants gathered in the islands of North Uist, Harris and Lewis during a botanical excursion in the month of August, 1841. *Trans. bot. Soc. Edinb.* **1**: 145–154.

Ball, D. F.—See No. 2575.

196 **Ballantine, W. J.** (1961) A biologically-defined exposure scale for the comparative description of rocky shores. *Fld Stud.* **1** (3): 1–19.

197 **Bamford, E.** [1871] *The complete guide to Dovedale, Ashbourn, and Ilam*. Ashbourne: E. Bamford. [Lichens p. 101.]

198 **Bannister, P.** (1965) Biological Flora of the British Isles: *Erica cinerea* L. *J. Ecol.* **53**: 527–542.

199 —— (1966) Biological Flora of the British Isles: *Erica tetralix* L. *J. Ecol.* **54**: 795–813.

200 **Banwell, A. D.** (1949) *Lophozia silvicola* new to Great Britain. *Trans. Br. bryol. Soc.* **1**: 194–198.

201 **Barber, D.,** ed. (1970) *Farming and Wildlife. A Study in Compromise*. Wisbech and London: R.S.P.B. [Lichens p. 89.]

202 **Barclay-Estrup, P.** (1966) *Interpretation of cyclical processes in a Scottish heath community*. Ph.D. thesis, University of Aberdeen.

203 **Barkham, J. P.** (1971) A report on the upland vegetation of Foula—August 1968. *Brathay Fld Stud. Rept* **11**: 25–47.

204 **Barkley, S. Y.** (1953) The vegetation of the Island of Soay, Inner Hebrides. *Trans. Proc. bot. Soc. Edinb.* **36**: 119–131.

205 **Barkman, J. J.** (1958) *Phytosociology and Ecology of Cryptogamic Epiphytes*. Assen, Netherlands: Van Gorcum. [Re-issued 1969.]

206 —— **Rose, F. and Westhoff, V.** (1969) Discussion in Section 5: The effects of air pollution on non-vascular plants. In *Air Pollution. Proceedings of the First European Congress on the Influence of Air Pollution on*

Plants and Animals, Wageningen 1968: 237–241. Wageningen: Centre for Agricultural Publishing and Documentation.

Barlow, F.—See No. 1867.

Barnes, J. A. G.—See No. 779.

207 **Barrett, J. H.** (1973) Life on the seashore. In *Pembrokeshire Coast* (D. Miles, ed.): 30–35. [National Park Guide no. 10.] London: H.M.S.O.

208 —— **and Yonge, C. M.** (1958) *Collins Pocket Guide to the Sea Shore.* London: Collins. [Also revised edition, 1964.]

209 **Barry, G.** (1805) *The History of the Orkney Islands.* Edinburgh: privately printed. [Lichens p. 280.]

210 **Bastin, H.** (1955) *Plants without Flowers.* London: Hutchinson. [Lichens pp. 40–47.]

211 **Bates, J. W.** (1975) A quantitative investigation of the saxicolous bryophyte and lichen vegetation of Cape Clear Island, County Cork. *J. Ecol.* **63**: 143–162.

212 **Batten, L.** (1921) Note on the occurrence of *Arthopyrenia foveolata* at Plymouth. *J. mar. biol. Ass. U.K., n.s.* **12**: 557.

213 **Baxter, W.** (1825, 1828) *Stirpes cryptogamae Oxoniensis.* 2 fasc., nos. 1–100. Oxford. [Exsiccata.]

Beattie, E. P.—See No. 1959.

Beckett, P. J.—See No. 1618.

214 **Beddows, A. R.** (1959) Biological Flora of the British Isles: *Dactylis glomerata* L. *J. Ecol.* **47**: 223–239.

215 **Bednar, T. W. and Smith, D. C.** (1966) Studies in the physiology of lichens. VI. Preliminary studies of photosynthesis and carbohydrate metabolism of the lichen *Xanthoria aureola. New Phytol.* **65**: 211–220.

216 **Bell, J. N. B. and Tallis, J. H.** (1973) Biological Flora of the British Isles: *Empetrum nigrum* L. *J. Ecol.* **61**: 289–305.

217 **Bellamy, D. J.** (1970) The vegetation. In *Durham County and City with Teesside* (J. C. Dewdney, ed.): 133–141. Durham: British Association.

218 **Bellerby, W.** (1907) Natural history of Thorne Waste. *Naturalist, Hull* **1907**: 316–324.

Benson-Evans, K.—See No. 2351.

219 **Berkeley, M. J.** (1856) [Letter.] *Gdnrs' Chron.* **1856**: 84.

220 —— (1856) [Letter.] *Gdnrs' Chron.* **1856**: 172.

221 —— (1857) *Introduction to Cryptogamic Botany.* London, New York, Paris, Madrid: H. Bailliere. [Lichens pp. 372–420.]

222 —— **and Broome, C. E.** (1852) Notices of British fungi. *Ann. Mag. nat. Hist., ser.* 2, **9**: 377–387.

223 —— —— (1859) Notes on British fungi. *Ann. Mag. nat. Hist., ser.* 3, **3**: 356–377.

224 **Berkenhout, J.** (1770) *Outlines of the Natural History of Great Britain and Ireland.* Vol. 2. London: P. Elmsly. [Lichens pp. 311–319.]

225 —— (1789) *Synopsis of the Natural History of Great Britain and Ireland* [being a second edition of *The Outlines*]. Vol. 2. London: T. Cadell. [Lichens pp. 328–341.]

226 **Bescoby, B. E.** (1971) Lichens. In *Holden Clough, The Natural History of a small Lancashire Valley* (L. N. Kidd, M. G. Fitton and D. C. McRae, eds.): 66. Oldham: Oldham Public Libraries, Art Gallery & Museum.

227 [**Bevan, R. J.**] (1975) *Lichen Survey of Air Pollution*. South Yorkshire County Council, Environment Department.

228 —— **and Greenhalgh, G. N.** (1976) *Rhytisma acerinum* as a biological indicator of pollution. *Envir. Poll.* **10**: 271–285.

—— —See also No. 2420a.

229 **Bevis, J. F. and Griffin, W. H.,** eds. (1909) Botany. The flora of Woolwich and West Kent. In *A Survey and Records of Woolwich and West Kent* (C. H. Grinling, T. A. Ingram and B. C. Polkinghorne, eds.): 31–230. Woolwich. [Lichens pp. 220–224; see also No. 842.]

230 **Bingley, K.** (1816) *Useful Knowledge*. Vol. 2. London: Baldwin, Cradock & Joy. [Lichens pp. 297–300.]

Birks, H. H.—See No. 232.

231 **Birks, H. J. B.** (1973) *Past and Present Vegetation of the Isle of Skye. A Palaeoecological Study*. London: Cambridge University Press.

232 —— **and Birks, H. H.** (1975) Studies on the bryophyte flora and vegetation of the Isle of Skye. I. Flora. *J. bryol.* **8**: 19–64.

—— —See also No. 10.

233 **Birks, T.** (1879) Goole Scientific Society. *Naturalist, Hull* **4**: 187–188.

Birnie, A. C.—See No. 2325.

234 **Birse, E. L.** (1974) Bioclimatic characteristics of Shetland. In *The Natural Environment of Shetland* (R. Goodier, ed.): 24–32. Edinburgh: Nature Conservancy Council.

—— —See also No. 1896.

235 **Birse, E. M.** (1958) Ecological studies on growth-form in bryophytes. III. The relationship between the growth-form of mosses and ground-water supply. *J. Ecol.* **46**: 9–27.

236 **Bishop, J. A., Cook, L. M., Muggleton, J. and Seaward, M. R. D.** (1975) Moths, lichens and air pollution along a transect from Manchester to North Wales. *J. appl. Ecol.* **12**: 83–98.

237 **Bitter, G.** (1901) Zur Morphologie und Systematik von *Parmelia*, Untergattung *Hypogymnia. Hedwigia* **40**: 171–274.

Blackler, M.—See No. 1322.

238 **Blackstone, J.** (1746) *Specimen Botanicum quo plantarum plurium rariorum Angliae indiginarum*. London. [Lichens pp. 45–46, 55–56.]

Blackwood, W.—See No. 33.

239 **Blight, J. T.** (1861) *A Week at the Land's End*. London: Longman etc.

240 —— (1876) *A Week at the Land's End*. [Ed. ?] Truro: Lake & Lake.

241 **Block, W.** (1966) Seasonal fluctuations and distribution of mite populations in moorland soils, with a note on biomass. *J. anim. Ecol.* **35**: 487–503.

242 **Bloom, J. H. and Frohawk, W.** (1924) *Hand List of Ferns, Mosses, Lichens, Etc. collected at Mickleham in Surrey*. Balham: privately printed.

243 **Bloomfield, E. N.** (1898) *The Natural History of Hastings and St. Leonards and the*

vicinity. Third Supplementary List. St Leonards-on-Sea: privately printed.

244 —— (1905) Lichens of Norfolk and Suffolk. *Trans. Norfolk Norwich Nat. Soc.*
8: 117–137.

245 —— (1908) Annual notes on the local fauna, flora etc. *Hastings E. Suss. Nat.*
1: 124–129.

246 —— (1910) *Lecanora mougeotioides* Schaer. in Britain. *J. Bot., Lond.* 48: 141.

247 —— (1911) Lichenes (Lichens). In *The Victoria History of the Counties of
England, Suffolk* (W. Page, ed.) 1: 79–81. London: Constable.

248 —— (1911) Annual notes on the local fauna and flora [lichens]. *Hastings E.
Suss. Nat.* 1: 309.

249 —— (1914) Annual notes on the local fauna and flora, etc., for 1913, lichens.
Hastings E. Suss. Nat. 2: 103.

—— —See also Nos. 351 and 352.

250 **Bloxam, A.** (18—) *Correspondence.* Letters in BM (1 vol.). [Not abstracted.]

251 [——] [*c.* 1854] *Leicestershire cellular cryptogamia.* Manuscript in BM. [See
also Hawksworth No. 950.]

252 —— (1860) Lichens. In *The Isle of Wight* (E. Venables *et al.*): 503–505.
London: E. Stanford. [With addenda to list by T. Salwey.]

—— —See also No. 1853.

253 **Boatman, D. J. and Armstrong, W.** (1968) A bog type in north-west Sutherland. *J. Ecol.* 56: 129–141.

254 **Bøcher, T. W.** (1954) Studies on European calcareous fixed dune communities. *Vegetatio* 6: 562–570.

255 **Bohler, J.** (1835–37) *Lichenes Britannici.* 16 fasc., nos. 1–128. Sheffield: G.
Ridge. [Includes some drawings and notes by R. Deakin; copy of
entire text with D. L. Hawksworth.]

—— —See also Nos. 122 and 1083.

255a **Bolam, G.** (1913) *Wild Life in Wales.* London: L. Palmer.

256 **Bolton, E. M.** (1960) *Lichens for Vegetable Dyeing.* London: Studio Books.

257 **Bolton, J. H.** (1788–91) *The History of the Fungusses growing about Halifax.*
4 vols. Huddersfield: Privately printed.

—— —See also No. 2477.

258 **Bond, G. and Scott, G. D.** (1955) An examination of some symbiotic systems
for fixation of nitrogen. *Ann. Bot., n.s.,* 19: 67–77.

259 **Boney, A. D.** (1961) A note on the intertidal lichen *Lichina pygmaea* Ag. *J.
mar. biol. Ass. U.K.* 41: 123–126.

260 **Borrer, W.** (1804–61) [*Correspondence.*] Letters in K (26 vols), LINN,
Trinity Coll. (Cambridge), Bangor, OXF, etc. [Incl. field and other
notes. See Fleming, *Schedule of William Borrer correspondence,*
manuscript in LINN.] [Not abstracted.]

261 —— (1846) Notices of north of England plants. *Phytologist* 2: 424–437.

—— —See also Nos. 427, 2220 and 2390.

262 **Boulger, G. S.** (1919) The cryptogams of Andrew's herbarium. *J. Bot., Lond.*
57: 337–340.

263 **Bouly de Lesdain, M.** (1906) Notes lichénologiques, V. *Bull. Soc. bot. Fr.* 53:
515–519.

264 **Bousfield, J. and Peat, A.** (1976) The ultrastructure of *Collema tenax* with particular reference to microtubule-like inclusions and vesicle production. *New Phytol.* **76**: 121–128.

265 **Bowen, H. J. M.** (1968) *The Flora of Berkshire*. Oxford: Privately printed. [Lichens pp. 70–77.]

266 —— (1970) Air pollution and its effects on plants. In *The Flora of a Changing Britain* (F. [H.] Perring, ed.): 119–127. [B.S.B.I. Conference Reports no. 11.] Hampton, Middx.: E. W. Classey.

267 —— (1970) Determination of sulphate ion by replacement of iodate in iodine-131 labelled barium iodate. *Analyst* **95**: 665–667.

268 —— (1970) Lichens in the Thames Valley. *Reading Nat.* **22**: 29–31.

269 —— (1976) The lichen flora of Dorset. *Lichenologist* **8**: 1–33.

270 —— **Rose, F. and Mason, J.** (1972) [*List of species from Wychwood, Oxon 1970 and* 1972]. Manuscript. Report to the Nature Conservancy; copy with BLS.

271 **Boyd, D. A.** (1904) West Kilbride and the North Ayrshire coast. *Trans. Edinb. Fld Nat. microsc. Soc.* **5**: 108–116.

272 —— (1905) Largs and its surroundings. *Trans Edinb. Fld Nat. microsc. Soc.* **5**: 208–218.

273 —— (1910) List of fungi, 1909. *A. Conf. crypt. Soc. Scotl.* **34–5**: 5–11.

274 —— (1911) List of fungi, 1911 *A. Conf. crypt. Soc. Scotl.* **36**: 5–12.

275 —— (1911) Additonal records of Scottish cryptogamic plants, 1911. *A. Conf. crypt. Soc. Scotl.* **36**: 13–16.

276 —— (1916) *Sphinctrina turbinata* (Pers.) Fr. *Glasg. Nat.* **8**: 63–64.

277 **Bradshaw, M. E.** (1970) The Teesdale flora. In *Durham County and City with Teesdale* (J. C. Dewdney, ed.): 141–152. Durham: British Association.

278 **Bramley, W. G.** (1950) The spring foray at Masham. *Naturalist, Hull* **1950**: 161–164.

279 —— (1951) The spring foray at Hebden Bridge. *Naturalist, Hull* **1951**: 201.

280 **Brebner, [J.]** (1882) Lichenes. In *The Flora of Arbroath and its Neighbourhood* (Arbroath Horticultural and Natural History Association, ed.): 50–51. Arbroath: T. Buncle.

281 —— (1912) The flora of Forfarshire. In *Handbook and Guide to Dundee and District* (A. W. Paton and A. H. Millar, eds.): 597–610. Dundee: British Association.

Brent, F.—See No. 2057.

282 **Brewer, S.** (1726–27) *Botanical journey through Wales in the year* 1726 [*and* 1727]. Manuscript in BM.
—— —See also No. 1114.

Brewster, D.—See Nos. 33 and 1650.

Brewster, J.—See No. 1030.

283 **Brightman, F. H.** (1959) Neglected plants – lichens. *New Biology* **29**: 75–94.

284 —— (1959) Some factors influencing lichen growth in towns. *Lichenologist* **1**: 104–108.

285 —— (1959) Lichen weekend at Flatford. *Lichenologist* **1**: 119.

286 —— (1959) *West Sutherland, V.C.* 108. Manuscript in BM. [Kept with Wade

(1957–59).]

287 —— (1960) Lichen notes. *Bull. Kent Fld Club* **5**: 26–28.

288 —— (1960) Field meeting at Flatford. *Lichenologist* **1**: 203–206.

289 —— (1961) Lichen notes. *Bull. Kent Fld Club* **6**: 7–8.

290 —— (1961) Reports of field meetings held during 1960. September 25th—Joyden's Wood. *Bull. Kent Fld Club* **6**: 33–44.

291 —— (1962) Lichen notes. *Bull. Kent Fld Club* **7**: 8.

292 —— (1962) Reports of field meetings held during 1961. April 15th—Ightham. *Bull. Kent Fld Club* **7**: 19–20.

293 —— (1962) Reports of field meetings held during 1961. October 1st—Angley Wood, Cranbrook. *Bull. Kent Fld Club* **7**: 44.

294 —— (1962) Field meeting at Arnside. *Lichenologist* **2**: 97–100.

295 —— (1963) Reports of recorders. Lichens. *Bull. Kent Fld Club* **8**: 6–7.

296 —— (1964) Reports of recorders. Lichen notes. *Bull. Kent Fld Club* **9**: 7.

297 —— (1964) Reports of field meetings held during 1963. April 21st—'wall tour' from Maidstone. *Bull. Kent Fld Club* **9**: 16.

298 —— (1964) Reports of field meetings held during 1963. October 26th-27th: Lichen Society meeting. *Bull. Kent Fld Club* **9**: 37.

299 —— (1964) *Cyphelium notarisii* in Britain. *Lichenologist* **2**: 283–284.

300 —— (1965) Lichens. *Bull. Br. mycol. Soc.* [old series] **24**: 11–13.

301 —— (1965) Reports of recorders. Lichens. *Bull. Kent Fld Club* **10**: 7.

302 —— (1965) Reports of field meetings held during 1964. 3rd May—'wall tour' in Tonbridge area. *Bull. Kent Fld Club* **10**: 10.

303 —— (1965) Insect on lichens. *Lichenologist* **3**: 154.

304 —— (1965) Field meeting at Folkestone. *Lichenologist* **3**: 173–174.

305 —— (1965) The lichens of Cambridge walls. *Nature Cambs.* **8**: 45–50.

306 —— (1965) Some patterns of distribution of lichens in Southern England. *SEast Nat.* **69**: 10–17.

307 —— (1966) Reports of recorders. Lichen notes. *Bull. Kent Fld Club* **11**: 8.

308 —— (1966) Reports of field meetings held in 1965. May 15th—'wall tour' from Tunbridge Wells. *Bull. Kent Fld Club* **11**: 13–14.

309 —— (1967) Reports of recorders. Lichen notes. *Bull. Kent Fld Club* **12**: 7–8.

310 —— (1967) Reports of field meetings held in 1966. 7th May—'wall tour' from Chilham. *Bull. Kent Fld Club* **12**: 9–10.

311 —— (1968) Reports of recorders. Lichen notes. *Bull. Kent Fld Club* **13**: 7–8.

312 —— (1969) Reports of recorders. Lichens. *Bull. Kent Fld Club* **14**: 6–7.

313 —— (1969) Reports of field meetings. 19th May—'wall tour' and 22nd June—'tree tour'. *Bull. Kent Fld Club* **14**: 24.

314 —— (1970) Reports of recorders. Lichens. *Bull. Kent Fld Club* **15**: 7–9.

315 —— (1970) Field meetings. 11th May—'wall tour' from Rye. *Bull. Kent Fld Club* **15**: 19.

316 —— (1971) Reports of recorders—1970. Lichens. *Bull. Kent Fld Club* **16**: 7–8.

317 —— (1971) Field meetings. 10th May—'wall tour' from Birchington. *Bull. Kent Fld Club* **16**: 20.

318 —— (1974) Reports of field meetings. 29th April—wall tour at Tunbridge

Wells. *Bull. Kent Fld Club* **19**: 12–13.

319 —— (1975) Reports of field meetings. 5th May—wall tour at Canterbury. *Bull. Kent Fld Club* **20**: 5–6.

320 —— (1976) Reports of field meetings 1975. 4th May—wall tour at Folkestone. *Bull. Kent Fld Club* **21**: 7.

321 —— **James, P. W. and Rose, F.** (1973) Distribution maps of lichens in Britain. Map 6. *Parmelia acetabulum* (Neck.) Duby. *Lichenologist* **5**: 476–477.

—— —See also Nos. 1659 and 2184.

322 **Britten, J.** (1893) Gilbert White's Selborne plants. *J. Bot., Lond.* **31**: 289–294.

323 [——] (1908) *Cladonia luteoalba. J. Bot., Lond.* **47**: 324.

Broad, K.—See No. 1586.

324 **Broadhead, E.** (1958) The psocid fauna of larch trees in northern England— an ecological study of mixed species populations exploiting a common resource. *J. anim. Ecol.* **27**: 217–263.

325 —— **and Thornton, I. W. B.** (1955) An ecological study of three closely related psocid species. *Oikos* **6**: 1–50.

326 **Brokenshire, F. A.** (1945) Flowerless plants. In *The Natural History of Ilfracombe*: 51–52. Ilfracombe: Chronicle Press.

—— —See Nos. 571, 572, 573, 1547, 1548, 2260 and 2261.

Broome, C. E.—See also Nos. 222, 223.

Brown, A. H. F.—See No. 363.

327 **Brown, D. H.** (1972) The effect of Kuwait crude oil and a solvent emulsifier on the metabolism of the marine lichen *Lichina pygmaea. Mar. Biol. Berlin* **12**: 309–315.

328 —— (1973) The lichen flora of the lead mines at Charterhouse, Mendip Hills. *Proc. Bristol Nat. Soc.* **32**: 267–274.

329 —— (1973) Toxicity studies on the components of an oil-spill emulsifier using *Lichina pygmaea* and *Xanthoria parietina. Mar. Biol. Berlin* **18**: 291–297.

330 —— (1974) Field and laboratory studies on detergent damage to lichens at the Lizard, Cornwall. *Cornish Stud.* **2**: 33–40.

331 —— (1976) Mineral uptake by lichens. In *Lichenology: Progress and Problems* (D. H. Brown, D.L. Hawksworth and R. H. Bailey, eds.): 419–439. [Systematics Association Special Volume No. 8.] London, New York and San Francisco: Academic Press.

332 —— **and Brown, R. M.** (1968) The lichens of Blakeney Point, Norfolk. *Lichenologist* **4**: 1–15.

333 —— —— (1969) Lichen communities at Blakeney Point, Norfolk. *Trans. Norfolk Norwich Nat. Soc.* **21**: 235–250.

334 —— **and DiMeo, J. A.** (1972) Influence of local maritime conditions on the distribution of two epiphytic lichens. *Lichenologist* **5**: 305–310.

335 —— **and Slingsby, D. R.** (1972) The cellular location of lead and potassium in the lichen *Cladonia rangiformis* (L.) Hoffm. *New Phytol.* **71**: 297–305.

—— —See also Nos. 111, 161, 331, 433, 703, 984, 1144, 2049 and 2146.

336 **Brown, E.** (1863) The fauna and flora of the district surrounding Tutbury and

Burton-on-Trent. In *The Natural History of Tutbury* (O. Mosley): 85–406. London. [Lichens pp. 317–327.]

337 **Brown, R.** (1793) *Observations made in a Botanical journey to the Highlands of Scotland in the year* 1793. Manuscript in BM; copy in E.

338 **Brown, R.** (1860) Sketches of Caithness and its botany, with a list of the phanerogamous plants and ferns. *Trans. bot. Soc. Edinb.* **6**: 328–329.

Brown, R. M. [née Cox]—See Nos. 332, 333 and 441.

339 **[Brown, T.]** (1834) List of plants discovered within the district, since the publication of Dr. Johnston's Flora of Berwick-upon-Tweed. *Hist. Berwicksh. Nat. Club* **1**: 31–32, 57, 62–63.

340 **Bryce, J.** (1845) Parish of Ardrossan. In *The New Statistical Account of Scotland*. Vol. 5. *Ayr*: 191–210. Edinburgh and London: W. Blackwood & Sons. [Botany by D. Landsborough.]

341 **Buck, G. W.** (1975) *An investigation into the use of various plants as indicators of metallic lead in the substrate soil, on the site of the ruins of Magpie Mine, Derbyshire and on industrial archaeological interpretation of the measured lead levels.* B.Sc. thesis, University of Salford. [Copy with D. H. Brown.]

342 **Bullock, P.** (1971) The soils of the Malham Tarn area. *Fld Stud.* **3**: 381–408.

343 **Bundy, H.** (1880) Do lichens injure forest trees? *Gdnrs' Fmrs' J., n.s.* **5**: 741.

Bunting, W.—See No. 1817.

344 **Burges, A.** (1951) The ecology of the Cairngorms. Part III. The *Empetrum-Vaccinium* zone. *J. Ecol.* **39**: 271–284.

345 **Burkitt, A., Lester, P. and Nickless, G.** (1972) Distribution of heavy metals in the vicinity of an industrial complex. *Nature, Lond.* **238**: 327–328.

346 **Burn, R.** (1942) A new Suffolk lichen. *Trans. Suffolk Nat. Soc.* **5**: 26.

Burn, R.—See No. 1662.

Burnett, J. H.—See Nos. 808, 809, 1236, 1502, 1503, 1504, 1505, 1506, 1911 and 1912.

347 **Burnip, M., Graham, G. G. and Moses, P.,** eds. (1975) *Witton-le-Wear Nature Reserve. Plant list with habitats.* Durham: Durham County Conservation Trust. [Lichens pp. 25–26 by G. G. Graham, J. F. Skinner and F. B. Stubbs.]

348 **Burns, R. and MacNair, R.** (1845) Town and parishes of Paisley. In *The New Statistical Account of Scotland*. Vol. 7. *Renfrew*: 135–306. Edinburgh and London: W. Blackwood & Sons. [Botany by A. R. Young.]

349 **Burrell, W. H.** (1926) Rare lichen in Yorks. *Naturalist, Hull* **1926**: 287.

350 **Burrows, R.** (1971) *The Naturalist in Devon and Cornwall.* Newton Abbot: David & Charles.

351 **Butler, E. A. and Bloomfield, E. N.** (1878) *The Natural History of Hastings and St. Leonards.* Hastings. [Lichens pp. 55–56.]

352 —— —— (1883) *The Natural History of Hastings and St. Leonards. First Supplement.* Hastings. [Lichens pp. 49–50.]

353 **Buxton, J. and Lockley, R. M.** (1950) *Island of Skomer.* London and New York: Staples Press.

Bywater, J.—See No. 11a.

354 **Cain, R. F.** (1957) *Synaptospora*, a new genus of amerosporous Ascohymeniales (Ascomycetes). *Sydowia, Beih.* **1**: 4–8.

355 **Calvert, M.** (1844) *The History of Knaresbrough.* Knaresbrough: W. Parr. [Lichens p. 166.]

356 **Cambridge, J. C.** (1975) *Cation analysis of lichens.* B.Sc. thesis, University of Bristol.

357 **Camden, W.** (1789) *Britannia.* 3 vols. London: T. Payne and Son. [Translated from the edition published by the author in 1607. Enlarged by R. Gough.] [Botany by B. M., E. and T. F. Forster.]

358 —— (1806) *Britannia.* 4 vols. London: J. Stockdale. [Translated from the edition published by the author in 1607. Enlarged by R. Gough. Ed. 2.]

359 **Cameron, J.** (1882) The gaelic names of plants. *Scott. Nat.* **6**: 297–305.

360 —— (1883) *Gaelic Names of Plants* (*Scottish and Irish*). Edinburgh and London: William Blackwood and Sons. [Lichens pp. 97–98.]

361 —— (1900) *The Gaelic Names of Plants* (*Scottish, Irish, and Manx*). Glasgow: John MacKay. [Lichens pp. 129–131.]

Campbell, B.—See No. 1670.

362 **Campbell, M. E.** (1936) Fungi and lichens. In: The natural history of Barra, Outer Hebrides (J. E. Forrest, A. R. Waterson, and E. V. Watson, eds.). *Proc. R. phys. Soc. Edinb.* **22**: 258–260.

Campbell, M. S.—See No. 1276.

363 **Carlisle, A. and Brown, A. H. F.** (1968) Biological Flora of the British Isles: *Pinus sylvestris* L. *J. Ecol.* **56**: 269–307.

—— —See also No. 2262.

364 **Carmichael, D.** (1829) *Cryptog* [*amiae*] *Appinens*[*is*]. Manuscript in K.

—— —See also No. 1481.

365 **Carpenter, G. C.** [1850] Address to the members of the Berwickshire Naturalists' Club, delivered at the anniversary meeting, held at Etal, September 12, 1849. *Hist. Berwicksh. Nat. Club* **2**: 341–348.

366 **Carr, A. A.** (1836) *A History of Coldingham Priory.* Edinburgh: Adam and Charles Black. [Lichens p. 194.]

367 **Carr, J. W.**, ed. (1893) *A Contribution to the Geology and Natural History of Nottinghamshire.* Nottingham: James Bell. [Lichens pp. 86–87.]

368 —— (1906) Botany. In *The Victoria History of the Counties of England, Nottinghamshire* (W. Page, ed.) **1**: 41–74. London: Constable. [Lichens pp. 67–68.]

369 **Carr, R.** (1845) Parish of Luss. In *The New Statistical Account of Scotland.* Vol. 8. *Dumbarton*: 155–168. Edinburgh and London: W. Blackwood & Sons.

340 **Carrington, [B.]** (1858) On the geological distribution of plants in some districts of Yorkshire. *Rep. Br. Ass. Advmt Sci.* **1858**, Notices and abstracts of miscellaneous communications to the sections: 115.

—— —See also No. 1597.

371 **Carroll, I.** (1865) Contributions to British lichenology; being notices of new or rare species observed since the publication of Mudd's "Manual". I. *J. Bot., Lond.* **3**: 286–293.

372 —— (1866) Contributions to British lichenology; being notices of new or rare species observed since the publication of Mudd's "Manual". II. *J. Bot., Lond.* **4**: 22–26.

373 —— (1867) Contributions to British lichenology. *J. Bot., Lond.* **5**: 254–260.

374 **Carter, G. S.** (1969) *Study of some plant antibiotics.* B.Sc. thesis, University of Bristol.

375 **[Carter, J.]** (1850) *A Visit to Sherwood Forest.* London: Longman etc.

376 **Carter, P. W.** (1955) Some account of the history of botanical exploration of Herefordshire. *Trans. Woolhope Nat. Fld Club* **34**: 232–267.

377 —— (1955) Some account of botanical exploration in Merionethshire. *Merioneth Miscellany extra publications, ser.* 1, **3**: 1–40.

378 —— (1956) Notes on the botanical exploration of Flintshire. *Flintshire Miscellany* [*Publs. Flintshire hist.* **16**] **1**: 1–21.

Carter, T.—See No. 1634.

Cartwright, E.—See No. 550.

Cathrall, W.—See No. 2083.

379 **Chadwick, M. J.** (1960) Biological Flora of the British Isles: *Nardus stricta* L. *J. Ecol.* **48**: 255–267.

Chamberlain, Y. M.—See No. 1855.

380 **Chandapillai, M. M.** (1970) Variation in fixed dune vegetation at Newborough Warren, Anglesey. *J. Ecol.* **58**: 193–201.

381 **Chandler, J. H.** (1967) The Uffington gravel pits: an ecological study. *Trans. Lincs. Nat. Un.* **16**: 203–215.

Chapman, D. S.—See No. 960.

382 **Chapman, S. B.** (1967) Nutrient budgets for a dry heath ecosystem in the south of England. *J. Ecol.* **55**: 677–689.

383 **Chapman, V. J.** (1947) Biological Flora of the British Isles: *Suaeda fruticosa* Forsk. *J. Ecol.* **35**: 303–310.

—— —See also No. 1109.

384 **Cheetham, C. A.** (1941) *Nowellia curvifolia* (Dicks.) Mitt. and its associates in a new locality. *Naturalist, Hull*, **1941**: 173.

385 —— (1943) Some Austwick lichens and suggested method of study. *Naturalist, Hull* **1943**: 71–72.

386 —— (1943) Yorkshire naturalists at Bolton Percy—Mosses and lichens. *Naturalist, Hull* **1943**: 89.

387 [——] (1943) Yorkshire naturalists at Market Weighton. *Naturalist, Hull* **1943**: 126.

388 —— (1946) Yorkshire Naturalists' Union excursions in 1946: Robin Hood's Bay, July 20th—Mosses and lichens. *Naturalist, Hull* **1946**: 166–168.

389 —— (1947) *Solorina spongiosa* (Sm.) Carroll. *Naturalist, Hull* **1947**: 58.

390 **Chevallier, F.-F.** (1824) *Histoire des Graphidées.* Paris: F. Didot & Son.

391 **Choisy, M.** (1927) *Atlas lichénographique d'Europe centrale et occidentale.* [Fasc. 2] Lyon: M. Choisy.

392 —— (1927) *Atlas lichénographique d'Europe centrale et occidentale.* Fasc. 3. Lyon: M. Choisy.

393 —— (1927) *Atlas lichénographique d'Europe centrale et occidentale.* [Fasc. 4]

Lyon: M. Choisy.

394 —— (1928) *Atlas lichénographique d'Europe centrale et occidentale*. Fasc. 5. Lyon: M. Choisy.

395 —— (1928) *Atlas lichénographique d'Europe centrale et occidentale*. Fasc. 6. Lyon: M. Choisy.

396 —— (1929) *Icones lichenum universalis*. Fasc. 4. Lyon: M. Choisy.

397 —— (1930) *Icones lichenum universalis, ser*. II, fasc. 1. Lyon: M. Choisy.

398 —— (1931) *Icones lichenum universalis, ser*. II, fasc. 2. Lyon: M. Choisy.

399 **Christy, M. and Worth, R. H.** (1922) The ancient dwarfed woods of Dartmoor. *Rep. Trans. Devon. Ass. Advmt Sci*. **54**: 291–342.

Clapham, A. R.—See No. 580.

400 **Clark, M. C. and Rotheroe, M.** (1976) The Warwickshire Fungus Survey. *Bull. Br. mycol. Soc*. **10**: 12–14.

Clarke, T.—See No. 57.

401 **Clauzade, G. and Roux, C.** (1975) Lichens recoltes par Jean Amic en Norvege et en Grande-Bretagne. *Bull. Soc. linn. Provence* **28**: 53–63.

402 —— (1976) *Les Champignons Lichénicoles non lichénises*. Montpellier: Institut de Botanique.

—— —See also No. 1740.

403 **C[leaver], R. D. M.** (1976) Lassington Wood. *Newsl. Gloucester. Trust Nat. Conserv*. **9**: 7–8.

404 **Cliffe, C. F.** (1850) *The Book of North Wales*. London: Longman, Brown, Green and Longmans. [Also Ed. 2 (1851).]

405 **Clouston, C.** (1845) Parish of Sandwick. In *The New Statistical Account of Scotland*. Vol. 15. *Orkney*: 41–67. Edinburgh and London: W. Blackwood & Sons.

Coker, A. M.—See No. 413.

406 **Coker, P. D.** (1966) Biological Flora of the British Isles: *Sibbaldia procumbens* L. *J. Ecol*. **54**: 823–831.

407 —— (1966) The destruction of bryophytes by lichens, fungi, myxomycetes and algae. *Trans. Br. bryol. Soc*. **5**: 142–143.

408 —— (1967) *Some aspects of the ecology and phytosociology of epiphytic bryophyte communities*. Ph.D. thesis, University of London.

409 —— (1967) Damage to lichens by gastropods. *Lichenologist* **3**: 428–429.

410 —— (1967) The effects of sulphur dioxide pollution on bark epiphytes. *Trans. Br. bryol. Soc*. **5**: 341–347.

411 —— (1968) The epiphytic communities of the Ruislip district. *J. Ruislip Distr. nat. Hist. Soc*. **17**: 26–37.

412 —— (1970) Some observations on the flora of The Cairnwell, Perthshire. I. The limestone areas. *Trans. Proc. bot. Soc. Edinb*. **40**: 592–603.

413 —— **and Coker, A. M.** (1973) Biological flora of the British Isles: *Phyllodoce caerulea* (L.) Bab. *J. Ecol*. **61**: 901–913.

414 **Cole, J.** (1829) *Cole's Scarborough Guide*. Scarborough: J. Cole. [Lichen p. 120.]

Cole, L. W.—See Nos. 2300 and 2301.

Cole, R. L.—See No. 645.

66

415 [**Coleman, W. H.**] (1863) Catalogue of Leicestershire lichens. In *History, Gazetteer, and Directory, of the Counties of Leicester and Rutland*, Ed. 2 (W. White, ed.): 9–10. Sheffield: W. White.

416 **Collingwood, C.** (1853) A day at Clova. *Naturalist*, old series **3**: 140–143.

417 —— (1859) On the British lichens: their character and uses. *Gdnrs' Chron.* **1859**: 21–22.

418 —— (1859) On the British lichens: their character and uses. *Pharmaceutical J.* **18**: 362–365.

419 **Colman, J.** (1933) The nature of the intertidal zonation of plants and animals. *J. mar. biol. Ass. U.K.* **18**: 435–476.

420 —— (1940) On the faunas inhabiting inter-tidal seaweeds. *J. mar. biol. Ass. U.K.* **24**: 129–183.

Conway, V. M.—See No. 818.

Cook, L. M.—See No. 236.

421 **Cooke, G. A.** [1810] *Topographical and Statistical Description of the County of Westmorland*. London: C. Cooke. [Lichens pp. 128–129.]

422 **Cooke, M. C.** (1871) Apple twig parasites. *Gdnrs' Chron.* **1871**: 1004–1006.

423 —— (1873) British fungi. *Grevillea* **1**: 155–156.

424 —— (1874) New British fungi. *Grevillea* **3**: 65–69.

425 —— **and Plowright, C. B.** (1879) British Sphaeriacei. *Grevillea* **7**: 77–89.

Cooper, C. W.—See No. 1634.

426 **Cooper, D.** (1837) *Flora metropolitana*. London: S. Highley.

Cooper, E. F.—See No. 1634.

427 **Cooper, T. H.** (1834) *The Botany of the County of Sussex*. Lewes: Sussex Press. [Lichens pp. 45–50 mainly from W. Borrer.]

—— —See also No. 1068.

428 **Coppins, B. J.** (1971) *A lichen flora of Looe Island, Cornwall*. Manuscript with M. R. D. Seaward.

429 —— (1971) Lichens of the Middle Torne Valley. *Naturalist, Hull* **1971**: 113.

430 —— (1971) *Report on the lichen flora of two woodland sites in South Devon*. Manuscript. Report to the Nature Conservancy; copy with BLS.

431 —— (1972) Field meeting at Richmond, Yorkshire. *Lichenologist* **5**: 326–336.

432 —— (1973) The "Drought Hypothesis". In *Air Pollution and Lichens* (B. W. Ferry, M. S. Baddeley and D. L. Hawksworth, eds.): 124–142. London: Athlone Press of the University of London.

433 —— (1976) Distribution patterns shown by epiphytic lichens in the British Isles. In *Lichenology: Progress and Problems* (D. H. Brown, D. L. Hawksworth and R. H. Bailey, eds.): 249–278. [Systematics Association Special Volume No. 8.] London, New York and San Francisco: Academic Press.

434 —— **and James, P. W.** (1974) Distribution maps of lichens in Britain. Map 9. *Cyphelium notarisii* (Tul.) Blomb. & Forss. *Lichenologist* **6**: 172–174.

435 —— —— (1974) Distribution maps of lichens in Britain. Map 10. *Cyphelium tigillare* (Ach.) Ach. *Lichenologist* **6**: 175–177.

436 —— **and Lambley, P. W.** (1974) Changes in the lichen flora of the parish of

Mendlesham, Suffolk, during the last fifty years. *Trans. Suffolk Nat. Soc.* **16**: 319–335.

437 —— **and Seaward, M. R. D.** (1976) A lichen collection from the Yorkshire Museum, York. *Naturalist, Hull* **1976**: 13–15.

438 —— **and Shimwell, D. W.** (1971) Cryptogam complement and biomass in dry *Calluna* heath of different ages. *Oikos* **22**: 204–209.

—— —See also Nos. 961, 962, 963, 970, 1574 and 2052.

439 **Corbet, S. A. and Lan, O. B.** (1974) Moss on a roof, and what lives in it. *J. Biol. Educ.* **8**: 153–160.

Cormack, E.—See No. 811.

Couch, R. C. E.—See Nos. 1607 and 1977.

440 **Courtney, J. S.** (1845) *A Guide to Penzance and its Neighbourhood, including the Isles of Scilly.* Penzance: E. Rowe. [Lichens pp. 17–18.]

441 **[Cox, R. M.]** (1960) Mosses, liverworts and lichens. *Rep. Lundy Fld Soc.* **13**: 49–53.

—— —See also Nos. 332 and 333.

Crabbe, G.—See No. 1658.

442 **Crampton, C. B. and MacGregor, M.** (1913) The plant ecology of Ben Armine, Sutherlandshire. *Scot. geogr. Mag.* **29**: 169–192.

443 **Craster, E.** (1970) *Some aspects of the zonation of the flora on a coastal outcrop of the Whin Sill.* B.Sc. thesis, University of Bristol.

444 **Crawford, R. M. M. and Wishart, D.** (1966) A multivariate analysis of the development of dune slack vegetation in relation to coastal accretion at Tentsmuir, Fife. *J. Ecol.* **54**: 729–743.

445 —— —— (1967) A rapid multivariate method for the detection and classification of groups of ecologically related species. *J. Ecol.* **55**: 505–524.

446 —— —— (1968) A rapid classification and ordination method and its application to vegetation mapping. *J. Ecol.* **56**: 385–404.

447 **Crawley, M. J., Robinson, H. D., Seaward, M. R. D. and Shaw, P. J.** (1975) *Plant communities of the Ardmeanach Peninsula on the Isle of Mull.* Bradford: University of Bradford.

448 **Crawshaw, A.** (1936) Bovey Heathfield triangle and its flora. *Trans. Proc. Torquay nat. Hist. Soc.* **7**: 33–40.

449 **Crittenden, P. D.** (1971) *Lichen growth in the urban environment.* B.Sc. thesis, University of Manchester.

450 **Croall, A.** (1857) *Plants of Braemar.* 4 fasc., nos. 1–602. Montrose. [Intact set of the lichens in BM.]

451 **Crocker, J.** *et. al.*, eds. (1962) *Bradgate Park and Cropston Reservoir margins.* [Surveys of Leicestershire Natural History No. 1.] Loughborough: Loughborough Naturalists Club. [Lichens by M. Walpole.]

452 **Crombe, D. E. and Frost, L. C.** (1956) The heaths of the Cornish serpentine. *J. Ecol.* **44**: 226–256.

453 **Crombie, J. M.** (18—) [*Manuscripts*]. 15 vols. and boxes. Manuscripts in BM. [Includes manuscripts re- nos. 515 and 2204, and some unpublished data; not abstracted.]

454 —— (1861) *Braemar: its Topography and Natural History.* Aberdeen: John

Smith. [Lichens pp. 64–65.]

455 —— (1868) *The Geological Relations of the Alpine Flora of Great Britain.* London: Geologists' Association.

456 —— (1869) Lichenes. In *Flora of Middlesex* (H. Trimen and W. T. Dyer, eds.): 405–407. London: Hardwicke.

457 —— (1869) New British lichens. No. I. *J. Bot., Lond.* **7**: 48–51.

458 —— (1869) New British lichens. No. II. *J. Bot., Lond.* **7**: 105–108.

459 —— (1869) New British lichens. No. III. *J. Bot., Lond.* **7**: 232–234.

460 —— (1870) Additions to the British lichen-flora. *J. Bot., Lond.* **8**: 95–99.

461 —— (1870) *Lichenes Britannici.* London: Reeve.

462 —— (1871) Additions to the British lichen flora. No. II. *J. Bot., Lond.* **9**: 177–179.

463 —— (1871) New lichens recently discovered in Great Britain. *J. Linn. Soc. (Bot.)* **11**: 481–490.

464 —— (1872) New British lichens. *Grevillea* **1**: 61–62.

465 —— (1872) Collecting and preserving. No. IX.—Lichens. *Sci. Gossip* **8**: 217–220.

466 —— (1872) Notes on the British Ramalinas in the herbarium of the British Museum. *J. Bot., Lond.* **10**: 70–75.

467 —— (1872) Notes on the lichens in Sowerby's herbarium. No. I. *Usnea— Solorina. J. Bot., Lond.* **10**: 231–235.

468 —— (1872) On a new erratic British *Parmelia. J. Bot., Lond.* **10**: 306–307.

469 —— (1872) Notes on the lichens in Sowerby's herbarium. No. II. *Stictina— Gyrophora. J. Bot., Lond.* **10**: 356–360.

470 —— (1873) New British lichens. *Grevillea* **1**: 141–142.

471 —— (1873) Lichens of Sowerby's herbarium. *Grevillea* **1**: 158–160.

472 —— (1873) On the rarer lichens of Blair Athole. *Grevillea* **1**: 170–174.

473 —— (1873) Note on *Lecanora ralfsii* (Salwey) Cromb. *Grevillea* **2**: 13–14.

474 —— (1873) Note on *Solorina bispora* (Nyl.). *Grevillea* **2**: 79–80.

475 —— (1873) New British lichens. *Grevillea* **2**: 89–91.

476 —— (1873) Additions to the British lichen-flora. *J. Bot., Lond.* **11**: 132–135.

477 —— (1874) New British lichens. *Grevillea* **3**: 22–24.

478 —— (1874) Lichenes Britannici Exsiccati. *Grevillea* **3**: 81–83

479 —— (1874) British Collemacei. *Grevillea* **3**: 92–95.

480 —— (1874) Recent additions to the British lichen-flora. IV. *J. Bot., Lond.* **12**: 146–149.

481 —— (1874) On *Ptychographa*, Nyl., a new genus of lichens. *J. Bot., Lond.* **12**: 257–258.

482 —— (1874) Revision of the British Collemacei. *J. Bot., Lond.* **12**: 330–337.

483 —— (1874–77) *Lichenes Britannici exsiccati.* 2 fasc., nos. 1–200. London. [Not abstracted; intact set not located.]

484 —— (1875) On a new British species of *Xylographa. Grevillea* **3**: 128.

485 —— (1875) Note on *Lecidea didymospora*, Strn., &c. *Grevillea* **3**: 142–143.

486 —— (1875) On two new British species of Collemacei. *Grevillea* **3**: 190–191.

487 —— (1875) Recent additions to the British lichen-flora. *J. Bot., Lond.* **13**: 140–142.

488 —— (1876) New British lichens. *Grevillea* **4**: 180–181.

489 —— (1876) New British lichens. *Grevillea* **5**: 25–30.

490 —— (1876) Conspectus of the genera of British lichens, according to the more recent Nylanderian arrangement. *Grevillea* **5**: 76–78.

491 —— (1876) Recent additions to the British lichen-flora. *J. Bot., Lond.* **14**: 359–363.

492 —— (1877) New British lichens. *Grevillea* **6**: 18–20.

493 —— (1877) New. British lichens. *Grevillea* **5**: 106–108.

494 —— (1877) Note on the British species of *Pterygium. Grevillea* **5**: 108–109.

495 —— (1877) Lichenes Britannici Exsiccati. *Grevillea* **6**: 20–22.

496 —— (1878) New British lichens. *Grevillea* **6**: 111–115.

497 —— (1879) Note on *Parmelia horrescens* Tayl. *Grevillea* **7**: 98.

498 —— (1879) Additions to the British *Ramalinei. Grevillea* **7**: 141–142.

499 —— (1879) Note on *Lecidea farinaria.* Borr. *Grevillea* **7**: 142.

500 —— (1879) New British lichens. *Grevillea* **8**: 28–30.

501 —— (1880) New British lichens. *Grevillea* **8**: 112–114.

502 —— (1880) On the lichens of Dillenius's '*Historia Muscorum*', as illustrated by his herbarium. *J. Linn. Soc. (Bot.)* **17**: 553–581.

503 —— (1881) New British lichens. *Grevillea* **10**: 22–24.

504 —— (1881) Observations on *Parmelia olivacea* and its British allies. *Grevillea* **10**: 24–26.

505 —— (1882) Recent additions to the British lichen-flora. *J. Bot., Lond.* **20**: 271–276.

506 —— (1883) Enumeration of the British *Cladoniei. Grevillea* **11**: 111–115.

507 —— (1883–84) On the lichens in Dr. Withering's herbarium. *Grevillea* **12**: 56–62, 70–76.

508 —— (1884) New British lichens. *Grevillea* **12**: 89–91.

509 —— (1884) Additions to the British *Cladoniei. Grevillea* **12**: 91–92.

510 —— (1885) Recent additions to the British lichen-flora. *J. Bot., Lond.* **23**: 194–196.

511 —— (1885) On the lichen-flora of Epping Forest, and the causes affecting its recent diminution. *Trans. Essex Fld Cl.* **4**: 54–75.

512 —— (1886–87) Index lichenum britannicorum. (According to the most recent Nylanderian arrangement) [Part I]. *Grevillea* **15**: 10–15, 44–49, 74–79.

513 —— (1889–91) Index lichenum britannicorum. Part II. *Grevillea* **18**: 43–47, 67–70; **19**: 57–60.

514 —— (1893–94) Index lichenum britannicorum. Part III. *Grevillea* **22**: 8–11, 57–60.

515 —— (1894) *A Monograph of Lichens found in Britain.* Vol. 1. London: British Museum (Natural History).

516 —— (1901) Lichenes (Lichens). In *The Victoria History of the Counties of England, Norfolk* (H. A. Doubleday, ed.) **1**: 70–72. Westminster: Constable.

517 **Crossland, C. and Needham, J.** (1902) Woodland studies: 1.—The flora of a boulder in March. *Halifax Nat.* **8**: 8–10.

70

518 —— —— (1904) The plants of Pecket Wood. *Naturalist, Hull* **1904**: 165–175.
—— —See also Nos. 520 and 1579.

519 **Crothers, J. H.** (1976) On the distribution of some common animals and plants along the rocky shores of West Somerset. *Fld Stud.* **4**: 369–389.

520 **Crump, W. B. and Crossland, C.** (1904) *The Flora of the Parish of Halifax.* Halifax: Halifax Scientific Society. [Lichens pp. 215–228.]

521 **Crundwell, A. C.** (1957) A lichen new to Scotland. *Glasgow Nat.* **17**: 279.

522 —— (1958) The meetings of the British Bryological Society, 1957. *Bryologist* **61**: 154–155.

523 —— (1970) *Herberta borealis,* a new species from Scotland and Norway. *Trans. Br. bryol. Soc.* **6**: 41–49.

524 **Cubbon, B. D.** (1970) Flora records of the Cave Research Group of Great Britain from 1939 to June 1969. *Trans. Cave Res. Grp Gt Br.* **12**: 57–74.

525 **Culberson, C. F.** (1965) Some constituents of the lichen *Ramalina siliquosa. Phytochemistry* **4**: 951–961.

526 —— (1967) Some microchemical tests for two new lichen substances, scrobiculin and 4-0-methylphysodic acid. *Bryologist* **70**: 70–75.

527 —— (1967) The structure of scrobiculin, a new lichen depside in *Lobaria scrobiculata* and *Lobaria amplissima. Phytochemistry* **6**: 719–725.

528 —— (1969) Chemical studies in the genus *Lobaria* and the occurrence of a new tridepside, 4-0-methylgyrophoric acid. *Bryologist* **72**: 19–27.

529 —— (1969) *Chemical and Botanical Guide to Lichen Products.* Chapel Hill, N. Carol.: The University of North Carolina Press.

530 —— (1970) Supplement to "Chemical and Botanical Guide to Lichen Products". *Bryologist* **73**: 177–377.
—— —See also Nos. 537 and 538.

531 **Culberson, W. L.** (1952) Étude systématique et phytogéographique de l'*Enterographa crassa* (DC) Fée. *Revue bryol. lichén.* **21**: 276–284.

532 —— [1964] A summary of the lichen genus *Haematomma* in North America. *Bryologist* **66**: 224–236.

533 —— (1966) Chimie et taxonomie des lichens du groupe *Ramalina farinacea* en Europe. *Revue bryol. lichén.* **34**: 841–851.

534 —— (1967) Analysis of chemical and morphological variation in the *Ramalina siliquosa* species complex. *Brittonia* **19**: 333–352.

535 —— (1969) The use of chemistry in the systematics of lichens. *Taxon* **18**: 152–166.

536 —— (1970) *Parmelia discordans,* lichen peu connu d'Europe. *Revue bryol. lichén.* **37**: 183–186.

537 —— **and Culberson, C. F.** (1967) Habitat selection by chemically differentiated races of lichens. *Science, N.Y.* **158**: 1195–1197.

538 —— —— (1968) The lichen genera *Cetrelia* and *Platismatia* (Parmeliaceae). *Contr. U.S. natn. Herb.* **34**: 449–558.

539 **Cullinane, J. P., McCarthy, P. and Fletcher, A.** (1975) The effect of oil pollution in Bantry Bay. *Mar. Pollut. Bull.* **6**: 173–176.

540 **Culpepper, N.** (1652) *The English Physitian.* London. [And over 100 later editions from 1653 onwards.]

541 **Curdie, J.** (1845) Parish of Eigha and Cava. In *The New Statistical Account of Scotland*. Vol. 7. *Argyle*: 394–407. Edinburgh and London: W. Blackwood & Sons.

542 **Curnow, W.** (1844) Rarer plants observed in the neighbourhood of Penzance. *Phytologist* **1**: 1143–1144.

543 —— (1872) The cryptogamia of the Scilly Isles. *Sci. Gossip* **8**: 210.

544 **Currie, A.** (1961) Notes on the lichen flora of Fair Isle, Shetland (V.C.112). *Trans. Proc. bot. Soc. Edinb*. **39**: 236–238.

545 **Dahl, E.** (1952) *Analytical Keys to British Macrolichens*. Cambridge: Botany School.

546 —— (1953) Notes on some British macrolichens. *Ann. Mag. nat. Hist., ser* 12, **6**: 425–431.

547 —— (1968) *Analytical Keys to British Macrolichens*. Ed. 2. London: British Lichen Society.

548 —— **and Krog, H.** (1970) On the distribution of *Cladonia luteoalba* Wils. & Wheld. *Nytt Mag. Bot*. **12**: 143–144.

549 **Dale, C. W.** (1878) *The History of Glanville's Wootton*. London: Hatchards. [Lichens pp. 379–380.]

550 **Dallaway, J. and Cartwright, E.** (1815) *A History of the Western Division of the County of Sussex*. London. [Not seen.]

551 **Dallman, A. A.** (1907) Notes on the flora of Flintshire. *J. Bot., Lond*. **45**: 138–153.

552 —— (1908) Notes on the flora of Flintshire. *J. Bot., Lond*. **46**: 222–230.

553 **Daniels, R. E. and Pearson, M. C.** (1974) Ecological studies at Roydon Common, Norfolk. *J. Ecol*. **62**: 127–150.

554 **Darbishire, O. V.** (1898) Monographia Roccelleorum. *Biblthca bot*. **9** (45): 1–103.

555 —— (1906) The wild fauna and flora of the Royal Botanic Gardens, Kew. 4. Lichenes. *Kew Bull., add. ser*. **5**: 102–103.

556 —— [*c.* 1906] Botany. In *The Victoria History of the Counties of England, Cheshire* (W. Page, ed.). Unpublished manuscript in BM. [Lichens pp. 144–146.]

557 —— (1914) Some remarks on the ecology of lichens. *J. Ecol*. **2**: 71–82.

558 —— (1926) The structure of *Peltigera* with special reference to *P. praetextata*. *Ann. Bot*. **40**: 727–758.

559 —— (1928) Roccellaceae Mass. (Nyl.). In *Die Pflanzenareale* (L. Diels, G. Samuelsson, E. Hannig and H. Winkler, eds.) **2** (1): 1–4, Karte 1–5. Jena: G. Fischer.

560 —— (1934) The *Lichen fucoides* of various authors and its fate. *Trans. Br. mycol. Soc*. **34**: 308–313.

561 **Darling, F. F.** (1947) *Natural History in the Highlands and Islands*. [New Naturalist No. 6.] London: Collins.

Darwin, E.—See No. 1835.

562 **Davies, E. G.** (1944) Figyn Blaen Brefi: A Welsh upland bog. *J. Ecol*. **32**: 147–166.

563 **Davies, G.** (1864) Notes on some new and rare British cryptogams. *Bot. Chron.* **5**: 37.

564 **Davies, H.** (1794) Descriptions of four new British lichens. *Trans. Linn. Soc. Lond.* **2**: 283–285.

565 —— (1813) Remarks on *Lichen scaber* and some of its allies. *Trans. Linn. Soc. Lond.* **11**: 79–85.

566 —— (1813) *Welsh Botanology*, I. *Systematic Catalogue of the native plants of the Isle of Anglesey.* London. [Lichens pp. 110–115.]

567 **Davies, J. B.** (1852) A day's botany in Roslin and Hawthornden, in October. *Naturalist, o.s.* **2**: 49–51.

568 **Davies, P.** (1972) Pollution in a National Park? *Peak Park News, February* 1972: sine pagin.

Davies, R. I.—See No. 1157.

569 **Davis, B. N. K. and Lambley, P. W.** (1972) *Sotterley Park.* Manuscript. Report to the Nature Conservancy; copy with BLS.

570 **Davis, J. W. and Lees, F. A.** (1878) *West Yorkshire: An Account of its Geology, Physical Geography, Climatology, and Botany.* London: Reeve.

Dawson, W. R.—See No. 2217.

571 **Day, F. M. and Brokenshire, F. A.** (1943) Thirty-fifth report on the botany of Devon. *Rep. Trans. Devon. Ass. Advmt Sci.* **75**: 57–64.

572 —— —— (1944) Thirty-sixth report on the botany of Devon. *Rep. Trans. Devon. Ass. Advmt Sci.* **76**: 41–55.

573 —— —— (1947) Thirty-seventh report on the botany of Devon. *Rep. Trans. Devon. Ass. Advmt Sci.* **77**: 57–67.

574 **Deakin, R.** [18—] [Drawings and MSS of lichens purchased from R. R. Hutchinson July 3rd. 1907]. 10 boxes. Manuscript in BM.

575 —— (1854) Description and illustrations of new species of *Verrucaria* and *Sagedia* found about Torquay, Devonshire. *Ann. Mag. nat. Hist., ser. 2,* **13**: 32–41.

—— —See also No. 255.

576 **Deering, C.** (1738) *Catalogus stirpium, &c, or A Catalogue of Plants naturally growing and commonly cultivated in divers parts of England, more especially about Nottingham.* Nottingham. [Lichens pp. 126–133.]

577 **Degelius, G.** (1935) Das ozeanische Element der Strauch- und Laubflechten-flora von Skandinavien. *Acta phytogeogr. suec.* **7**: i–xii, 1–411.

578 —— (1954) The lichen genus *Collema* in Europe. *Symb. bot. upsal.* **13** (2): 1–499.

579 —— (1974) The lichen genus *Collema* with special reference to the extra-European species. *Symb. bot. upsal.* **20** (2): 1–215.

580 **Deighton, F. C. and Clapham, A. R.** (1926) The vegetation of Scolt Head Island. A preliminary account. *Trans. Norfolk Norwich Nat. Soc.* **12**: 86–111.

Delhanty, J. E.—See No. 2351.

581 **Dennis, N.** (1966) Records of non-vascular cryptogams in Skye. *Trans. Proc. bot. Soc. Edinb.* **40**: 204–231.

Dewdney, J. C.—See Nos. 217 and 277.

582 **Dibben, M. J.** (1974) *The chemosystematics of the lichen genus* Pertusaria *in North America north of Mexico.* Ph.D. thesis, Duke University, North Carolina. [Copy in BM.]

Dickens, R. F.—See No. 2567.

583 **Dickie, G.** (1860) *The Botanist's Guide to the Counties of Aberdeen, Banff and Kincardine.* Aberdeen: A. Brown. [Lichens pp. 268–280.]

584 **Dickinson, C. I.** (1934) The lichens. In *Scolt Head Island. The Story of its Origin: The Plant and Animal Life of the Dunes and Marshes* (J. A. Steers, ed.): 151–160. Cambridge: Norfolk and Norwich Naturalists' Society.

585 **Dickson, J.** (1785–1801) *Fasciculus [secundus, etc.] plantarum cryptogamicarum britanniae.* 4 fasc. London. [I, 1785; II, 1790; III, 1793; IV, 1801; reprinted (1976) Richmond: Richmond Publishing.] [Original water-coloured drawings by J. Sowerby in BM.]

586 —— (1789–91) *A Collection of Dried Plants.* 4 fasc., nos. 1–100. London. [Intact copy in BM.]

587 —— (1793–1802) *Hortus siccus brittanicus.* 19 fasc. [Specimens in each fascicle separately numbered.] London. [Intact copy in BM.]

Dickson, J. H.—See No. 2194.

588 **Dickson, W. E. C.** (1898) Ben Lawers. *J. Cairngorm Club* **2**: 192–205.

Diels, L.—See No. 559.

589 **Dillenius, J. J.** (17—) *[Correspondence; manuscripts; notes; drawings.]* Manuscripts in OXF (Bodlein as Sherard MSS 202–211). [Not seen and not abstracted; see *Catalogue of manuscripts belonging to Oxford University Department of Botany deposited in the Bodlein Library* (1958) for details.]

590 —— (1726) *Plants observed in a journey from London in North Wales by the way of Wiltshire, Somersetshire, Worcestershire, & Shropshire.* Manuscript in OXF (Bodlein). [Reproduced in Druce and Vines (1907); original not seen.]

591 —— [1742] (1741) *Historia muscorum.* Oxford: Sheldonian Theatre. [Lichens pp. 56–228.] [Dated '1741' but first issued in March 1742; original water-coloured drawings in BM; copy annotated by Dillenius in OXF (Bodlein).] [Re-issued in 1811 by C. Stewart, Edinburgh, with an Appendix giving binomials.]

592 —— (1763) *Historia muscorum.* London: J. Millan. [Plates of the 1742 edition (some re-numbered) with a much shortened text.] [Re-issued by the same publisher in 1768.]

—— —See also Nos. 502, 613, 1927 and 2530.

593 **Dillwyn, L. W.** (1809) *British Confervae.* London: W. Phillips.

—— —See also No. 2391.

DiMeo, J. A.—See No. 334.

Ditchfield, P. H.—See No. 611.

594 **Dix, H. M.** (1954) Coventry Nature Reserve records. Lichens. *Proc. Coventry Distr. nat. Hist. scient. Soc.* **2**: 232.

595 —— (1959) Coventry Nature Reserve records. Lichens. *Proc. Coventry Distr. nat. Hist. scient. Soc.* **3**: 69.

596 —— and H[ughes], V. R. (1953) Miscellaneous Reserve records. *Proc. Coventry Distr. nat. Hist. scient. Soc.* **2**: 201–202.

597 **Dixon, H. N.** (1902) Lichenes (Lichens). In *The Victoria History of the Counties of England, Northamptonshire* (H. A. Doubleday, ed.) **1**: 84. Westminster: Constable.

598 **Dixon, J. H.** (1886) *Gairloch in North-West Ross-shire.* Edinburgh: Co-operative Printing. [Lichens pp. 234–235 by C. F. Newcombe.]

Döbbeler, P.—See No. 1845.

599 **Dobson, F. S. and Hawksworth, D. L.** (1976) *Parmelia pastillifera* (Harm.) Schub. & Klem. and *P. tiliacea* (Hoffm.) Ach. in the British Isles. *Lichenologist* **8**: 47–59.

600 **Dodge, C. W.** (1929) A synopsis of *Stereocaulon* with notes on some exotic species. *Annls Cryptog. exot.* **2**: 93–153.

601 **Doherty, A. J.** (1884) *Solorina saccata. J. Micr. & nat. Sci.* **3**: 28–31.

602 **Don, G.** (1804–12) *Herbarium Britannicum.* 9 fasc., nos. 1–225. Edinburgh. [Intact copy in BM.]

603 —— (1813) Account of the native plants in the county of Forfar, and the animals to be found there. In [Appendix to] *General View of the Agriculture of the county of Angus, or Forfarshire* (J. Headrick, ed.): 11–59. Edinburgh.

Don. G. [jr.]—See No. 1445.

604 **Doody, S.** (1690) Stirpes & observationes. In *Synopsis methodica stirpium Britannicarum* (J. Ray): 243–246. London.

605 —— (1696) Species strirpium novae & observationes. In *Synopsis methodica stirpium Britannicarum,* Ed. 2 (J. Ray): 327–346. London.

Doubleday, H. A.—See Nos. 134, 135, 516, 597, 1045, 1046, 1065, 2168 and 2634.

606 **Douglas, J. A.** (1880) Bradford Naturalists' Society. *Naturalist, Hull* **5**: 170–171.

607 **Douglas, R.** (1845) Parish of Kilbarchan. In *The New Statistical Account of Scotland.* Vol. 7. *Renfrew*: 353–382. Edinburgh and London: W. Blackwood & Sons.

608 **Drive Publications** (1973) *Book of the British Countryside.* London: Drive Publications.

609 **Druce, G. C.** (1886) *The Flora of Oxfordshire.* Oxford: Parker. [Lichens pp. 441–442.]

610 —— (1905) Botany. In *The Victoria History of the Counties of England, Buckinghamshire* (W. Page, ed.) **1** (2): 27–68. London: Constable. [Lichens pp. 63–64, list provided by E. M. Holmes.]

611 —— (1906) Botany. In *The Victoria History of the Counties of England, Berkshire* (P. H. Ditchfield and W. Page, eds.) **1** (2): 27–68. London: Constable. [Lichens p. 67.]

612 —— (1930) *The Flora of Northamptonshire.* Arbroath: T. Buncle.

613 —— **and Vines, S. H.** (1907) *The Dillenian Herbaria.* Oxford: Clarendon Press.

—— —See also No. 2422.

614 **Druery, J. H.** (1829) *Historical and Topographical notices of Great Yarmouth.* London: Nichols. [Lichens pp. 379–381 by D. Turner.]

615 **Drummond, J.** (1844) Lichens eaten by squirrels. *Gdnrs' Chron.* **1844**: 28.

616 **Duffey, E.** (1955) Notes on the natural history of Eynhallow, Orkney. *Scott. Nat.* **67**: 40–51.

617 —— (1962) A population study of spiders in limestone grassland. *J. anim. Ecol.* **31**: 571–599.

618 —— (1968) An ecological analysis of the spider fauna of sand dunes. *J. anim. Ecol.* **37**: 641–674.

619 **Duncan, J.** (1832) *Flora of Jedburgh, Roxburghshire.* Manuscript in K.

620 **Duncan, U. K.** [1958] *Additions to Census Catalogue (up to Dec. 31st. 1957).* [Manuscript in BM with Wade (1957–59).]

621 —— [1959] *Additions to Census Catalogue (since Jan. 1st. 1958).* [Manuscript in BM with Wade (1957–59).]

622 —— (1959) *A Guide to the Study of Lichens.* Arbroath: T. Buncle.

623 —— (1959) *Lecanora andrewii* B. de Lesd. in east Scotland. *Lichenologist* **1**: 111–112.

624 —— (1960) A survey of the bryophytes and lichens of 'The Burn', Kincardine. *Trans. Proc. bot. Soc. Edinb.* **39**: 62–84.

625 —— (1961) A visit to the Shetland Isles. *Lichenologist* **1**: 267–268.

626 —— (1962) The bryophytes and lichens of the Loch Tay area. *Rep. Scott. Fld Stud. Ass.* **1962**: 20–31.

627 —— (1963) *Lichen Illustrations.* Arbroath: T. Buncle.

628 —— (1963) A list of Fair Isle lichens. *Lichenologist* **2**: 171–178.

629 —— (1966) Additions and corrections to the bryophyte and lichen flora of 'The Burn', Kincardine. *Trans. Proc. bot. Soc. Edinb.* **40**: 232–234.

630 —— [**assisted by P. W. James**] (1970) *Introduction to British Lichens.* Arbroath: T. Buncle.

631 —— (1975) Kirkton of Auchterhouse, 26th April 1975. *News bot. Soc. Edinb.* **17**: 8.

632 **Duncumb, J.** (1804) *Collections towards the History and Antiquities of the County of Hereford.* Vol. 1. Hereford. [Lichens p. 186.]

633 —— (1805) *A General View of the Agriculture of the County of Hereford.* Hereford & London. [Lichens pp. 169–171.]

—— —See also No. 105.

634 **Dunn, M. D.** (1941) The marine algal associations of St. Andrews district. Part I: The dominant associations of the spray and littoral regions. *Trans. Proc. bot. Soc. Edinb.* **33**: 83–93.

Dunston, A. E. A.—See No. 636.

635 **Dunston, T. F. [G. W.]** (1903) A list of plants found in div. IX of Preston's "Flowering Plants of Wilts." *Rep. Marlboro. Coll. nat. Hist. Soc.* **51**: 98–107.

636 —— **and Dunston, A. E. A.** [**revised W. Watson**] (1940) Notes on some of the lichens found in south-west Wiltshire and especially round Donhead St. Mary. *Wilts. archaeol. nat. Hist. Mag.* **49**: 526–533.

637 **Du Rietz, G. E.** (1922) Flechtensystematische Studien. II. *Leptogium sernanderi* n. sp. und einige verwandte Arten. *Bot. Notiser* [**75**]: 317–322.

638 —— (1935) Glacial survival of plants in Scandinavia and the British Isles. *Proc. R. Soc., B*, **118**: 226–229.

Dyer, W. T.—See No. 456.

639 **E[——], R.** (1842) Lungs of the oak. *Gdnrs' Chron.* **1842**: 84.

640 **Earland-Bennett, P. M.** (1971) Yorkshire Naturalists' Union excursions in 1971: Hardcastle Crags—Lichens. *Naturalist, Hull* **1971**: 145–146.

641 —— (1973) Lichens from the Tanfield and Masham areas of North Yorkshire. *Naturalist, Hull* **1973**: 51–57.

642 —— (1973) Yorkshire Naturalists' Union excursions in 1973: Mickletown, near Castleford—Lichens. *Naturalist, Hull* **1973**: 148–149.

643 —— (1973) Spring foray, Hebden Bridge—Lichens. *Naturalist, Hull* **1973**: 154–155.

644 —— (1974) *Umbilicaria spadochroa* (Hoffm.) DC. from Ireland. *Lichenologist* **6**: 128–129.

645 —— (1974) The lichen flora of the Basildon area. In *The Natural History of Basildon*, Ed. 2 (R. L. Cole, H. J. Peck and G. Reid, eds.): 35–44. Basildon: Basildon Natural History Society.

646 —— (1975) *Lecanora subaurea* Zahlbr., new to the British Isles. *Lichenologist* **7**: 162–167.

647 —— (1975) Lichens from the Leeds region. *Naturalist, Hull* **1975**: 101–106.

648 —— (1975) Some preliminary notes on Manx lichens. *Peregrine* **4**: 160–162.

649 —— (1976) The lichen flora of the Basildon area. *Essex Nat.* **33**: 139–154.

650 —— (1976) The lichen flora of Water Newton. *Rep. Huntingdon Fauna Flora Soc.* **28**: 15–24.

651 —— (1976) Yorkshire Naturalists' Union, Mycological Section, Spring Foray 1975. *Bull. Br. mycol. Soc.* **10**: 14.

652 —— **and Seaward, M. R. D.** (1974) Field meeting in the Hebden Valley, Yorkshire. *Lichenologist* **6**: 115–121.

—— —See also No. 990.

653 **Eddy, A., Welch, D. and Rawes, M.** (1969) The vegetation of the Moor House National Nature Reserve in the Northern Pennines, England. *Vegetatio* **16**: 239–284.

654 **Edgell, M. C. R.** (1969) Vegetation of an upland ecosystem: Cader Idris, Merionethshire. *J. Ecol.* **57**: 335–359.

655 **[Edmondston, T.]** (1842) On the native dyes of the Shetland Islands. *Trans. bot. Soc. Edinb.* **1**: 123–126.

656 —— (1844) Note on *Cetraria sepincola. Phytologist* **1**: 905–906.

657 —— (1845) *A Flora of Shetland.* Aberdeen: G. Clark & Son. [Lichens pp. 48–52.]

658 **Edwards, K. C. [assisted by H. H. Swinnerton and R. H. Hall]** (1962) *The Peak District.* [New Naturalist No. 44.] London: Collins. [Lichens pp. 107–109.]

Edwards, P. J.—See No. 1970.

Edwards, R. W.—See No. 773.

659 Eggeling, W. J. (1960) *The Isle of May: A Scottish Nature Reserve*. Edinburgh and London: Oliver & Boyd. [Lichens p. 181.]

660 —— (1965) Check list of the plants of Rhum, Inner Hebrides (VC 104, North Ebudes). Part II. Lichens, liverworts and mosses. *Trans. Proc. bot. Soc. Edinb.* **40**: 60–99.

661 Elgee, F. (1912) *The Moorlands of North-Eastern Yorkshire. Their Natural History and Origin*. London, Hull and York: A. Brown & Sons.

662 Elkington, T. T. (1963) Biological Flora of the British Isles: *Gentiana verna* L. *J. Ecol.* **51**: 755–767.

663 —— (1964) Biological Flora of the British Isles: *Myosotis alpestris* F. W. Schmidt *J. Ecol.* **52**: 709–722.

664 —— (1971) Biological Flora of the British Isles: *Dryas octopetala* L. *J. Ecol.* **59**: 887–905.

665 Eller, I. (1841) *The Natural History of Belvoir Castle*. London: Tyas & Groombridge. [Lichens pp. 402–3.]

666 Elliot, G. F. S. (1901) Lichens. In *Fauna, Flora and Geology of the Clyde area* (G. F. S. Elliot, M. Laurie and J. B. Murdoch, eds.): 50–60. Glasgow: British Association.

667 —— (1907) Notes on the Trap-flora of Renfrewshire. *Ann. Anderson. Nat. Soc.* **3**: 1–10.

668 Ellis, C. J. (1975) *Initial vegetation survey of Ubley Warren, Mendip*. B.Sc. thesis, University of Bristol.

669 Ellis, E. A. (1934) Wheatfen Broad, Surlingham. *Trans. Norfolk Norwich Nat. Soc.* **13**: 422–451. [Lichens p. 444.]

670 —— (1939) The Norfolk sea floods. 4. Detailed observations. *Trans. Norfolk Norwich Nat. Soc.* **14**: 373–390. [Lichens p. 382.]

671 —— (1939) Miscellaneous observations. Lichenes. *Trans. Norfolk Norwich Nat. Soc.* **14**: 478.

672 —— (1960) The lichens. In *Scolt Head Island* (J. A. Steers, ed.): 177–178. Cambridge: Heffer.

673 —— (1965) *The Broads*. [New Naturalist No. 46.] London: Collins. [Lichens pp. 123–124.]

—— —See also No. 1661.

674 Elmes, E. V. (1916) *Catalogue of Parfitt Collection of Devon Lichens*. Manuscript in BM.

675 Erichsen, C. F. E. (1935–36) Pertusariaceae. *Rabenh. Krypt.-Fl.* **9**, 5 (1): 319–728.

676 —— (1940) Neue Pertusarien nebst Mitteilungen über die geographische Verbreitung der europäischen Arten. *Annls mycol.* **38**: 16–55.

677 Evans, A. H. (1911) A short flora of Cambridgeshire. *Proc. Camb. phil. Soc. math. phys. Sci.* **16**: 198–284. [Also issued separately by Cambridge University Press.] [Lichens pp. 280–284 by P. G. M. Rhodes.]

678 Evans, A. L. (1971) *The Naturalists' Lake District including maritime Cumbria*. Clapham: Dalesman.

679 Evans, E. P. (1932) Cader Idris: a study of plant communities in south west

Merionethshire. *J. Ecol.* **20**: 1–52.

680 **Evans, J.** (1800) *A Tour through North Wales, in the year* 1798, *and at other times.* London: J. White.

681 **Evans, R. G.** (1947) The intertidal ecology of Cardigan Bay. *J. Ecol.* **34**: 273–309.

682 —— (1947) The intertidal ecology of selected localities in the Plymouth neighbourhood. *J. mar. biol. Ass. U.K.* **27**: 173–218.

683 —— (1949) The intertidal ecology of rocky shores in south Pembrokeshire. *J. Ecol.* **37**: 120–139.

684 **Ewen, A. H. and Prime, C. T.** (1975) *Ray's Flora of Cambridgeshire* (*Catalogus plantarum circa Cantabrigiam nascentium*). Hitchin: Wheldon & Wesley. [Lichens p. 79.]

685 **Ewing, P.** (1912) The summit-flora of the Breadalbane Range. *Glasg. Nat.* **4**: 48–62.

686 **Farrar, J. F.** (1974) A method for investigating lichen growth rates and succession. *Lichenologist* **6**: 151–155.

687 —— (1976) Ecological physiology of the lichen *Hypogymnia physodes*. I. Some effects of constant water saturation. *New Phytol.* **77**: 93–103.

688 **Farrow, E. P.** (1916) On the ecology of the vegetation of Breckland I. General description of Breckland and its vegetation. *J. Ecol.* **3**: 211–228.

689 —— (1916) On the ecology of the vegetation of Breckland II. Factors relating to the relative distributions of *Calluna*-heath and grass-heath in Breckland. *J. Ecol.* **4**: 57–64.

690 —— (1918) On the ecology of the vegetation of Breckland. IV. Characteristic bare areas and sand hummocks. *J. Ecol.* **6**: 144–152.

691 —— (1925) *Plant life on East Anglian heaths.* Cambridge: Cambridge University Press.

692 **Fearn, G. M.** (1973) Biological Flora of the British Isles: *Hippocrepis comosa* L. *J. Ecol.* **61**: 915–926.

693 **Fenton, A. F.** (1962) *Les lichéns en tant qu'indicateurs du degre de pollution atmospherique: Observations faites à Belfast, Irlande du Nord.* Geneva: World Health Organization Report WHO/EBL/3.

694 **Fenton, E. W.** (1939) The vegetation of "screes" in certain hill grazing districts of Scotland. *J. Ecol.* **27**: 502–512.

695 —— (1940) The influence of rabbits on the vegetation of certain hill-grazing districts of Scotland. *J. Ecol.* **28**: 438–449.

696 **Ferguson, D.** (1860) *The Natural History of Redcar and its Neighbourhood.* London: Simpkin Marshall. [Lichens p. 128.]

697 **Fergusson, J.** (1881) Mosses and lichens of the county. In *Angus or Forfarshire, the land and people, descriptive and historical* (A. J. Warden) **2**: 190–193. Dundee: C. Alexander.

698 **Ferreira, R. E. C.** (1958) *A comparative ecological and floristic study of the vegetation of Ben Hope, Ben Loyal, Ben Lui and Glas Maol in relation to the geology.* Ph.D. thesis, University of Aberdeen.

699 —— (1959) Scottish mountain vegetation in relation to geology. *Trans. Proc.*

bot. Soc. Edinb. **37**: 229–250.

700 —— **and Roger, J. G.** (1957) *Saxifraga hypnoides* L. on the coast of Banffshire. *Trans. Proc. bot. Soc. Edinb.* **37**: 133–136.

701 **Ferry, B. W.** (1971) The lichens of the Dale Peninsula and other nearby localities. *Fld Stud.* **3**: 481–496.

702 —— (1973) Lichens. In *The Natural History of Cape Clear Island* (J. T. R. Sharrock, ed.): 180–182. Berkhamstead: Poyser.

703 —— **and Baddeley, M. S.** (1976) Sulphur dioxide uptake in lichens. In *Lichenology: Progress and Problems* (D. H. Brown, D. L. Hawksworth and R. H. Bailey, eds.): 407–418. [Systematics Association Special Volume No. 8.] London, New York and San Francisco: Academic Press.

704 —— **and Sheard, J. W.** (1969) Zonation of supralittoral lichens on rocky shores around the Dale Peninsula, Pembrokeshire. *Fld Stud.* **3**: 41–67.

—— —See also Nos. 127, 128, 129, 432, 787, 941, 970, 1139, 1315, 1622, 2037, 2164, 2165 and 2166.

Finch, J. E. M.—See No. 1634.

705 **Findlay, T.** (1845) Parish of West Kilbride. In *The New Statistical Account of Scotland*. Vol. 5. *Ayr*: 243–273. Edinburgh and London: W. Blackwood & Sons. [Botany by D. Landsborough.]

Finegan, E. J.—See Nos. 127, 128 and 129.

706 **Fisher, G. C.** (1975) Some aspects of the phytosociology of heathland and related communities in the New Forest, Hampshire, England. *J. Biogeogr.* **2**: 103–116.

707 **Fisher, H.** (1906) Botany. In *The Victoria History of the Counties of England, Lancashire* (W. Page, ed.) **1**(2): 37–86. London: Constable. [Lichens pp. 82–85.]

—— —See also No. 1074.

Fisher, J.—See No. 1855.

Fisk, D.—See No. 2351.

Fitton, M. G.—See No. 226.

Fleming, J.—See No. 1481.

708 **Fletcher, A.** (1972) *The ecology of marine and maritime lichens of Anglesey.* Ph.D. thesis, University of Wales.

709 —— (1973) The ecology of marine (littoral) lichens on some rocky shores of Anglesey. *Lichenologist* **5**: 368–400.

710 —— (1973) The ecology of maritime (supralittoral) lichens on some rocky shores of Anglesey. *Lichenologist* **5**: 401–422.

711 —— (1975) Key for the identification of British marine and maritime lichens I. Siliceous rocky shore species. *Lichenologist* **7**: 1–52.

712 —— (1975) Key for the identification of British marine and maritime lichens II. Calcareous and terricolous species. *Lichenologist* **7**: 73–115.

—— —See also Nos. 539 and 1802.

713 **Foister, C. E.** (1948) Report of the annual conference of the Cryptogamic Section, 1946. *Trans. Proc. bot. Soc. Edinb.* **34**: 392–396.

714 —— (1949) Report on the annual conference of the Cryptogamic Section,

80

1947. *Trans. Proc. bot. Soc. Edinb.* **35**: 103–107.

715 —— (1950) Report of the annual conference of the Cryptogamic Section, 1948. *Trans. Proc. bot. Soc. Edinb.* **35**: 203–206.

716 **Folan, A.** (1965) *The lichen flora of Derryclare Wood.* Honours thesis, University College, Galway. [Not seen.]

Folkes, B. F.—See No. 2640.

717 **Follmann, G. and Huneck, S.** (1969) Mitteilungen über Flechteninhaltsstoffe LXXI. Zur Phytochemie und Chemotaxonomie der Diploschistaceae. *Bot. Jb.* **89**: 344–352.

718 **Forrest, G. I.** (1971) Structure and production of North Pennine blanket bog vegetation. *J. Ecol.* **59**: 453–479.

Forrest, J. E.—See No. 362.

719 **Forssell, K. B. J.** (1885) Beiträge zur Kenntnis der Anatomie und Systematik der Gloeolichenen. *Nova Acta R. Soc. Scient. upsal., ser.* 3, **13**(6): 1–118.

Forster, B. M.—See Nos. 357 and 720.

720 **Forster, E.** (1781–1817) [*Botanical Notes.*] 6 vols. Manuscripts in BM. [Includes notes by B. M. Forster and possibly others.]
—— —See also No. 357.

721 **Forster, T. F.** (1816) *Flora Tonbrigensis.* London. [Lichens pp. 142–161.]

722 —— (1842) *Flora Tonbrigensis.* Ed. 2. Tonbridge: J. Clifford. [Lichens pp. 142–161.]
—— —See also No. 357.

723 **Fowler, W.** (1881) Notes on the flora of Hodder-Dale. *Naturalist, Hull, o.s.* **7**: 15–16.

724 —— (1895) Lichens at Flamborough. *Naturalist, Hull* **1895**: 242.

725 **Fraser, G. T.** (1938) Thirtieth botany report. *Rep. Trans. Devon. Ass. Advmt Sci.* **70**: 67–82.

726 —— (1939) Thirty-first botany report. *Rep. Trans. Devon. Ass. Advmt Sci.* **71**: 77–90.

727 —— (1940) Thirty-second report on the botany of Devon. *Rep. Trans. Devon. Ass. Advmt Sci.* **72**: 67–80.

728 —— (1941) Thirty-third report on the botany of Devon. *Rep. Trans. Devon. Ass. Advmt Sci.* **73**: 63–72.

729 —— (1942) Thirty-fourth report on the botany of Devon. *Rep. Trans. Devon. Ass. Advmt Sci.* **74**: 69–83.

730 **Fraymouth, J.** (1928) The moisture relations of terrestrial algae. III. The respiration of certain lower plants, including terrestrial algae, with special reference to the influence of drought. *Ann. Bot.* **42**: 75–100.

731 **Friend, H.** (1890) A mud-capped dyke. *Sci. Gossip* **26**: 135–136.

732 **Fries, E. M.** (1831) *Lichenographia Europaea reformata.* Lund.

733 —— (1831) *Primitiae geographiae Lichenum.* Lund.

734 **Fries, Th. M.** (1858) Monographia Stereocaulorum et Pilophororum. *Nova Acta R. Soc. Scient. upsal., ser.* 3, **2**(1): 1–76.

735 **Fritsch, F. E. and Salisbury, E. J.** (1915) Further observations on the heath association on Hindhead Common. *New Phytol.* **14**: 116–138.

Frohawk, W.—See No. 242.

Frost, L. C.—See No. 452.

736 **Fry, E. J.** (1922) Some types of endolithic limestone lichens. *Ann. Bot.* **36**: 541–562.

737 —— (1923–28) *Expeditions note book*. Manuscript in LINN no. 8° 40.

738 —— (1927) The mechanical action of crustaceous lichens on substrata of shale, schist, gneiss, limestone and obsidian. *Ann. Bot.* **41**: 437–460.

739 —— (1928) The penetration of lichen gonidia by the fungal component. *Ann. Bot.* **42**: 141–148.

Futty, D. W.—See No. 1896.

740 **Galinou, M.-A.** (1954) Contributions à la connaissance des lichens du Massif Armoricains. *Bull. Soc. scient. Bretagne* **29**: 49–56.

—— —See also No. 1386.

741 **Galpine, J.** (1819) *A Synoptical Compend of British Botany*. Ed. 2. Liverpool. [Some copies dated 1820; later Eds. not seen.]

742 **Galt, W.** (1872) New British lichen, *Lecidea fossarum. Grevillea* **1**: 75.

743 **Gardiner, M.** (1845) Parish of Bothwell. In *The New Statistical Account of Scotland*. Vol. 6. *Lanark*: 765–804. Edinburgh and London: W. Blackwood & Sons.

744 **[Gardiner, W.]** (1841) Notice of some rare plants, chiefly cryptogamic, found on the Sands of Barrie and elsewhere, in May and June 1841. *Proc. bot. Soc. Edinb.* **1840–41**: 58.

745 [——] (1841) Notice of the occurrence of several rare cryptogamic plants on the Sidlaw Hills. *Trans. bot. Soc. Edinb.* **1**: 62.

746 —— (1842) Sketch of an excursion to the Clova Mountains, in July and August, 1840. *Phytologist* **1**: 212–217.

747 —— (1842) Notice of the occurrence of several rare cryptogamic plants on the Sidlaw Hills. *Phytologist* **1**: 321.

748 —— (1843) Contributions towards a flora of the Breadalbane mountains. *Phytologist* **1**: 468–476.

749 —— (1843) Account of two botanical visits to the Reeky Linn and Den of Airly, Forfarshire, in April and June, 1842. *Trans. bot. Soc. Edinb.* **1**: 182–184.

750 —— (1844) Two botanical visits to the Reeky Linn and Den of Airley, in April and June, 1842. *Phytologist* **1**: 898–901.

751 —— (1844) List of a few botanical rarities collected in Scotland in 1843. *Phytologist* **1**: 915–918.

752 —— (1844) *Alectoria jubata. Phytologist* **1**: 939–940.

753 —— (1844) Corrections of some errors in Mr. Gardiner's list of plants. *Phytologist* **1**: 972.

754 —— (1844) The Sidlaw-hills. *Phytologist* **1**: 972.

755 —— (1845) *Botanical Rambles in Braemar in 1844; with an Appendix on Forfarshire Botany*. Dundee.

756 —— (1848) *The Flora of Forfarshire*. London: Longman, Brown, Green and Longmans. [Lichens pp. 268–287.]

756a —— (1850) *Illustrations of British Botany* [*to accompany the Flora of Forfarshire*]. Dundee. [Bound volume with printed title label, hand-written subtitle, and specimens with hand-written labels, in BM.]

756b —— (1851) *Illustrations of British Botany*. Vol. 15. *Cellulares. Musci, Hepaticae, Lichenes, Fungi*. Dundee. [One of 16 bound vols. with manuscript labels in BM; contents different from No. 756a.]

757 **Garms, H.** (1967) *The Natural History of Europe*. London: Paul Hamlyn.

758 **Garner, R.** (1844) *The Natural History of the County of Stafford*. London. [Lichens pp. 428–433.]

759 **Garnett, T.** (1800) *Observations on a tour through the Highlands and part of the Western Isles of Scotland*. Vol. 1. London: T. Cadell.

760 **Garrad, L. S.** (1972) The wildlife of the Ayres. *Peregrine* **4**: 21–41.

761 **Garrett, R. M.** (1967) *Studies on lichen ascospore dispersal*. Science Long Study Dissertation, St Paul's College of Education, Cheltenham. [Copy with BLS.]

762 **Garrod, G.** (1971) Ringmere. *Trans. Norfolk Norwich Nat. Soc.* **22**: 73–82.

Gemmell, A. R.—See No. 812.

763 **Gentlemen, Several Literary** (1830) *The Teignmouth, Dawlish, and Torquay Guide*. Teignmouth: E. Croydon. [See also No. 2395.]

764 **Gerard, J.** (1597) *The Herball or Generall Historie of Plantes*. London.

765 —— (1633) *The Herball or Generall Historie of Plantes* [Edition enlarged and amended by T. Johnson]. London. [Lichens pp. 1557–1566 lichen section little changed in 1638 and later issues.]

766 **Gerson, U.** (1973) Lichen-arthropod associations. *Lichenologist* **5**: 434–443.

767 **Gibbs, A. E.** [*c.* 1900] *The lichen flora of Hertfordshire with the localities in which the species occur in the different river basins*. Vol. 1. Manuscript in the City Museum, St Albans. [Location of vol. 2 unknown.]
—— —See also No. 1065.

768 **Gibbs, T.** (1902) Yorkshire Naturalists' Union at Masham. *Naturalist, Hull* **1902**: 15–20.

769 **Gibson, S.** (1842) Additions to the list of Wharfedale mosses. *Phytologist* **1**: 291–292.

770 **Gilbert, J. L.** (1959) *Parmelia glomellifera* new to British Isles. *Lichenologist* **1**: 84–85.

771 —— (1959) Huntingdonshire lichens II. *Rep. Huntingdon. Fauna Flora Soc.* **11**: 14–16. [For part I see No. 1301.]

772 —— (1972) Kew's lichens. *J. Kew Guild* **9**: 38–40.

773 **Gilbert, O. L.** (1965) Lichens as indicators of air pollution in the Tyne Valley. In *Ecology and the Industrial Society* (G. T. Goodman, R. W. Edwards and J. M. Lambert, eds.): 35–47. [Symposium of the British Ecological Society. No. 5.] Oxford: Blackwell Scientific Publications.

774 —— (1966) Lichen pathogens on *Lecanora conizaeoides* Nyl. ex Cromb. *Lichenologist* **3**: 275.

775 —— (1968) *Biological indicators of air pollution*. Ph.D. thesis, University of Newcastle upon Tyne.

776 —— (1968) Bryophytes as indicators of air pollution in the Tyne Valley. *New Phytol.* **67**: 15–30.

777 —— (1968) Biological estimation of air pollution. In *Plant Pathologist's Pocketbook* (Commonwealth Mycological Institute, ed.): 206–207. Kew: Commonwealth Mycological Institute.

778 —— (1969) The effect of SO₂ on lichens and bryophytes around Newcastle upon Tyne. In *Air Pollution. Proceedings of the First European Congress on the Influence of Air Pollution on Plants and Animals, Wageningen 1968*: 223–235. Wageningen: Centre for Agricultural Publishing and Documentation.

779 —— (1970) Lichens. In *Natural History of the Lake District* (G. A. K. Hervey and J. A. G. Barnes, eds.): 72–75. London: Warne.

780 —— (1970) Further studies on the effect of sulphur dioxide on lichens and bryophytes. *New Phytol.* **69**: 605–627.

781 —— (1970) A biological scale for the estimation of sulphur dioxide pollution. *New Phytol.* **69**: 629–634.

782 —— (1970) New tasks for lowly plants. *New Scient.* **46**: 288–289.

783 —— (1971) Some indirect effects of air pollution on bark-living invertebrates. *J. appl. Ecol.* **8**: 77–84.

784 —— (1971) Studies along the edge of a lichen desert. *Lichenologist* **5**: 11–17.

785 —— (1971) The effect of airborne fluorides on lichens. *Lichenologist* **5**: 26–32.

786 —— (1972) Field meeting in Northumberland. *Lichenologist* **5**: 337–341.

787 —— (1973) The effect of airborne fluorides. In *Air Pollution and Lichens* (B. W. Ferry, M. S. Baddeley and D. L. Hawksworth, eds.): 176–191. London: Athlone Press of the University of London.

788 —— [1974] (1973) Lichens and air pollution. In *The Lichens* (V. Ahmadjian and M. E. Hale, eds.): 443–472. New York and London: Academic Press.

789 —— (1974) An air pollution survey by school children. *Envir. Poll.* **6**: 175–180.

790 —— (1974) Reindeer grazing in Britain. *Lichenologist* **6**: 165–167.

791 —— (1975) Distribution maps of lichens in Britain. Map 19. *Solorina saccata* (L.) Ach. *Lichenologist* **7**: 181–183.

792 —— (1975) Distribution maps of lichens in Britain. Map 20. *Solorina spongiosa* (Sm.) Anzi. *Lichenologist* **7**: 184–185.

793 —— (1975) Distribution maps of lichens in Britain. Map 21. *Solorina bispora* Nyl. *Lichenologist* **7**: 186–188.

794 —— (1975) Distribution maps of lichens in Britain. Map 22. *Solorina crocea* (L.) Ach. *Lichenologist* **7**: 190–192.

795 —— (1975) Lichens. In *Bedford Purlieus: its history, ecology and management* (G. F. Peterken and R. C. Welch, eds.): 125–129. [Monks Wood Symposium No. 7.] Huntingdon: Institute of Terrestrial Ecology.

796 —— (1975) *Wildlife Conservation and Lichens.* Exeter: Devon Trust for Nature Conservation.

797 —— (1976) A lichen-arthropod community. *Lichenologist* **8**: 96.

798 —— (1976) An alkaline dust effect on epiphytic lichens. *Lichenologist* **8**: 173–178.

84

799 —— (1976) The construction, interpretation and use of lichen/air pollution maps. In *Proceedings of the Kuopio Meeting on Plant Damages Caused by Air Pollution* (W. Kärenlampi, ed.): 83–92. Kuopio: University of Kuopio and Kuopio Naturalists' Society.

800 —— **Holligan, P. M. and Holligan, M. S.** (1973) The flora of North Rona 1972. *Trans. Proc. bot. Soc. Edinb.* **42**: 43–68.

801 —— **and Wathern, P.** (1976) The flora of the Flannan Isles. *Trans. Proc. bot. Soc. Edinb.* **42**: 487–503.

—— —See also No. 1897.

Gilchrist, J.—See No. 2216.

802 **Gillham, M. E.** (1953) An ecological account of the vegetation of Grassholm Island, Pembrokeshire. *J. Ecol.* **41**: 84–99.

803 —— (1954) An annotated list of the bryophytes and lichens of Skokholm Island, Pembrokeshire. *NWest. Nat.* **25**: 37–48.

804 —— (1954) The marine algae of Skokholm and Grassholm Islands, Pembrokeshire. *NWest Nat.* **25**: 204–225.

805 —— (1956) Ecology of the Pembrokeshire Islands. V. Manuring by the colonial seabirds and mammals, with a note on seed distribution by gulls. *J. Ecol.* **44**: 429–454.

806 **Gilmour, J. L. S.,** ed. (1972) *Thomas Johnson, Botanical Journeys in Kent and Hampstead. A Facsimile Reprint with Introduction and Translation of his Iter-Plantarum 1629, Descriptio Itineris Plantarum 1632.* Pittsburgh: The Hunt Botanical Library.

807 **Gimingham, C. H.** (1960) Biological Flora of the British Isles: *Calluna vulgaris* (L.) Hull. *J. Ecol.* **48**: 455–483.

808 —— (1964) Maritime and sub-maritime communities. In *The Vegetation of Scotland* (J. H. Burnett, ed.): 67–142. Edinburgh and London: Oliver and Boyd.

809 —— (1964) Dwarf-shrub heaths. In *The Vegetation of Scotland* (J. H. Burnett, ed.): 232–288. Edinburgh and London: Oliver and Boyd.

810 —— (1972) *Ecology of Heathlands.* London: Chapman & Hall.

811 —— **and Cormack, E.** (1964) Plant distribution and growth in relation to aspect on hill slopes in North Scotland. *Trans. Proc. bot. Soc. Edinb.* **39**: 525–538.

812 —— **Gemmell, A. R. and Greig-Smith, P.** (1949) The vegetation of a sand-dune system in the Outer Hebrides. *Trans. Proc. bot. Soc. Edinb.* **35**: 82–96.

813 —— **Miller, G. R., Sleigh, L. M. and Milne, L. M.** (1961) The ecology of a small bog in Kinlochewe Forest, Wester Ross. *Trans. Proc. bot. Soc. Edinb.* **39**: 125–147.

—— —See also No. 1972.

814 **Gittins, R.** (1965) Multivariate approaches to a limestone grassland community. I. A stand ordination. *J. Ecol.* **53**: 385–401.

815 —— (1965) Multivariate approaches to a limestone grassland community. II. A direct species ordination. *J. Ecol.* **53**: 403–409.

Gliddon, W. A.—See Nos. 93 and 94.

816 **Glover, S.** (1829) *The History of the County of Derby.* Vol. 1. Derby: H. Mosley & Son. [Lichens p. 128.]

817 —— (1831) *The History of the County of Derby.* Vol. 1 [Ed. 2]. Derby: H. Mosley. [Lichens pp. 110–111.]

818 **Godwin, H. and Conway, V. M.** (1939) The ecology of a raised bog near Tregaron, Cardiganshire. *J. Ecol.* **27**: 313–363.

819 **Goode, D. A.** (1972) A note on the bog flora of Wigtownshire. *Trans. Proc. bot. Soc. Edinb.* **41**: 541–545.

820 —— (1974) The flora and vegetation of Shetland. In *The Natural Environment of Shetland* (R. Goodier, ed.): 50–72. Edinburgh: Nature Conservancy Council.

Goodier, R.—See Nos. 234, 820, 1120 and 2248.

Goodman, G. T.—See No. 773.

Gordon, V.—See No. 2099.

Gough, R.—See Nos. 357 and 358.

821 **Gowland T. S.** [*c.* 1872] *The Guide to Eastbourne.* Ed. 9. Eastbourne: Gowland. [Earlier editions not seen; many later editions with changed pagination to at least Ed. 19 (*c.* 1880).] [Lichens p. 82.]

822 **Graham, G. G.** (1971) *Phytosociological studies of relict woodlands in the north east of England.* M.Sc. thesis, University of Durham.
—— —See also No. 347.

823 **Graham, R.** (1830) Notice of plants observed in an excursion made by Dr. Graham with part of his botanical pupils, accompanied by a few friends, in August last. *Edinb. N. phil. J.* **9**: 360–363.

824 **Gray, J.** [pseud. H. Speight] (1891) *Through Airedale from Goole to Malham.* Leeds: Walker and Laycock.
—— —See also No. 2249.

825 **Gray, P.** (1887) A gossip about lichens. *Trans. Dumfries. Galloway nat. hist. antiq. Soc., ser.* 2, **4**: 142–143.
—— —See also Nos. 1053 and 1054.

826 **Gray, S. F.** (1821) *A Natural Arrangement of British Plants.* Vol. 1. London: Baldwin, Cradock & Joy. [Lichens pp. 394–507.]

Green, B. H.—See No. 1953.

Green, H. E.—See No. 845.

827 **Green, T. G. A. and Smith, D. C.** (1974) Lichen physiology XIV. Differences between lichen algae in symbiosis and in isolation. *New Phytol.* **73**: 753–766.

Greenhalgh, G. N.—See No. 228.

828 **Greenwood, A.** (1846) *Borrera flavicans* discovered in fruit near Penzance. *Phytologist* **2**: 496.

829 **Gregory, P. H.** (1951) The fungi of Hertfordshire. *Trans. Herts. nat. Hist. Soc. Fld Club* **23**: 135–208. [Lichens pp. 160–168.]

830 **Gregson, C. S.** (1887) A day's 'scientific' insect-hunting on the Isle of Man in June. *Young Nat.* **8**: 153–155.

Greig-Smith, P.—See Nos. 812 and 1399.

831 **Grelet, l'Abbé L.-J.** (1927) Discomycètes nouveaux (2e série). *Bull. Soc. mycol. Fr.* **42**: 203–207.

832 **Greville, R. K.** (1824) *Flora Edinensis.* Edinburgh: William Blackwood. [Lichens pp. 324–354.] [Copy in E with annotations by Greville.]

833 —— (1826) *Scottish Cryptogamic Flora.* Vol. 4. Edinburgh: MacLachlan and Stewart. [Lichens Pls. 186, 219 and 221.]

834 —— (1830) *Algae Britannicae.* Edinburgh: MacLachlan and Stewart. [Lichens pp. 21–23.]

835 **Grieve, S.** (1882) How we spent the 30th July 1879 in the wilds of Kilmonivaig and north-west Badenoch. *Trans. Edinb. nat. Fld Club.* **1**: 3–8.

Griffin, W. H.—See No. 229.

836 **Griffith, J. E.** [1895] *The Flora of Anglesey and Carnarvonshire.* Bangor: Nixon and Jarvis. [Lichens pp. 174–218.]

837 **Griffith, J. L.** (1966) *Some aspects of the effect of atmospheric pollution on the lichen flora to the west of Consett, Co. Durham.* M.Sc. thesis, University of Durham.

Grime, A.—See No. 1780.

838 **Grimshaw, J.** (1976) Chemical differentiation of lichens. *Educ. Chem.* **13**(3): 74–75.

839 **Grindon, L. H.** (1859) *The Manchester Flora.* London: W. White. [Lichens pp. 510–519.]

840 —— (1866) *Summer Rambles in Cheshire, Derbyshire, Lancashire, and Yorkshire.* Manchester & London: Palmer & Howe.

841 —— (1868) The Manchester peat-bogs, locally 'mosses'. *Rep. Proc. Manchr Fld Nat. Archaeol. Soc.* **1867**: 28–36.

842 **Grinling, C. H., Ingram, T. A. and Polkinghorne, B. C.,** eds. (1909) *A Survey and Record of Woolwich and West Kent.* Woolwich [Lichens pp. 220–224; see also No. 229.]

843 **Grove, W. B.** (1893) The fungi of Abbot's "Flora Bedfordiensis" (1798). *Midland Nat.* **16**: 212–216, 235–240.

—— —See also Nos. 1628 and 1629.

844 **Groves, E. W.** (1965) Three new vice-county records for Middlesex from the Ruislip Local Nature Reserve. *J. Ruislip Distr. nat. Hist. Soc.* **14**: 45–46.

845 **Grubb, P. J., Green, H. E. and Merrifield, R. C. J.** (1969) The ecology of chalk heath: its relevance to the calcicole-calcifuge and soil acidification problems. *J. Ecol.* **57**: 175–212.

846 **Grummann, V. [J.]** (1963) *Catalogus Lichenum Germaniae.* Stuttgart: G. Fischer.

847 **Guillaumot, l'Abbé M.** (1951) *Flora des Lichens de France et de Grande-Bretagne.* [*Encycl. Biol.* **42**: 1–605.] Paris: Lechevalier.

848 **Gulliver, G.** (1841) *A Catalogue of Plants collected in the neighbourhood of Banbury.* London, etc. [Lichens pp. 31–35.]

849 **Gunther, W. T.,** ed. (1928) *Further Correspondence of John Ray.* London: Ray Society.

Guse, K.—See No. 1391.

850 **Gutch, J. W. G.** (1842) A list of plants met with in the neighbourhood of Swansea, Glamorgan. *Phytologist* **1**: 180–187.

851 —— (1842) Additions to the list of plants met with in the neighbourhood of Swansea. *Phytologist* **1**: 377–380.

852 **Gyelnik, V.** (1932) *Nephroma*-Studien. *Hedwigia* **72**: 1–30.

853 —— (1935) Conspectus Bryopogonum. *Reprium Spec. nov. Regni veg.* **38**: 219–255.

854 —— (1940) Lichinaceae. *Rabenh. Krypt.-Fl.* **9**, 2 (2): 1–110.

855 —— (1940) Pannariaceae. *Rabenh. Krypt.-Fl.* **9**, 2 (2): 135–272.

856 **Hafellner, J. and Poelt, J.** (1976) *Rhizocarpon schedomyces* spec. nov., eine fast delichenisierte parasitische Flechte und seine Verwandten. *Herzogia* **4**: 5–14.

857 **Haggart, D. A.** (1915) Botanical notes: Appin, Fortingall, Schiehallion, and Ben Lawers. *Trans. Proc. Perthsh. Soc. nat. Sci.* **6**: 44–55.

Hailstone, S.—See No. 2620.

858 **Hale, M. E.** (1956) Chemical strains of the lichen *Parmelia furfuracea*. *Am. J. Bot.* **43**: 456–459.

859 —— (1964) The *Parmelia conspersa* group in North America and Europe. *Bryologist* **67**: 462–473.

860 —— (1965) A monograph of *Parmelia* subgenus *Amphigymnia*. *Contr. U.S. natn. Herb.* **36**: 193–357.

861 —— (1965) Studies on the *Parmelia borreri* group. *Svensk bot. Tidskr.* **59**: 37–48.

862 —— (1968) A synopsis of the lichen genus *Pseudevernia*. *Bryologist* **71**: 1–11.

863 —— **and Kurokawa, S.** (1962) *Parmelia* species first described from the British Isles. *Lichenologist* **2**: 1–5.

—— —See also Nos. 788, 1219 and 1890.

Hall, R. H.—See No. 658.

Hannig, E.—See Nos. 559 and 2088a.

864 **Hardy, J.** (1863) New British cryptogamia. *J. Bot., Lond.* **1**: 307–308.

865 —— (1863) The lichen flora of the Eastern Borders. *Hist. Berwicksh. Nat. Club* **4**: 396–428.

866 —— (1872) On Langleyford Vale and the Cheviots. *Hist. Berwicksh. Nat. Club* **6**: 353–375.

867 —— [1875] Miscellanea; extracts from correspondence, &c. II.—Botanical. *Hist. Berwicksh. Nat. Club* **7**: 299–300.

868 —— [1885] Report of meetings of the Berwickshire Naturalists' Club for the year 1884. Prestonkirk, for Whittingham and Presmennan. *Hist. Berwicksh. Nat. Club* **10**: 461–476.

869 —— [1885] The tomb of Cockburn of Henderland, and its vicinity. *Hist. Berwicksh. Nat. Club* **10**: 604–607.

870 —— [1885] Miscellanea. *Hist. Berwicksh. Nat. Club* **10**: 607–610.

871 —— [1886] Report of meetings of Berwickshire Naturalists' Club, for the year 1885. Jedburgh for Oxnam. Supplement. *Hist. Berwicksh. Nat. Club* **11**: 26–31.

872 —— [1886] Report of meetings of Berwickshire Naturalists' Club, for the year 1885. Rothbury. Supplement. *Hist. Berwicksh. Nat. Club* **11**: 37–50.

873 —— [1886] Report of meetings of Berwickshire Naturalists' Club, for the year 1885. Westruther and Wedderlie. *Hist. Berwicksh. Nat. Club* **11**: 64–75.

874 —— [1886] Report of meetings of Berwickshire Naturalists' Club, for the year 1885. Cockburnspath for Aikengall. *Hist. Berwicksh. Nat. Club* **11**: 75–91.

875 —— [1887] Report of meetings of the Berwickshire Naturalists' Club, for the year 1886. Lauder, etc. *Hist. Berwicksh. Nat. Club* **11**: 350–361.

876 —— [1887] Report of meetings of the Berwickshire Naturalists' Club, for the year 1886. Peebles. *Hist. Berwicksh. Nat. Club* **11**: 361–386.

877 —— [1888] Report of the meetings of the Berwickshire Naturalists' Club, for the year 1887. Alwinton, the Drake Stone, Harbottle Lough, Harbottle, Holystone, Biddleston. *Hist. Berwicksh. Nat. Club* **12**: 38–55.

878 —— [1888] Report of the meetings of the Berwickshire Naturalists' Club, for the year 1887. Rule Water, Jedburgh, Weens, Bonchester Bridge, Wells, Bedrule, the Dunion. *Hist. Berwicksh. Nat. Club* **12**: 68–76.

879 —— [1889] Report of the meetings of the Berwickshire Naturalists' Club, for the year 1888. Jedburgh, Minto, Chesters, Ancrum, Monteviot. *Hist. Berwicksh. Nat. Club* **12**: 185–194.

880 —— [1890] Report of the meetings of the Berwickshire Naturalists' Club for the year 1889. *Hist. Berwicksh. Nat. Club* **12**: 436–453, 469–497.

881 —— [1891] Report of the meetings of the Berwickshire Naturalists' Club for the year 1890. Beanley. *Hist. Berwicksh. Nat. Club* **13**: 22–30.

882 —— [1891] Report of the meetings of the Berwickshire Naturalists' Club for the year 1890. Callaly Camps. *Hist. Berwicksh. Nat. Club* **13**: 41–50.

883 —— [1892] Report of the meetings of the Berwickshire Naturalists' Club for the year 1891. Dunbar, Aikengall, Shippath, Thurston. *Hist. Berwicksh. Nat. Club* **13**: 312–320.

884 —— [1892] Remarks on some new localities for rare plants. *Hist. Berwicksh. Nat. Club* **13**: 410–411.

885 —— [1893] Report of the meetings of the Berwickshire Naturalists' Society for the year 1892. Wauchope meeting. Additional particulars. *Hist. Berwicksh. Nat. Club* **14**: 13–18.
—— —See also Nos. 1203 and 1929.

886 **Hardy, M.** (1905) *Esquisse de la Géographie et de la Végétation des Highlands d'Ecosse.* Paris: Imprimerie Générale Lahure.

887 **Harriman, J.** (1806) [Letter to Shute Barrington, Bishop of Durham]. Manuscript in LINN (Smith MSS no. 38).

888 **Harris, G. P.** (1969) *A study of the ecology of corticolous lichens.* Ph.D. thesis, University of London.

889 —— (1971) The ecology of corticolous lichens. I. The zonation on oak and birch in South Devon. *J. Ecol.* **59**: 431–439.

890 —— (1971) The ecology of corticolous lichens. II. The relationship between physiology and the environment. *J. Ecol.* **59**: 441–452.

891 —— (1972) The ecology of corticolous lichens. III. A simulation model of productivity as a function of light intensity and water availability. *J. Ecol.* **60**: 19–40.

—— —See also Nos. 1233 and 1234.

892 **Harris, G. T.** (1926) Eighteenth botany report. *Rep. Trans. Devon. Ass. Advmt Sci.* **58**: 121–132.

893 —— (1927) Nineteenth botany report. *Rep. Trans. Devon. Ass. Advmt Sci.* **59**: 73–85.

894 —— (1927) An ecological reconnaissance of East Devon. *Rep. Trans. Devon. Ass. Advmt Sci.* **58**: 299–335.

895 —— (1929) Twenty-first botany report. *Rep. Trans. Devon. Ass. Advmt Sci.* **61**: 89–107.

896 —— (1930) Twenty-second botany report. *Rep. Trans. Devon. Ass. Advmt Sci.* **62**: 125–133.

897 —— (1933) Twenty-fifth botany report. *Rep. Trans. Devon. Ass. Advmt Sci.* **65**: 89–102.

898 —— (1934) Twenty-sixth botany report. *Rep. Trans. Devon. Ass. Advmt Sci.* **66**: 45–52.

899 —— (1935) Twenty-seventh botany report. *Rep. Trans. Devon. Ass. Advmt Sci.* **67**: 83–96.

900 —— (1936) Twenty-eighth botany report. *Rep. Trans. Devon. Ass. Advmt Sci.* **68**: 55–62.

901 —— (1937) Twenty-ninth botany report. *Rep. Trans. Devon. Ass. Advmt Sci.* **69**: 115–121.

902 —— (1938) An ecological reconnaissance of Dartmoor. *Rep. Trans. Devon. Ass. Advmt Sci.* **70**: 37–55.

903 **Harris, T. M.** (1946) Zinc poisoning of wild plants from wire netting. *New Phytol.* **45**: 50–55.

904 **Hartley, J. W. and Wheldon, J. A.** (1914) The Manx sand-dune flora. *J. Bot., Lond.* **52**: 170–175.

905 —— —— (1927) The lichens of the Isle of Man. *NWest. Nat.* **2** *Suppl.*: 1–38. [Issued in 3 parts, all in 1927.]

Harvey, G. A. K.—See No. 779.

906 **Harvey, L. B. and Russell, B.** (1976) A preliminary survey of the distribution of bryophytes and lichens, epiphytic on oaks, on a transect from Wolverhampton to the Welsh border. *Proc. Bgham nat. Hist. Soc.* **23**: 109–116.

Harvey, R.—See No. 1893.

907 **Harvey, W. H.** (1841) *A Manual of British Algae*. London: J. van Voorst. [Lichens pp. 22–23; unchanged in Ed. 2 (1849).]

908 —— (1854) *The Sea-side Book*. London: J. van Voorst.

909 **Hasselrot, T. E.** (1953) Nordliga lavar i syd-och mellansverige. *Acta phytogeogr. suec.* **33**: 1–200.

910 **Hasted, E.** (1782) *The History and Topographical Survey of the County of*

Kent. Vol. 2. Canterbury: privately printed. [Lichens p. 343.]

Hastings, S.—See Nos. 1772a and 1773.

911 **Hawksworth, D. L.** (1966) The lichen flora of Foula (Shetland). *Lichenologist* **3**: 218–223.

912 —— (1966) Foula—Botanical studies & preliminary report on the peat deposits. *A. Rept Brathay Expl. Gp* **1965**: 38–42.

913 —— (1967) Foula—Botanical studies: 2. *A. Rept Brathay Expl. Gp* **1966**: 139–146.

914 —— (1967) Notes on Shetland lichens: 1. *Trans. Proc. bot. Soc. Edinb.* **40**: 283–287.

915 —— (1967) A note on Suffolk lichens. *Trans. Suffolk Nat. Soc.* **13**: 323.

916 —— (1967) Lichens collected by Jonathan Salt between 1795 and 1807 now in the herbarium of Sheffield Museum. *Naturalist, Hull* **1967**: 47–50.

917 —— (1968) A note on the chemical strains of the lichen *Ramalina subfarinacea*. *Bot. Notiser* **121**: 317–320.

918 —— (1969) The scanning electron microscope. An aid to the study of cortical hyphal orientation in the lichen genera *Alectoria* and *Cornicularia*. *J. microsc., Paris* **8**: 753–760.

919 —— (1969) Chemical and nomenclatural notes on *Alectoria* (Lichenes) I. *Taxon* **18**: 393–399.

920 —— (1969) The lichen flora of Derbyshire. *Lichenologist* **4**: 105–193.

921 —— (1969) Notes on the flora and vegetation of Foula, Zetland (v.c. 112). *Proc. bot. Soc. Br. Isl.* **7**: 537–547.

922 —— (1969) The bryophyte flora of Foula (Shetland). *Revue bryol. lichén.* **36**: 213–218.

923 —— (1970) *Systematic studies on the lichen genus* Alectoria *Ach. with particular reference to the British species*. 2 vols. Ph.D. thesis, University of Leicester.

924 —— (1970) The chemical constituents of *Haematomma ventosum* (L.) Massal. in the British Isles. *Lichenologist* **4**: 248–255.

925 —— (1970) Guide to the literature for the identification of British lichens. *Bull. Br. mycol. Soc.* **4**: 73–95.

926 —— (1970) A nomenclatural note on *Racodium* Pers. *Trans. Br. mycol. Soc.* **54**: 323–326.

927 —— (1971) Lichens as litmus for air pollution: a historical review. *Int. J. Environ. Studies* **1**: 281–296.

928 —— (1971) *Lobaria pulmonaria* (L.) Hoffm. transplanted into Dovedale, Derbyshire. *Naturalist, Hull* **1971**: 127–128.

929 —— (1971) Field meeting at Leicester. *Lichenologist* **5**: 170–174.

930 —— (1971) *An Introduction to the Lichens of Devonshire*. Kew: Commonwealth Mycological Institute.

931 —— (1972) Regional studies in *Alectoria* (Lichenes) II. The British species. *Lichenologist* **5**: 181–261.

932 —— (1972) The natural history of Slapton Ley Nature Reserve IV. Lichens. *Fld Stud.* **3**: 535–578.

933 —— (1972) *Leicestershire & Rutland. Report on the Lichen Flora and Vegeta-*

tion of Sites of Scientific Importance for lichens together with notes on other notified NNR and SSSI localities. Manuscript. Report to the Nature Conservancy; copy with BLS.

934 —— (1972) *Berry Head (including briefer observations on Chudleigh Rocks & Buckfastleigh Caves). Report on the Lichen Flora and Vegetation.* Manuscript. Report to the Nature Conservancy; copy with BLS.

935 —— (1972) *Dartmoor Oak Woodlands (Piles Copse, Sampford Spiney, Wistmans Wood, Holne Chase, Black Tor Copse & Yarner Wood). Report on the Lichen Flora and Vegetation.* Manuscript. Report to the Nature Conservancy; copy with BLS.

936 —— (1972) *Axmouth-Lyme Regis Undercliffs National Nature Reserve. Report on the Lichen Flora and Vegetation.* Manuscript. Report to the Nature Conservancy; copy with BLS.

937 —— (1973) *Watersmeet, North Devon. Report on the Lichen Flora and Vegetation.* Manuscript. Report to the Nature Conservancy; copy with BLS.

938 —— (1973) The lichen flora and vegetation of Berry Head, South Devonshire. *Trans. Torquay nat. hist. Soc.* **16**: 55–66.

939 —— (1973) Ecological factors and species delimitation in the lichens. In *Taxonomy and Ecology* (V. H. Heywood, ed.): 31–69. [Systematics Association Special Volume No. 5.] London and New York: Academic Press.

940 —— (1973) Some advances in the study of lichens since the time of E. M. Holmes. *Bot. J. Linn. Soc.* **67**: 3–31.

941 —— (1973) Mapping studies. In *Air Pollution and Lichens* (B. W. Ferry, M. S. Baddeley and D. L. Hawksworth, eds.): 38–72. London: Athlone Press of the University of London.

942 —— (1974) Lichens and indicators of environmental change. *Environment and Change* **2**: 381–386.

943 —— (1974) *Mycologist's Handbook.* Kew: Commonwealth Mycological Institute.

944 —— (1974) *Report on the Lichen Flora of the Peak District National Park (with emphasis on Sites of Special Scientific Interest).* Shrewsbury: Nature Conservancy Council.

945 —— (1974) Man's impact on the British fauna and flora. *Outl. Agric.* **8**: 23–28.

946 —— (1974) The lichen flora of Derbyshire—Supplement I. *Naturalist, Hull* **1974**: 57–64.

947 —— (1974) Progress in the study of the Devonshire lichen flora. *Bull. Br. Lichen Soc.* **35**: 5.

948 —— (1975) Notes on British lichenicolous fungi, I. *Kew Bull.* **30**: 183–203.

949 —— (1975) A revision of lichenicolous fungi accepted by Keissler in *Coniothecium. Trans. Br. mycol. Soc.* **65**: 219–238.

950 —— (1975) The changing lichen flora of Leicestershire. *Trans. Leicester lit. phil. Soc.* **68**: 32–56.

951 —— (1975) *Chudleigh Rocks SSSI. Report on the Lichen Flora and Vegetation.* Manuscript. Report to the Nature Conservancy; copy with BLS.

952 —— (1975) *Investigations into the occurrence and status of rare epiphytic*

92

lichens in South Devonshire. Manuscript. Report to the World Wildlife Fund; copy with BLS.

953 —— (1975) Autumn field meeting at Leek, Staffordshire, 1975. *Bull. Br. Lichen Soc.* **36**: 2.

954 —— (1975) *Report on the Lichen Flora of the Peak District National Park. Supplement I. Further Studies in the Cheshire and Staffordshire parts of the National Park.* Manuscript. Report to the Nature Conservancy; copy with BLS.

955 —— (1976) Looking at lichens. *Nat. Hist. Book Rev.* **1**: 8–15.

956 —— (1976) *Dartmoor National Park. Report on the Lichen Flora and Vegetation I. Epiphytic species.* Manuscript. Report to the Nature Conservancy; copy with BLS.

957 —— (1976) The natural history of Slapton Ley Nature Reserve X. Fungi. *Fld Stud.* **4**: 391–439.

958 —— (1976) Field meeting in north Staffordshire. *Lichenologist* **8**: 189–196.

959 —— (1976) New and interesting microfungi from Slapton, South Devonshire: Deuteromycotina III. *Trans. Br. mycol. Soc.* **67**: 51–59.

960 —— **and Chapman, D. S.** (1971) *Pseudevernia furfuracea* (L.) Zopf and its chemical races in the British Isles. *Lichenologist* **5**: 51–58.

961 —— **and Coppins, B. J.** (1973) A new species of *Tryblidium* from *Fraxinus* in Britain. *Trans. Br. mycol. Soc.* **61**: 597–599.

962 —— —— (1976) *Alectoria nidulifera* Norrl. discovered in Scotland. *Lichenologist* **8**: 95–96.

963 —— —— **and Rose, F.** (1974) Changes in the British lichen flora. In *The Changing Flora and Fauna of Britain* (D. L. Hawksworth, ed.): 47–78. [Systematics Association Special Vol. 6.] London and New York: Academic Press.

964 —— **and James, P. W.** (1973) *Alectoria capillaris* in Ireland. *Ir. Nat. J.* **17**: 425.

965 —— —— (1974) Distribution maps of lichens in Britain. Map 16. *Schismatomma virgineum* D. Hawksw. & P. James sp. nov. *Lichenologist* **6**: 194–196.

966 —— **and Punithalingham, E.** (1973) New and interesting microfungi from Slapton, South Devonshire: Deuteromycotina. *Trans. Br. mycol. Soc.* **61**: 57–69.

967 —— **and Rose, F.** (1969) A note on the lichens and bryophytes of the Wyre Forest. *Proc. Bgham nat. Hist. Soc.* **21**: 191–197.

968 —— —— (1970) Qualitative scale for estimating sulphur dioxide air pollution in England and Wales using epiphytic lichens. *Nature, Lond.* **227**: 145–148.

969 —— —— (1976) *Lichens as Pollution Monitors.* [Studies in Biology No. 66.] London: Arnold.

970 —— —— **and Coppins, B. J.** (1973) Changes in the lichen flora of England and Wales attributable to pollution of the air by sulphur dioxide. In *Air Pollution and Lichens* (B. W. Ferry, M. S. Baddeley and D. L. Hawksworth, eds.): 330–367. London: Athlone Press of the University of London.

971 —— and **Skinner, J. F.** (1974) The lichen flora and vegetation of Black Head, Ilsham, Torquay. *Trans. Torquay nat. hist. Soc.* **16**: 121–136.

972 —— and **Sowter, F. A.** (1969) Leicestershire and Rutland lichens, 1950–1969. *Trans. Leicester lit. phil. Soc.* **63**: 50–61.

973 —— and **Walpole, P. R.** (1966) The lichens of Bradgate Park, Leicestershire. *Trans. Leicester lit. phil. Soc.* **60**: 48–56.

—— —See also Nos. 111, 129, 161, 331, 432, 433, 599, 703, 787, 984, 1139, 1143, 1144, 1315, 1622, 1664, 1673, 2037, 2049, 2052, 2146, 2246, 2462 and 2690.

974 **Haworth, F. [M.]** (1927) [Lichen dyes]. *Proc. Linn. Soc. Lond.* **1926–1927**: 35–36.

—— —See also No. 1448.

975 **Haynes, F. N.** (1964) Lichens. *Viewpts Biol.* **3**: 64–115.

976 —— and **Morgan-Huws, D. I.** (1970) The importance of field studies in determining the factors influencing the occurrence and growth of lichens. *Lichenologist* **4**: 362–368.

—— —See also No. 1622.

Headrick, J.—See No. 603.

977 **Hebden, T.** (1916) The lichen flora of Harden Beck Valley. *Naturalist, Hull* **1916**: 132–134, 159–162.

—— —See also Nos. 2129, 2156 and 2157.

978 **Hedlund, T.** (1892) Kritische Bemerkungen über einige Arten der Flechtengattungen *Lecanora* (Ach.), *Lecidea* (Ach.) und *Micarea* (Fr.). *Bih. K. svenska VetenskAkad. Handl.*, III, **18** (3): 1–104.

979 **Henderson, A.** and **Seaward, M. R. D.** (1976) The lichens of Harewood. *Naturalist, Hull* **1976**: 61–71.

980 **Henderson, D. M.** (1958) New and interesting Scottish fungi: I. *Notes R. bot. Gdn Edinb.* **22**: 593–597.

Henderson, M. E. K.—See No. 2569.

981 **Henslow, J. S.** and **Skepper, E.** (1860) *Flora of Suffolk*. London: Simpkins and Marshall. [Lichens pp. 109–113.]

982 **Henssen, A.** (1963) Eine Revision der Flechtenfamilien Lichinaceae und Ephebaceae. *Symb. bot. upsal.* **18** (1): 1–123.

983 —— (1974) New or interesting cyanophilic lichens II. *Lichenologist* **6**: 106–111.

984 —— (1976) Studies in the developmental morphology of lichenized Ascomycetes. In *Lichenology: Progress and Problems* (D. H. Brown, D. L. Hawksworth and R. H. Bailey, eds.): 107–138. [Systematics Association Special Volume No. 8.] London, New York and San Francisco: Academic Press.

—— —See also Nos. 1144 and 1611.

985 **Hepburn, A.** [1852] Notes on some of the mammalia and birds found at St. Abb's Head. *Hist. Berwicksh. Nat. Club* **3**: 70–77.

986 **Hepburn, I.** (1943) A study of the vegetation of sea-cliffs in North Cornwall. *J. Ecol.* **31**: 30–39.

987 —— (1952) *Flowers of the Coast*. [New Naturalist No. 24.] London: Collins.

94

988 **Hepp, P.** (1853–67) *Die Flechten Europas.* 32 fasc., nos. 1–962. Zurich. [Exsiccata; see his nos. 423, 428, 665, 715–6, 869 and 927.]

989 **Hepper, F. N.** (1956) Biological Flora of the British Isles: *Silene nutans* L. *J. Ecol.* **44**: 693–700.

990 **Hering, T. F.** (1976) Spring fungus foray, Pickering, 29 May to 2 June 1975. *Naturalist, Hull* **1976**: 105 [Lichens by P. M. Earland-Bennett.]

991 **Heron-Allen E.,** (1911) *Selsey Bill: Historic and Prehistoric.* London: Duckworth.

992 **Hertel, H.** (1967) Revision einiger calciphiler Formenkreis der Flechtengattung *Lecidea. Nova Hedwigia Beih.* **24**: 1–155.

993 —— (1968) Beitrage zur Kenntnis der Flechtenfamilie Lecideaceae I. *Herzogia* **1**: 25–39.

994 —— (1969) Die Flechtengattung *Trapelia* Choisy. *Herzogia* **1**; 111–130.

995 —— (1969) Beiträge zur Kenntnis der Flechtenfamilie Lecideaceae II. *Herzogia* **1**: 321–329.

996 —— (1970) Trapeliaceae—eine neue Flechtenfamilie. *Votr. GesGeb. Bot.* [*Dtsch. bot. Ges.*], *n.f.* **4**: 171–185.

997 —— (1970) Parasitische lichenisierte Arten der Sammelgattung *Lecidea* in Europa. *Herzogia* **1**: 405–438.

998 —— (1970) Beiträge zur Kenntnis der Flechtenfamilie Lecideaceae III. *Herzogia* **2**: 27–62.

999 —— (1971) Über holarctische Krustenflechten aus den venezuelanischen Anden. *Willdenowia* **6**: 225–272.

1000 —— (1971) Beiträge zur Kenntnis der Flechtenfamilie Lecideaceae IV. *Herzogia* **2**: 231–261.

1001 —— (1973) Beiträge zur Kenntnis der Flechtenfamilie Lecideaceae V. *Herzogia* **2**: 479–515.

1002 —— (1975) Über einige gesteinsbewohnende Krustenflechten aus der Umgebung von Finse (Norwegen, Hordaland). *Mitt. bot. StSamml., Münch.* **12**: 113–152.

1003 —— (1975) Beiträge zur Kenntnis der Flechtenfamilie Lecideaceae V [sic] [VI]. *Herzogia* **3**: 365–406.

1004 —— (1975) Ein vorlaufiger Bestimmungsschlüssel für die kryptothallinen, schwarzfrüchtigen, saxicolen Arten der Sammelgattung *Lecidea* (Lichenes) in der Holarktis. *Decheniana* **127**: 37–78.

1005 —— **and Leuckert, C.** (1969) Über Flechtenstoffe und Systematik einiger Arten der Gattung *Lecidea, Placopsis* und *Trapelia* mit C+ rot reagierendem Thallus. *Willdenowia* **5**: 369–383.

1006 **Hewitt, C. G.** (1907) A contribution to the flora of St. Kilda: being a list of certain lichens, mosses, Hepaticae and fresh-water algae. *Ann. Scot. nat. Hist.* **1907**: 239–241.

Heywood, V. H.—See Nos. 939 and 2099.

1007 **Hiern, W. P.** (1897) Isle of Man plants. *J. Bot., Lond.* **35**: 11–15.

1008 —— (1910) Second report of the botany committee. *Rep. Trans. Devon. Ass. Advmt Sci.* **42**: 112–139.

1009 —— (1912) Fourth report of the botany committee. *Rep. Trans. Devon. Ass. Advmt Sci.* **44**: 126–135.

1010 —— (1913) Fifth report of the botany committee. *Rep. Trans. Devon. Ass. Advmt Sci.* **45**: 117–126.

1011 —— (1915) Seventh report of the botany committee. *Rep. Trans. Devon. Ass. Advmt Sci.* **47**: 160–170.

1012 —— (1917) Ninth report of the botany committee. *Rep. Trans. Devon. Ass. Advmt Sci.* **49**: 99–108.

1013 —— (1917) Address of the President. *Rep. Trans. Devon. Ass. Advmt Sci.* **49**: 27–59.

1014 —— (1918) Tenth report of the botany committee. *Rep. Trans. Devon. Ass. Advmt Sci.* **50**: 219–227.

1015 —— (1919) Eleventh report of the botany committee. *Rep. Trans. Devon. Ass. Advmt Sci.* **51**: 114–129.

1016 **Hill, D. J.** (1971) Experimental study of the effect of sulphite on lichens with reference to atmospheric pollution. *New Phytol.* **70**: 831–836.

1017 —— (1972) The movement of carbohydrate from the alga to the fungus in the lichen *Peltigera polydactyla*. *New Phytol.* **71**: 31–39.

1018 —— (1974) Some effects of sulphite on photosynthesis in lichens. *New Phytol.* **73**: 1193–1205.

1019 —— **and Smith, D. C.** (1972) Lichen physiology XII. The 'inhibition technique'. *New Phytol.* **71**: 15–30.

1020 —— **and Woolhouse, H. W.** (1966) Aspects of the autecology of *Xanthoria parietina* agg. *Lichenologist* **3**: 207–214.
     —— —See also Nos. 1126, 1816 and 1954.

1021 **Hill, J.** (1751) *A General Natural History.* Vol. 2. *A History of Plants.* London: T. Osborne.

1022 —— (1760) *Flora Britannica.* Vol. I. London. [Some copies dated 1759.] [Lichens pp. 580–593.]

1023 **Hillmann, J.** (1936) Parmeliaceae. *Rabenh. Krypt.-Fl.* **9**, 5(3): 1–309, *1–10*.
     Hine, R. L.—See No. 1438.

1024 **Hitch, C. J. B. and Millbank, J. W.** (1975) Nitrogen metabolism in lichens. VI. The blue-green phycobiont content, heterocyst frequency and nitrogenase activity in *Peltigera* species. *New Phytol.* **74**: 473–476.

1025 —— —— (1975) Nitrogen metabolism in lichens VII. Nitrogenase activity and heterocyst frequency in lichens with blue-green phycobionts. *New Phytol.* **75**: 239–244.

1026 —— **and Stewart, W. D. P.** (1973) Nitrogen fixation by lichens in Scotland. *New Phytol.* **72**: 509–524.

1027 **Hobkirk, C. P.** (1868) *Huddersfield: Its history and natural history.* Ed. 2. Huddersfield: G. Tindall. [Lichens pp. 203–204.] [No lichens in Ed. 1 (1859).]
     Hodson, N. L.—See No. 1298.

1028 **Hoffmann, G. F.** (1790–1801) *Descriptio et adumbratio plantarum e classe cryptogamica Linnaei quae Lichenes dicunter.* 3 vols. Leipzig.

1029 **Hogan, F. E., Hogan, J. and MacErlean, J. C.** (1900) *Luibhleabhrán: Irish and Scottish Gaelic Names of Herbs, Plants, Trees etc.* Dublin: M. H. Gill & Son.

Hogan, J.—See No. 1029.

1030 **Hogg, J.** (1827) *On the Natural History of the Vicinity of Stockton-on-Tees.* Stockton: T. Jennett. [Separately printed from *The History of Stockton* (J. Brewster).]

1031 **Holden, M.** (1968) Spring Foray, Pickering, 18th-22nd May 1967. *Bull. Br. mycol. Soc.* **2**: 2–5.

1032 **Holl, H. B.** (1870) Additions and corrections to the list of lichens in the "Botany of the Malvern Hills". *Trans. Malvern Nat. Fld Cl.* **3**: 119–129.

Holligan, M. S.—See No. 800.

Holligan, P. M.—See No. 800.

1033 **Holmes, E. M.** [18—-19—] [*Correspondence etc.*]. Letters in LINN (from 128 people; 3 boxes); various notes 1876–1920 in the Wellcome Institute of the History of Medicine Library, London. [Not abstracted.]

1034 —— [1872] Lichens of Devon and Cornwall. *A. Rep. Plymouth Inst.* **3**: 11–39. [The date of publication has been variously cited but 1872 is now considered most probable (see No. 940); additional support for this is that H. B. Holl wrote to thank Holmes for sending him a copy on 2 February 1872 (letter in LINN).]

1035 —— (1873) *Physcia intricata. J. Bot., Lond.* **11**: 206.

1036 —— (1874) Lichenological notes. *Grevillea* **3**: 91–92.

1037 —— (1878) The cryptogamic flora of Kent. Lichens. *J. Bot., Lond.* **16**: 117–120, 209–212, 329–345, 373–376.

1038 —— (1881) Chemical tests for lichens. *J. Bot., Lond.* **19**: 115–116.

1039 —— (1882) A botanical ramble around Weymouth. *Sci. Gossip* **18**: 78–80.

1040 —— (1882) [On *Placodium cesatii* of Leighton's 'Lichen Flora'.] *Proc. Linn. Soc. Lond.* **1875–1880**: 1.

1041 —— (1883) [British lichens.] *Proc. Linn. Soc. Lond.* **1880–1882**: 10.

1042 —— (1883) [Lichens] Field Meeting, 28th October, 1883. Watford. *Trans. Herts. nat. Hist. Soc. Fld Club* **2**: lxvi-lxviii.

1043 —— (1886) [Lichens] Field Meeting, 17th October, 1885. Bricket Wood, Watford. *Trans. Herts. nat. Hist. Soc. Fld Club* **3**; lxxi-lxxii.

1044 —— (1891) Lists of mosses, lichens and Hepaticae gathered in Hatfield Forest. *Essex Nat.* **4**: 220–221.

1045 —— (1902) Lichens (Lichenes). In *The Victoria History of the Counties of England, Surrey* (H. A. Doubleday, ed.) **1**: 60–63. Westminster: Constable.

1046 —— (1904) The lichens (Lichenes). In *The Victoria History of the Counties of England, Bedfordshire* (H. A. Doubleday and W. Page, eds.) **1**(2): 60–61. London: Constable.

1047 —— (1906) Lichens (Lichenes). In *The Victoria History of the Counties of England, Cornwall* (W. Page, ed.) **1**: 98–106. London: Constable.

1048 —— (1906) Lichenes (Lichens). In *The Victoria History of the Counties of England, Devonshire* (W. Page, ed.) **1**: 117–123. London: Constable.

1049 —— [*c.* 1906] Lichens (Lichenes). In *The Victoria History of the Counties of England, Dorset* (W. Page, ed.). Unpublished manuscript in BM.

1050 —— (1906) Lichens (Lichenes). In *The Victoria History of the Counties of*

England, Somerset (W. Page, ed.) **1**: 56–58. London: Constable.
1051 —— (1908) Lichens (Lichenes). In *The Victoria History of the Counties of England, Kent* (W. Page, ed.) **1**: 77–79. London: Constable.
1052 —— (1910) [*Parmelia rugosa* var. *concentrica* Cromb. exhibited.] *Proc. Linn. Soc. Lond.* **1909-1910**: 57.
1052a —— (1923) Lichens in the British Isles. *Pharm. J.* **110**: 67.
1053 —— **and Gray, P.** (1886) *British Fungi, Lichens and Mosses Including Scale Mosses and Liverworts.* [The Young Collector.] London: Swan Sonnenschein, Lowrey. [Lichens by P. Gray, sections individually paged.]
1054 —— —— (1907) *British Fungi, Lichens and Mosses Including Scale Mosses and Liverworts.* [The Young Collector.] Ed. 2. London: Swan Sonnenschein. [Lichens by P. Gray, sections individually paged.]
—— See also Nos. 610, 940 and 1868.
Holt, A.—See Nos. 1961, 2612 and 2613.
1055 **Home D. M.** [1862] Address delivered at Berwick, on the 27th of September, 1861. *Hist. Berwicksh. Nat. Club* **4**: 219–260.
1056 **Honnor, K. J. and Honnor, M.** (1966) Lichens. In *Northaw Great Wood. Its history and natural history* (B. L. Sage, ed.): 90–91. Hertford: Hertfordshire County Council Education Department.
Honnor, M.—See No. 1056.
1057 **Hooker, W. J.** (18—) [*Correspondence*]. Letters and notes in K (*c.* 15 vols.). [Not abstracted.]
1058 —— (1821) *Flora Scotica.* Vol. 2. London: Constable. [Lichens pp. 35–73, 96.]
1059 —— (1833) *The English Flora of Sir James Edward Smith.* Vol. 5(1). London: Longman, etc. [Lichens pp. 129–241, 270.]
1060 —— (1844) *The English Flora of Sir James Edward Smith.* Vol. 5(1). [Ed. 2.] London: Longman, etc. [Lichens pp. 133–245; as in the 1833 edition but pagination changed.]
—— —See also No. 2220.
1061 **Hope-Simpson, J. F.** (1938) A chalk-flora on the Lower Greensand: its use in interpreting the calcicole habit. *J. Ecol.* **26**: 218–235.
1062 —— (1941) Studies of the vegetation of the English chalk. VII. Bryophyte and lichens in chalk grassland, with a comparison of their occurrence in other calcareous grasslands. *J. Ecol.* **29**: 107–116.
1063 —— (1941) Studies of the vegetation of the English chalk. VIII. A second survey of the chalk grasslands of the South Downs. *J. Ecol.* **29**: 217–267.
—— —See also No. 2640.
1064 **Hopkins, B.** (1957) Pattern in the plant community. *J. Ecol.* **45**: 451–463.
1065 **Hopkinson, J., Gibbs, A. E. and Saunders, J.** (1902) Botany. In *The Victoria History of the Counties of England, Hertfordshire* (H. A. Doubleday, ed.) **1**(2): 43–80. Westminster: Constable. [Lichens pp. 69–70.]
1066 **Hopkirk, T.** (1813) *Flora Glottiana.* Glasgow.
Hornby, R.—See No. 2568.

1067 **Horrill, A. B., Sykes, J. M. and Idle, E. T.** (1975) The woodland vegetation of Inchcailloch, Loch Lomond. *Trans. Proc. bot. Soc. Edinb.* **42**: 307–334.

1068 **Horsfield, T. W.** (1835) *The History, Antiquities, and Topography of the County of Sussex.* Vol. 2. Lewes: Sussex Press. [Botany by T. H. Cooper; lichens pp. 9, 20–22 (appendix 2).]

1069 **[Horton, R. J.]** (1962) *Scarba Expedition* 1961. Oundle: Oundle School. [Lichens p. 24.]

1070 **Horwood, A. R.** (1904) Leicestershire lichens, 1886–1903. *J. Bot., Lond.* **42**: 47–49.

1071 —— (1907) A proposed exchange club for lichens. *J. Bot., Lond.* **45**: 412.

1072 —— (1907) On the disappearance of cryptogamic plants from Charnwood Forest, Leicestershire, within historic times. *J. Bot., Lond.* **45**: 334–339.

1073 —— (1907) *A Guide to Leicester and District.* Leicester: British Association.

1074 —— (1907). Lichens. In *The Victoria History of the Counties of England, Leicestershire* (H. Fisher, ed.): 53–55. London: Constable.

1075 —— (1909) The cryptogamic flora of Leicestershire. *Trans. Leicester lit. phil. Soc.* **13**: 15–86. [Lichens pp. 60–69.]

1076 —— (1910) Report of the Secretary for 1909. *Rep. Lichen Exch. Club* **1909**: 2–5.

1077 —— (1910) The extinction of cryptogamic plants. *SEast. Nat.* [**15**]: 56–86.

1078 —— (1911) Report of the Secretary for 1910. *Rep. Lichen Exch. Club* **1910**: 2(*18*)–5(*21*).

1079 —— (1912) Report of the Secretary, 1911. *Rep. Lichen Exch. Club* **1911**: 2(*26*)–11(*35*).

1080 —— [1913] *A Hand-List of the Lichens of Great Britain, Ireland and the Channel Islands.* Leicester: Lichen Exchange Club.

1081 —— (1928) Lichen dyeing today: the revival of an ancient industry. *Sci. Progr., Lond.* **23**: 279–283.

—— —See also No. 2425.

1082 **How, W.** (1650) *Phytologia Britannica, natales exhibens indigenarum stirpium sponte emergentium.* London: R. Cotes.

1083 **Howitt, G.** (1839) *Flora of Nottinghamshire.* London: A. Hamilton. [Lichens pp. 102–113, ? by J. Bohler.]

1084 **Hubbard, J. C. E.** (1970) The shingle vegetation of Southern England: A general survey of Dungeness, Kent and Sussex. *J. Ecol.* **58**: 713–722.

1085 **Hubbertsy, N.** [*c.* 1800] *Derbyshire Plants.* Manuscript in Derby Borough Reference Library.

1086 **Hudson, W.** (1762) *Flora Anglica.* London: C. Moran. [Lichens pp. 441–463.]

1087 —— (1778) *Flora Anglica.* Ed. 2. Vol. 2. London. [Lichens pp. 523–562.]

1088 —— (1798) *Flora Anglica.* Ed. 3. London. [Identical to Ed. 2 but with slightly altered preface; pagination unaltered.]

1089 **Hue, A. [l'Abbé]** (1911) Monographiam generis *Solorina* Ach. morphologice et anatomice addito de genre *Psoromaria* Nyl. appendice. *Mém. Soc. natn. Sci. nat. math. Cherbourg* **38**: 1–56.

1090 —— (1924) Monographia Crocyniarum. *Bull. Soc. bot. Fr.* **71**: 311–402.

Hueck-van der Plas, E. H.—See No. 1441.

1091 **Hughes, S. J.** (1951) Studies on micro-fungi. VIII. *Mycol. Pap.* **42**: 1–27.

Hughes, V. R.—See No. 596.

1092 **Hull, J.** (1799) *The British Flora.* Manchester. [Lichens pp. 284–308, 324.]

1093 **Huneck, S., Mathey, A. and Trotet, G.** (1967) Mitteilung über Flechtenin-haltsstoffe. 46. Über die Inhaltsstoffe von *Roccella fuciformis* DC. *Z. Naturf.* **22b**: 1367–1368.

—— —See also No. 717.

Huntley, B.—See No. 10.

1094 **Hurst, C. P.** (1920) Lichenological report. *Rep. Marlboro. Coll. nat. Hist. Soc.* **68**: 26–29.

1095 —— (1920) East Wiltshire mosses, hepatics, and lichens. *Wilts. archaeol. nat. Hist. Mag.* **41**: 40–52.

1096 [——] (1922) Notes. *Rep. Marlboro. Coll. nat. Hist. Soc.* **70**: 48–52.

1097 —— (1922) East Wiltshire lichens. *Wilts. archaeol. nat. Hist. Mag.* **42**: 1–10.

1098 [——] (1923) Lichens. *Rep. Marlboro. Coll. nat. Hist. Soc.* **71**: 61.

1099 —— (1924) The Wiltshire lichens in the Department of Botany at the British Museum. *Wilts. archaeol. nat. Hist. Mag.* **42**: 427–430.

1100 [——] (1924) Lichens. *Rep. Marlboro. Coll. nat. Hist. Soc.* **72**: 103.

1101 —— (1926) Lichens. *Rep. Marlboro. Coll. nat. Hist. Soc.* **74**: 116.

1102 —— (1928) Lichens. *Rep. Marlboro. Coll. nat. Hist. Soc.* **76**: 78–80. [Includes list by H. H. Knight.]

1103 —— (1928) Natural history notes round Great Bedwyn. II. *Wilts. archaeol. nat. Hist. Mag.* **44**: 128–137. [Lichens partly by H. H. Knight.]

1104 —— (1929) Natural history notes round Great Bedwyn (III.). *Wilts. archaeol. nat. Hist. Mag.* **44**: 401–406.

1105 —— (1929) Lichens. *Rep. Marlboro. Coll. nat. Hist. Soc.* **77**: 71.

1106 [——] (1930) Lichens. *Rep. Marlboro. Coll. nat. Hist. Soc.* **78**: 105.

1107 —— (1931) Natural history notes round Great Bedwyn (IV.). *Wilts. archaeol. nat. Hist. Mag.* **45**: 279–290.

1108 —— (1931) Natural history notes round Great Bedwyn (V.). *Wilts. archaeol. nat. Hist. Mag.* **45**: 407–417.

1109 [—— **and Chapman, V. J.**] (1927) Lichens. *Rep. Marlboro. Coll. nat. Hist. Soc.* **75**: 78.

—— —See also No. 1256.

1110 **Hurst, D. E.** (1868) *Horsham: Its History and Antiquities.* London: W. Macintosh. [Lichens p. 244.]

1111 —— (1889) *The History and Antiquities of Horsham* [Ed.2.]. Lewes: Farncombe. [Lichens pp. 225–226.]

Hutchins, J.—See Nos. 1876 and 1877.

1112 **Hutchinson, T. C.** (1966) The occurrence of living and sub-fossil remains of *Betula nana* L. in Upper Teesdale. *New Phytol.* **65**: 351–357.

1113 **Hutchinson, W.** (1794) *The History and Antiquities of the County Palatine of Durham.* Vol. 3. Carlisle: privately printed. [Lichens p. 516 by E. Robson.]

1114 **Hyde, H. A.** (1931) Samuel Brewer's diary. A chapter in the history of botanical exploration in North Wales. *Rep. botl Soc. Exch. Club Br. Isl.* **1930**, *Suppl.*: 1–30.
—— —See also No. 1483.

1115 **Idle, E. T.** (1974) Botany. In *A Natural History of Loch Lomond*: 25–35. Glasgow: University of Glasgow Press.
—— —See also No. 1067.

1116 **Ing, B.** (1966) Studies on bark epiflora. *Sch. Sci. Rev.* **164**: 37–44.

1117 **Ingold, C. T.** (1972) The advance of mycology. *Trans. Br. mycol. Soc.* **58**, *Suppl.*: 5–14.

1118 **Ingram, H. A. P. and Roger, J. G.** (1968) Vegetation and flora. In *Dundee and District* (S. J. Jones, ed.): 62–81. Dundee: British Association.

1119 **Ingram, M.** (1958) The ecology of the Cairngorms. IV. The *Juncus* zone: *Juncus trifidus* communities. *J. Ecol.* **46**: 707–737.
Ingram, T. A.—See Nos. 229 and 842.

1120 **Irvine, D. E. G.** (1974) The marine vegetation of the Shetland Isles. In *The Natural Environment of Shetland* (R. Goodier, ed.): 107–113. Edinburgh: Nature Conservancy Council.
Irvine, L. M.—See No. 2474.
Ismay, J.—See No. 2568.
Isoviita, P.—See No. 2691.

1121 **Ivimey-Cook, R. B.** (1959) The lichens of the Ewenny Downs, Glamorganshire. *Lichenologist* **1**: 97–103.

1122 —— (1959) Biological Flora of the British Isles: *Agrostis setacea* Curt. *J. Ecol.* **47**: 697–706.

1123 —— (1963) Biological Flora of the British Isles: *Hypericum linarifolium* Vahl. *J. Ecol.* **51**: 727–732.

1124 **Jackson, B. and Young, G.** (1973) Watching what we breath. *Sunday Times Magazine, 28th January* 1973: 40–46.
Jackson, B. D.—See No. 1868.

1125 **Jackson, G.** (1974) Lichens collected on Foula, August 1973. *Brathay Fld Stud. Rept* **27**: 9–12.

1126 **Jaggard, S. J., Seviour, J. A. and Hill, D. J.** (1974) The distribution of some ornithocoprophilous lichen species on carboniferous limestone in the Malham area. *Trans. nat. Hist. Soc. Northumb.* **41**: 194–203.

1127 **Jahns, H. M.** (1970) Remarks on the taxonomy of the European and North American species of *Pilophorus* Th. Fr. *Lichenologist* **4**: 199–213.

1128 **James, P. W.** (1958) *Gyrophora hirsuta* new to the British Isles. *Lichenologist* **1**: 39–40.

1129 —— (1959) New British records. *Lichenologist* **1**: 113–114.

1130 —— (1960) Notes on angiocarpous lichens in the British Isles: 1. *Lichenologist* **1**: 145–158.

1131 —— (1961) Field meeting at Dolgellau. *Lichenologist* **1**: 269–274.

1132 —— (1962) Angiocarpous lichens in the British Isles: 2. *Lichenologist* **2**: 86–94.

1133 —— (1965) A new check-list of British lichens. *Lichenologist* **3**: 95–153.

1134 —— (1965) Field meeting in Scotland. *Lichenologist* **3**: 155–172.

1135 —— (1966) A new check-list of British lichens: Additions and corrections – 1. *Lichenologist* **3**: 242–247.

1136 —— (1966) Lichens collected at the Pengelly Cave Research Centre. In *Minutes of the Ecological Meeting held at the Pengelly Centre at 10.30 a.m. on Sunday 15th May 1966*: 9–10. London: The Association of the William Pengelly Cave Research Centre. [Copy with D. L. Hawksworth.]

1137 —— (1970) The lichen flora of shaded acid rock crevices and overhangs in Britain. *Lichenologist* **4**: 309–322.

1138 —— (1971) New or interesting British lichens: 1. *Lichenologist* **5**: 114–148.

1139 —— (1973) The effect of air pollutants other than hydrogen fluoride and sulphur dioxide on lichens. In *Air Pollution and Lichens* (B. W. Ferry, M. S. Baddeley and D. L. Hawksworth, eds.): 143–175. London: Athlone Press of the University of London.

1140 —— (1974) Distribution maps of lichens in Britain. Map 17. *Sclerophyton circumscriptum* (Tayl.) Zahlbr. *Lichenologist* **6**: 197–199.

1141 —— (1975) The genus *Gyalideopsis* Vězda in Britain. *Lichenologist* **7**: 155–161.

1142 —— (1976) *A lichen survey of north-west Sutherland*. Manuscript. Report to the World Wildlife Fund; copy with BLS.

1143 —— **and Hawksworth, D. L.** (1974) Distribution maps of lichens in Britain. Map 11. *Parmelia britannica* D. Hawksw. & P. James. *Lichenologist* **6**: 178–180.

1144 —— **and Henssen, A.** (1976) The morphological and taxonomic significance of cephalodia. In *Lichenology: Progress and Problems* (D. H. Brown, D. L. Hawksworth and R. H. Bailey, eds.): 27–77. [Systematics Association Special Volume No. 8.] London, New York and San Francisco: Academic Press.

1145 —— **and Rose, F.** (1973) Distribution maps of lichens in Britain. Map 2. *Anaptychia ciliaris* (L.) Körb. *Lichenologist* **5**: 467–469.

1146 —— —— (1973) Distribution maps of lichens in Britain. Map 3. *Caloplaca luteoalba* (Turn.) Th. Fr. *Lichenologist* **5**: 470–471.

1147 —— —— (1973) Distribution maps of lichens in Britain. Map 4. *Opegrapha lyncea* (Sm.) Borr. ex Hook. *Lichenologist* **5**: 472–473.

1148 —— —— (1973) Distribution maps of lichens in Britain. Map 5. *Opegrapha prosodea* Ach. *Lichenologist* **5**: 474–475.

1149 —— —— (1973) Distribution maps of lichens in Britain. Map 7. *Parmelia soredians* Nyl. *Lichenologist* **5**: 478–480.

1150 —— —— (1974) Distribution maps of lichens in Britain. Map 12. *Parmelia elegantula* (Zahlbr.) Szat. *Lichenologist* **6**: 181–184.

1151 —— —— (1974) Distribution maps of lichens in Britain. Map 13. *Parmelia exasperatula* Nyl. *Lichenologist* **6**: 185–187.

1152 —— —— (1974) Distribution maps of lichens in Britain. Map 14. *Parmelia laciniatula* (Flag. ex Oliv.) Zahlbr. *Lichenologist* **6**: 188–190.

1153 —— —— (1974) Distribution maps of lichens in Britain. Map 15. *Pertusaria coccodes* (Ach.) Nyl. *Lichenologist* **6**: 191–193.

—— —See also Nos. 321, 434, 435, 630, 964, 965, 1779, 2053 and 2167.

1154 **Jane, F. W. and White, D. J. B.** (1971) The botany and plant ecology of Blakeney Point and Scolt Head Island. In *Blakeney Point and Scolt Head Island*, Ed. 3 (J. A. Steers, ed.): 26–43. [Norwich.]

1155 **Jeffrey, D. W. and Pigott, C. D.** (1973) The response of grasslands on sugar-limestone in Teesdale to application of phosphorus and nitrogen. *J. Ecol.* **61**: 85–92.

1156 **Jenkins, D. A.** (1964) *Trace element studies on some Snowdonian rocks, their minerals and related soils.* Ph.D. thesis, University of Wales.

1157 —— **and Davies, R. I.** (1966) Trace element content of organic accumulations. *Nature, Lond.* **210**: 1296–1297.

Jenkins, M.—See No. 2351.

1158 **Jenner, E.** (1845) *A Flora of Tunbridge Wells.* Tunbridge Wells & London. [Lichens pp. 75–87.]

Jensen, H. A. P.—See No. 2257.

1159 **Jerdon, A.** [1872] Note of a lichen new to the border flora. *Hist. Berwicksh. Nat. Club* **6**: 436.

1160 **Jermy, A. C. and Stott, P. A.** (1973) *Chalk Grassland: Studies on its Conservation and Management.* Maidstone: Kent Trust for Nature Conservation.

Johnson, C.—See No. 2221.

1161 **Johnson, T.** (1629) *Iter plantarum investigationes ergo suscepti in agrum Cantianum anno* 1629 *Julii* 13. London.

1162 —— (1632) *Enumeratio plantarum in Ericeto Hampsteadiano locisque vicinis crescentium.* London.

1163 —— (1632) *Decriptio itineris plantarum investigationis ergo suscepti in agrum Cantianum anno dom.* 1632. London.

1164 —— (1634) *Mercurius botanicus.* London. [Lichens p. 47.]

1165 —— (1638) *Catalogus plantarum juxte Tottenham lectorum anno dom.* 1638. Manuscript in K.

—— —See also Nos. 765 and 806.

1166 **Johnson, W.** (1879) Lichens, and a polluted atmosphere. *Sci. Gossip* **15**: 219.

1167 —— (1881) New British lichens. *J. Bot., Lond.* **19**: 113–114.

1168 —— (1881) The lichens of Cumberland. *Trans. Cumberland Ass. Advanc. Sci.* **6**: 129–157.

1169 —— (1881) An introduction to the study of lichens. *Nth. Microscopist* **1**: 85–90, 101–110.

1170 —— (1882) Additions to the British lichen flora. *J. Bot., Lond.* **20**: 184–185.

1171 —— (1883) Additions to the lichen-flora of Great Britain. *Grevillea* **11**: 94.

1172 —— (1884) A new lichen found in Northumberland. *Nat. Hist. Trans. Northumb.* **8**: 184.

1173 —— (1884) *Sirosiphon saxicola* (Naeg). *Grevillea* **12**: 76.

1174 —— (1886) Lichen memorabilia, 1884. *Nat. Hist. Trans. Northumb.* **8**: 217–219.

1175 —— (1886) A new British lichen. *Grevillea* **14**: 91.

1176 —— (1887) Lichenology: *Peltigerae. Wesley Nat.* **1**: 138–139.

1177 —— (1887) Lichenology. *Wesley Nat.* **1**: 114–115.
1178 —— (1887) Lichenology. *Wesley Nat.* **1**: 174–176.
1179 —— (1887) Lichenology: *Stictei. Wesley Nat.* **1**: 238–239.
1180 —— (1887) Lichenology: *Evernia. Wesley Nat.* **1**: 281–282.
1181 —— (1887) Lichenology: *Parmelia. Wesley Nat.* **1**: 314–316.
1182 —— (1888) Lichenology: *Parmelia. Wesley Nat.* **1**: 338–340.
1183 —— (1888) Lichenology: *Physcia. Wesley Nat.* **1**: 370–372.
1184 —— (1888) Lichenology: *Gyrophorei. Wesley Nat.* **2**: 19–22.
1185 —— (1888) Lichenology: *Baeomyces. Wesley Nat.* **2**: 53–55.
1186 —— (1888) Lichenology: Reply to queries. *Wesley Nat.* **2**: 116–118.
1187 —— (1888) Lichenology: *Cladonia. Wesley Nat.* **2**: 148–151.
1188 —— (1888) Lichenology: *Cladonia. Wesley Nat.* **2**: 247–249.
1189 —— (1888) Lichenology: *Cladina. Wesley Nat.* **2**: 275.
1190 —— (1888) Lichenology: *Roccella. Wesley Nat.* **2**: 313–315.
1191 —— (1889) Lichenology: *Ramalina. Wesley Nat.* **2**: 341–342.
1192 —— (1889) Lichenology: *Usnea. Wesley Nat.* **2**: 365–367.
1193 —— (1889) Lichenology: *Alectoria. Wesley Nat.* **3**: 15–16.
1194 —— (1889) Lichenology: *Cetraria. Wesley Nat.* **3**: 90–91.
1195 —— (1889) A catalogue of Mr. N. J. Winch's lichens, now in the museum of the Natural History Society, Newcastle-upon-Tyne. *Trans. nat. Hist. Soc. Northumb.* **8**: 307–325.
1196 —— (1894–1918) *The North of England Lichen-Herbarium.* 13 fasc., nos. 1–520. Newcastle/Darlington/Manchester. [Fasc. 1–11 intact in BM; 1–13 intact in HAMU; fasc. 13 not seen.]
1197 —— (1898) Lichens found in the neighbourhood of Darlington. In *Zig-Zag Ramblings of a Naturalist*, Ed. 2 (R. T. Manson, ed.): 207. Darlington: W. Dresser & Sons. [Ed. 1 (1884) not seen.]
1198 —— [1903] *Nature and Naturalists.* London: H. R. Allenson.
1199 —— (1912) *Wimbledon Common. Its geology, antiquities and natural history.* London: T. Fisher Unwin. [Lichens pp. 276–277.]
1200 —— (1917) A new British lichen. *Naturalist, Hull* **1917**: 88.
1201 —— (1918) An addition to the British lichen-flora. *Naturalist, Hull* **1918**: 103.
—— —See also Nos. 1852 and 2133.
1202 **Johnston, G.** (1831) *A Flora of Berwick-upon-Tweed.* Vol. 2. *Cryptogamous Plants.* Edinburgh: J. Carfrae & Son. [Lichens pp. 69–105.]
1203 —— (1835) *The lichens of Berwickshire and North Durham.* [Bound set of specimens in E including a few of T. Brown and J. Hardy; annotations by Hardy and Leighton.]
1204 —— (1853) *The Natural History of the Eastern Borders.* Vol. 1. *The Botany.* London: van Voorst. [Lichens pp. 264–270.]
1205 **Johnston-Lavis, H. T.** (1878) A glass-eating lichen. *Sci. Gossip* **14**: 128–130.
Jones, A. D.—See No. 2469.
Jones, D.—See Nos. 1996 and 2087.
1206 **[Jones, D. A.]** (1898) *A handbook of the botany of Merionethshire, North Wales.* Manuscript in NMW. [Lichens pp. 468–568.]

1207 —— (1925) Lichen flora of the Ingleton district. *Naturalist, Hull* **1925**: 241–244.

1208 **Jones, E. W.** (1952) Some observations on the lichen flora of tree boles, with special reference to the effect of smoke. *Revue bryol. lichén.* **21**: 96–115.

1209 **Jones, J. P.** (1820) *A Botanical Tour through various parts of the Counties of Devon and Cornwall.* Exeter. [Lichens pp. 71–74.]

1210 —— **and Kingston, J. F.** (1829) *Flora Devoniensis.* Part II. London: Longman etc. [Lichens pp. 25–48.]

Jones, S. J.—See No. 1118.

1211 **Jones, V. and Richards, P. W.** (1962) Biological Flora of the British Isles: *Silene acaulis* (L.) Jacq. *J. Ecol.* **50**: 475–487.

1211a **Jones, W. E.** (1955) The littoral and sublittoral marine algae of Bardsey. *Rep. Bardsey Bird Fld Obs.* **3**: 40–51.

1211b —— (1959) The effects of exposure on the zonation of algae on Bardsey Island. *Rep. Bardsey Bird Fld Obs.* **7**: 41–46.

1212 **Jørgensen, P. M.** (1969) *Sticta dufourii* Del. and its parasymbiont *Arthonia abelonae* P.M. Jörg. n.sp. in Norway. *Nova Hedwigia* **18**: 331–340.

1213 —— (1971) On some *Leptogium* species with short *Mallotium* hairs. *Svensk bot. Tidskr.* **67**: 53–58.

1214 —— (1973) Über einige *Leptogium*-Arten vom *Mallotium*-typ. *Herzogia* **2**: 453–468.

1215 —— (1975) Contributions to a monograph of the *Mallotium*–hairy *Leptogium* species. *Herzogia* **3**: 433–460.

1216 **Joshua, W.** (1875) On the Collemei of the Cirencester or Cotteswold District. *Grevillea* **4**: 42–43.

1217 —— (1879) *Microscopical slides of British Lichens.* Nos. 1–48. Cirencester. [Not seen; a set is said to have been purchased by BM in 1880 but has not been located there.]

1218 [——] (1879) Preparations of lichens for the microscope. *Grevillea* **7**: 145–146.

1219 **Kappen, L.** [1974] (1973) Response to extreme environments. In *The Lichens* (V. Ahmadjian and M. E. Hale, eds.): 311–380. New York and London: Academic Press.

Kärenlampi, L.—See Nos. 799 and 2150.

1220 [**Keddie, W.**] [1850] *Staffa and Iona.* Glasgow, etc.: Blackie & Son. [Lichen p. 148.]

1221 —— (1863) Notice of a botanical trip to Ben Lawers and Schihallion in September 1860. *Trans. bot. Soc. Edinb.* **7**: 202–207.

1222 **Keegan, P. Q.** (1907) The chemistry of some common plants. *Naturalist, Hull* **1907**: 24–25.

1223 **Keissler, K. von** (1928) Systematische Untersuchungen über Flechtenparasiten und lichenoide Pilze (VI. Teil, Nr. 51–60). *Annln Naturhist. Mus. Wien* **42**: 99–106.

1224 —— (1930) Die Flechtenparasiten. *Rabenh. Krypt.-Fl.* **8**: i–xi, 1–712.

1225 —— (1936–38) Pyrenulaceae bis Mycoporaceae, Coniocarpineae. *Rabenh. Krypt.-Fl.* **9**, 1(2): i–xii, 1–846.

1226 —— (1958–60). Familie Usneaceae. *Rabenh. Krypt.-Fl.* **9**, 5(4): i-xi, 1–755.

1227 **Kenworthy, J. B.,** ed. (1976) *John Anthony's Flora of Sutherland.* Edinburgh: Botanical Society of Edinburgh.

1228 **Kermode, P. M. C.** (1886) Introduction to the study of lichens in the Isle of Man. *Lioar Manninagh* **1**: 85–87.

Kerr, J. G.—See No. 2264.

1229 **Kershaw, K. A.** (1960) The genus *Stereocaulon* Schreb. in the British Isles. *Lichenologist* **1**: 184–203.

1230 —— (1961) The genus *Umbilicaria* in the British Isles. *Lichenologist* **1**: 251–265.

1231 —— (1963) Lichens. *Endeavour* **22**: 65–69.

1232 —— (1964) Preliminary observations on the distribution and ecology of epiphytic lichens in Wales. *Lichenologist* **2**: 263–276.

1233 —— **and Harris, G. P.** (1971) Simulation studies and ecology: a simple defined system model. In *Statistical Ecology* (G. P. Patil, E. C. Pielou and W. E. Waters, eds.) **3**: 1–21. University Park, Penna.: Pennsylvania State University Press.

1234 —— —— (1971) Simulation studies and ecology: use of the model. In *Statistical Ecology* (G. P. Patil, E. C. Pielou and W. E. Waters, eds.) **3**: 23–42. University Park, Penna.: Pennsylvania State University Press.

—— —See also Nos. 11a, 27, 1604 and 1605.

1234a **Ketchen, J.** (1965) Foulney Island: some of the lesser known flora & fauna. *Fld. Nat.* **10**: 22–24.

Kidd, L. N.—See No. 226.

1235 **King, J.** (1962) The *Festuca-Agrostis* grassland complex in south-east Scotland. *J. Ecol.* **50**: 321–355.

1236 —— **and Nicholson, I. A.** (1964) Grasslands of the forest and sub-alpine zones. In *The Vegetation of Scotland* (J. H. Burnett, ed.): 67–142. Edinburgh and London: Oliver and Boyd.

1237 **King, R. W.** (1934) *Cephalodia of* Solorina. B.Sc. thesis, University of Bristol.

Kingston, J. F.—See Nos. 1210 and 2395.

1238 **Knight, H. H.** (1921) The lichens of Minehead district. *Trans. Br. mycol. Soc.* **7**: 16–18.

1239 —— (1922) Lichens of Haslemere district. *Trans. Br. mycol. Soc.* **7**: 225.

1240 —— (1922) Lichens found during the Worcester foray. *Trans. Br. mycol. Soc.* **8**: 10.

1241 —— (1924) Keswick lichens. *Trans. Br. mycol. Soc.* **9**: 10–12.

1242 —— (1924) Lichens of the Windsor foray. *Trans. Br. mycol. Soc.* **10**: 9.

1243 —— (1925) Lichens of the Matlock foray. *Trans. Br. mycol. Soc.* **10**: 132–133.

1244 —— (1925) Bettws-y-Coed lichens. *Trans. Br. mycol. Soc.* **10**: 242–244.

1245 —— (1926) Lichens of the Tintern foray. *Trans. Br. mycol. Soc.* **11**: 9–10.

1246 —— (1927) Lichens of Aviemore district. *A. Conf. crypt. Soc. Scotl.* **46**: 14–15.

1247 —— (1928) Lichens of the Marlborough foray. *Trans. Br. mycol. Soc.* **13**: 149–150.

1248 —— (1928) Lichens of Aviemore district. *Trans. Br. mycol. Soc.* **13**: 314–316.

1249 —— (1929) Sussex lichens. *Trans. Br. mycol. Soc.* **14**: 191–193.

1250 —— (1931) Lichens of the Whitby foray. *Trans. Br. mycol. Soc.* **16**: 17–18.

1251 —— (1933) The lichens of the Isle of Wight. *Proc. Isle Wight nat. Hist. archaeol. Soc.* **2**: 221–232.

1252 —— (1933) Lichens of the Ludlow district. *Trans. Br. mycol. Soc.* **18**: 5–6.

1253 —— (1933) Lichens of the Haslemere foray. *Trans. Br. mycol. Soc.* **18**: 16–17.

1254 —— (1935) Norfolk lichens. *Trans. Br. mycol. Soc.* **20**: 15–16.

1255 —— [**revised W. Watson**] (1950) The lichens of Gloucestershire. *Proc. Cotteswold Nat. Fld Club* **30**: 70–96.

1256 [—— **and Hurst, C. P.**] (1931) Lichens. *Rep. Marlboro. Coll. nat. Hist. Club* **79**: 124.

—— —See also Nos. 105, 1102 and 1103.

1257 **Knight, T. A.** (1811) *Pomona Herefordiensis; containing engravings of the Old Cider and Perry Fruits of Herefordshire.* London: Agricultural Society of Herefordshire. [Plates II, III, X, XII, XIX, XXI, XXII, XXVIII, XXIX and XXX show lichens on twigs.]

1258 **Knowles, M. C.** (1926) Rare lichen, *Biatorina bouteillei* Arn., in Ireland. *Ir. Nat. J.* **1**: 97.

1259 **Knowlton, D.** (1973) *The Naturalist in Central Southern England.* [The Regional Naturalist Series.] Newton Abbot: David and Charles.

1260 —— (1974) *The Naturalist in Scotland.* Newton Abbot & London: David and Charles. [The Regional Naturalist Series.]

1261 **Knox, R. B.** (1960) Flora of Jura II: List of lichens. *Trans. Proc. bot. Soc. Edinb.* **39**: 114–115.

Köfaragó-Gyelnik, V.—See Gyelnik, V.

1262 **Kok, A.** (1966) A short history of the orchil dyes. *Lichenologist* **3**: 248–272.

1263 **Krempelhuber, A. von** (1867–72) *Geschichte und Litteratur der Lichenologie.* 3 vols. Munich: Wolf.

1264 **Krinos, M. H. G.** (1970) Some marine algae from the Mull of Kintyre region of Argyll. *Trans. Proc. bot. Soc. Edinb.* **40**: 545–556.

1265 **Krog, H.** (1951) Microchemical studies on *Parmelia. Nytt Mag. Naturvid.* **88**: 57–85.

1266 —— (1970) The Scandinavian members of the *Parmelia borreri* group. *Nytt Mag. Bot.* **17**: 11–15.

—— —See also Nos. 548 and 2323.

Krüger, U.—See No. 1846.

Kupczyk, B.—See No. 2363a.

Kurokawa, S.—See No. 863.

1267 [**Kynoch, J.**] (1910) *Wild Flowers of Barmouth and Neighbourhood.* Ed. 4. Brighton: J. Kynoch. [Lichens pp. 37–42 from T. Salwey (1863).]

1268 **Laird, [Rev. 'and others']** (1810) *Flora Orcadensis.* Manuscript in K.

1269 **Lamb, I. M.** (1936) Lichenological notes from the British Museum herbarium.—I. *J. Bot., Lond.* **74**: 174–178.

1270 —— (1938) Lichenological notes from the British Museum herbarium.—II. *J. Bot., Lond.* **76**: 153–165.

1271 —— (1938–39) *Collections*. 2 vols. Manuscript in BM.

1272 —— (1939) Lichenological notes from the British Museum herbarium.—III. *J. Bot., Lond.* **77**: 72–80.

1273 —— (1940) Lichenological notes from the British Museum herbarium.—IV. *Rhizocarpon* sect. *Catocarpon* in the British Isles. *J. Bot., Lond.* **78**: 129–138.

1274 —— (1941) Lichenological notes from the British Museum herbarium.—V. *J. Bot., Lond.* **79**: 89–97. [Index to I-V in this series pp. 95–97.]

1275 —— (1942) A lichenological excursion to the west of Scotland. *Trans. Proc. bot. Soc. Edinb.* **33**: 295–326.

1276 —— (1945) Lichens. In *The Flora of Uig* (M. S. Campbell, ed.): 57–58. Arbroath: T. Buncle.

1277 —— (1947) A monograph of the lichen genus *Placopsis* Nyl. *Lilloa* **13**: 151–288.

1278 —— (1951) On the morphology, phylogeny, and taxonomy of the lichen genus *Stereocaulon. Can. J. Bot.* **29**: 522–584.

1279 —— **and Ward, S.** (1974) A preliminary conspectus of the species attributed to the imperfect lichen genus *Leprocaulon* Nyl. *J. Hattori bot. Lab.* **38**: 499–553.

1280 **Lamb, W.** (1845) Parish of Carmichael. In *The New Statistical Account of Scotland.* Vol. 6. *Lanark*: 514–534. Edinburgh and London: W. Blackwood & Sons.

Lambert, J. M.—See No. 773.

1281 **Lambley, P. W.** (1973) Lichens on Weeting Heath. *Trans. Norfolk Norwich Nat. Soc.* **22**: 439.

1282 —— (1973) Lichens on Marsham Heath. *Trans. Norfolk Norwich Nat. Soc.* **22**: 439.

1283 —— (1973) Hull Wood. *Trans. Norfolk Norwich Nat. Soc.* **22**: 439–440.

1284 —— (1976) Some notable hollies *Ilex aquifolium* in Norfolk woodlands. *Trans. Norfolk Norwich Nat. Soc.* **23**: 269.

—— —See also Nos. 436, 569 and 2568.

Lan, O. B.—See No. 439.

1285 **Landsborough, D.** (1847) *Arran: a poem in six cantos and Excursions to Arran.* Edinburgh: J. Johnstone. [Lichens pp. 356–358.]

1286 —— (1849) *A Popular History of British Sea-Weeds.* London: Reeve, Benham and Reeve. [Lichens pp. 96–98.] [And later editions to Ed. 3 (1857).]

1287 —— (1851) *Excursions to Arran, Ailsa Craig, and the two Cumbraes, with Reference to the Natural History of the Islands.* Edinburgh: Johnstone and Hunter. [Lichens pp. 356–358.]

1288 —— **and Landsborough [D.], the Younger** (1875) *Arran: its Topography, Natural History and Antiquities.* Ardrossan: Arthur Guthrie. [Lichens pp. 482–483.]

—— —See also Nos. 340 and 705.

1289 **Landsburgh, S. Y.** (1955) *The morphology and vegetation of the Sands of Forvie.* Ph.D. thesis, University of Aberdeen.

Langdale, E.—See No. 34.

Lankester, E.—See No. 1480.

1290 **Larbalestier, C. du B.** (1867–72) *Lichenes caesarienses et sargienses exsiccati.* 6 fasc., nos. 1–280. Perrot, Jersey. [Intact set not seen.]

1291 —— (1879–81). *Lichen-Herbarium.* 9 fasc., nos. 1–360. London and St Aubin's, Jersey. [Intact set not seen; not abstracted.]

1292 —— (1896) *Lichenes exsiccati circa cantabrigiam collecti a C. du Bois Larbalestier, M.A.* Nos. 1–35. Cambridge. [Intact set in BM.]

—— —See also No. 103.

1293 **Larter, C. E.** (1914) Sixth report of the botany committee. *Rep. Trans. Devon. Ass. Advmt Sci.* **46**: 95–102.

1294 —— (1925) Seventeenth botany report. *Rep. Trans. Devon. Ass. Advmt Sci.* **57**: 75–89.

Latham, R. G.—See Nos. 103 and 104.

1295 **Laundon, J. R.** (1954) The survey of Bookham Common. Progress report. Lichens. *Lond. Nat.* **33**: 23.

1296 —— (1955) The survey of Bookham Common. Progress report. Lichens. *Lond. Nat.* **34**: 14.

1297 —— (1956) The lichen ecology of Northamptonshire. In *The First Fifty Years. A History of Kettering & District Naturalists Society & Field Club*: 89–96. Kettering: Kettering and District Naturalists Society and Field Club.

1298 —— (1957) Lichens. In *A Survey of Thoroughsale and Hazel Woods, Corby, Northants.* (N. L. Hodson *et al.*, eds.): [9 pp.; unnumbered]. Kettering.

1299 —— (1958) Lichens new to the British flora: 1. *Lichenologist* **1**: 31–38.

1300 —— (1958) The lichen vegetation of Bookham Common. *Lond. Nat.* **37**: 66–79.

1301 —— (1958) Huntingdonshire lichens. *Rep. Huntingdon. Fauna Flora Soc.* **10**: 14–17.

1302 —— (1960) Lichens new to the British flora: 2. *Lichenologist* **1**: 158–168.

1303 —— (1962) The taxonomy of sterile crustaceous lichens in the British Isles. 1. Terricolous lichens. *Lichenologist* **2**: 57–67.

1304 —— (1963) The taxonomy of sterile crustaceous lichens in the British Isles. 2. Corticolous and lignicolous species. *Lichenologist* **2**: 101–151.

1305 —— (1964) Semi-natural vegetation in Northamptonshire. *J. Northampt. nat. Hist. Soc.* **34**: 268–281.

1306 —— (1965) Lichens new to the British flora: 3. *Lichenologist* **3**: 65–71.

1307 —— (1966) Hudson's *Lichen siliquosus* from Wiltshire. *Lichenologist* **3**: 236–241.

1308 —— (1966) Frost damage to *Parmelia caperata. Lichenologist* **3**: 273.

1309 —— (1967) A study of the lichen flora of London. *Lichenologist* **3**: 277–327.

1310 —— (1970) Lichens new to the British flora: 4. *Lichenologist* **4**: 297–308.

1311 —— (1970) London's lichens. *Lond. Nat.* **49**: 20–69.

1312 —— (1971) Fumarprotocetraric acid in *Cladonia rangiformis. Lichenologist* **5**: 175–176.

1313 —— (1971) Lichen communities destroyed by psocids. *Lichenologist* **5**: 177.

1314 —— (1972) The lichen flora of Wood Walton Fen. *Rep. Huntingdon. Fauna Flora Soc.* **24**: 15–19.

1315 —— (1973) Urban lichen studies. In *Air Pollution and Lichens* (B. W. Ferry, M. S. Baddeley and D. L. Hawksworth, eds.): 109–123. London: Athlone Press of the University of London.

1316 —— (1973) *Lichens of Wicken Fen.* [Guides to Wicken Fen no. 10.] Wicken Fen: The National Trust.

1317 —— (1973) Lichens. In *Monks Wood. A Nature Reserve Record.* (R. C. Steele and R. C. Welch, eds.): 95–100. Huntingdon: The Nature Conservancy.

1318 —— (1974) *Leproplaca* in the British Isles. *Lichenologist* **6**: 102–105.

1319 —— (1976) Lichens new to the British flora: 5. *Lichenologist* **8**: 139–150.

1320 [——] (1976) Sarsen lichens at risk. *Bull. Br. Lichen Soc.* **38**: 1–2.

1321 [——] (1976) Lichens on stone: beauty or blight. *Bull. Br. Lichen Soc.* **39**: 1–2.

Laurie, A. E.—See No. 2172.

Laurie, M.—See No. 666.

1322 **Laverack, M. S. and Blackler, M.,** eds. (1974) *Fauna and Fauna of St. Andrews Bay.* Edinburgh & London: Scottish Academic Press. [Lichens pp. 283–284.]

1323 **Lawson, G.** (1848) The Dundee Naturalists' Association. *Phytologist* **3**: 126–128.

1324 **Lees, E.** (1842) *The Botanical Looker-out among the wild flowers of the fields, woods and mountains of England and Wales.* London: Tilt & Bogue.

1325 —— (1842) On the flora of the Malvern Hills. Part 3: Being a sketch of the cryptogamic vegetation indigenous to the chain. *Phytologist* **1**: 271–272.

1326 —— (1843) *The Botany of the Malvern Hills.* London: Tilt & Bogue. [Lichens pp. 56–61.]

1327 —— (1852) Notices of the flowering time and localities of some plants observed during an excursion through a portion of South Devon, in June, 1851. *Phytologist* **4**: 530–541.

1328 —— (1853) Account of the mosses and lichens of the Malvern Hills. *Phytologist* **4**: 863–866.

1329 —— (1855) On the plants that more particularly flourish on the Silurian limestones, with remarks incidental to the subject. *Trans. Malvern Nat. Fld Club* **1**: 15–28.

1330 —— (1856) *Pictures of Nature in the Silurian Region around the Malvern Hills and Vale of Severn.* Malvern: H. W. Lamb.

1331 —— (1868) *The Botany of the Malvern Hills.* [Ed. 2.] London: Simpkin & Marshall. [Lichens pp. 139–151.]

1332 **Lees, F. A.** (1878) A gossip about two new Lincolnshire cryptogams (*Hypnum salebrosum* and *Cetraria islandica*). *Naturalist, Hull* **3**: 97–102.

1333 —— (1879) The autumn flora of Whernside: An account of an excursion in search of mosses. *Naturalist, Hull* **4**: 136–137.

1334 —— [*c.* 1880–90, revised 1918] *MS. Florula Wetherby District (Wharfe & Nidd) of West Yorks.* Manuscript in possession of W. A. Sledge (Leeds).

1335 —— (1885) *Botany and Natural History of Upper Wensleydale*. Hawes: Routh.

1336 —— (1887) Botanical Section Report. *In* Yorkshire and Westmorland Naturalists at Sedbergh (Anon). *Naturalist, Hull* **1887**: 279–282.

1337 —— (1888) The flora of West Yorkshire. [*Trans. Yorks. Nat. Un., ser. Bot.,* **2**: 1–843.] London: L. Reeve. [Lichens pp. 631–660, 806–811.]

1338 —— (1892) Botany and outline flora of Lincolnshire. In *History, Gazetteer and Directory of the County of Lincolnshire* (W. White, ed.): 37–63. Sheffield: W. White.

1339 —— (1894) The botany of Nidderdale and the Garden of the Nidd. In *Nidderdale and the Garden of the Nidd: A Yorkshire Rhineland* (H. Speight, ed.): 17–63. London: Elliot Stock. [Lichens pp. 62–63.]

1340 —— **and West, W.** (1881) A local ramble and fungus hunt. *Naturalist, Hull* **7**: 61–63.

—— —See also Nos. 570, 2121, 2125 and 2137.

1341 **Leighton, W. A.** (1837) *Catalogue of the Cellulares or Flowering Plants of Great Britain*. London.

1342 —— (1849–53) [*Correspondence*]. Letters in K (1 vol.). [Not abstracted.]

1343 —— (1851) *The British Species of Angiocarpous Lichens, elucidated by their sporidia*. London: Ray Society.

1344 —— (1851–67) *Lichenes Britannici exsiccati*. 13 fasc., nos. 1–410. London: W. Pamplin; Shrewsbury. [Intact set in BM.]

1345 —— (1854) Monograph of the British *Graphideae*. *Ann. Mag. nat. Hist., ser.* 2, **13**: 81–97, 202–212, 264–279, 387–395, 436–446. [Also separately printed.]

1346 —— (1856) Monograph of the British *Umbilicariae*. *Ann. Mag. nat. Hist., ser.* 2, **18**: 273–297.

1347 —— (1856) New British *Arthoniae*. *Ann. Mag. nat. Hist., ser.* 2, **18**: 330–333.

1348 —— (1857) New British lichens. *Ann. Mag. nat. Hist., ser.* 2, **19**: 129–133.

1349 —— (1864) New British lichens. *Ann. Mag. nat. Hist., ser.* 3, **14**: 401–405.

1350 —— (1865) Notes on British lichens. *Ann. Mag. nat. Hist., ser.* 3, **16**: 8–12.

1351 —— (1866) Notulae lichenologicae. No. II. New British lichens described by Nylander. *Ann. Mag. nat. Hist., ser.* 3, **17**: 59–65.

1352 —— (1866) Notulae lichenologicae. No. V. New British lichens, described by Nylander. *Ann. Mag. nat. Hist., ser.* 3, **17**: 348–351.

1353 —— (1866) Notulae lichenologicae. No. VII. On three European species of *Thelocarpon. Ann. Mag. nat. Hist., ser.* 3, **18**: 23–24.

1354 —— (1866) Notulae lichenologicae. No. VIII. On new British lichens. *Ann. Mag. nat. Hist., ser.* 3, **18**: 103–106.

1355 —— (1866) Notulae lichenologicae. No. XI. On the examination and re-arrangement of the *Cladoniae* as tested by hydrate of potash. *Ann. Mag. nat. Hist., ser.* 3, **18**: 405–420.

1356 —— (1867) Notulae lichenologicae. No. XII. On the *Cladoniae* in the Hookerian herbarium of Kew. *Ann. Mag. nat. Hist., ser.* 3, **19**: 99–124.

1357 —— (1867) Notulae lichenologicae. No. XIV. On new British lichens, described by Nylander. *Ann. Mag. nat. Hist., ser.* 3, **19**: 330–334.

1358 —— (1867) Notulae lichenologicae. No. XV. Notes on the lichens of Cader Idris, North Wales. *Ann. Mag. nat. Hist.*, *ser.* 3, **19**: 402–409.

1359 —— (1867) Notulae lichenologicae. No. XVII. On new British lichens. *Ann. Mag. nat. Hist.*, *ser.* 3, **20**: 256–260.

1360 —— (1868) Notulae lichenologicae. No. XXI. Rev. E. William's list of Shropshire lichens. *Ann. Mag. nat. Hist.*, *ser.* 4, **1**: 183–188.

1361 —— (1868) Notulae lichenologicae. No. XXII. Dr. W. Nylander on new British lichens. *Ann. Mag. nat. Hist.*, *ser.* 4, **1**: 482–483.

1362 —— (1869) Notulae lichenologicae. No. XXVII. Nylander on new British lichens. *Ann. Mag. nat. Hist.*, *ser.* 4, **3**: 264–270.

1363 —— (1869) Notulae lichenologicae. No. XXX. Further notes on the lichens of Cader Idris, North Wales. *Ann. Mag. nat. Hist.*, *ser.* 4, **4**: 198–202.

1364 —— (1870) Notulae lichenologicae. No. XXXI. On certain new characters in the species of the genera *Nephroma* (Ach.) and *Nephromium*, Nyl. *Ann. Mag. nat. Hist.*, *ser.* 4, **5**: 37–41.

1365 —— (1870) Notulae lichenologicae. No. XXXIV. Notes on the chemical reaction in the British species of *Pertusaria*. *Ann. Mag. nat. Hist.*, *ser.* 4, **6**: 473–474.

1366 —— (1871) *The Lichen-flora of Great Britain, Ireland and the Channel Islands.* Shrewsbury: privately printed.

1367 —— (1872) Notulae lichenologicae. No. XXXV. On the genus *Ramalina*, by Nylander. *Ann. Mag. nat. Hist.*, *ser.* 4, **9**: 122–132.

1368 —— (1872) Lichenological memorabilia.—No. 1. *Grevillea* **1**: 8–9.

1369 —— (1872) Lichenological memorabilia.—No. 2. The lichens of Bettws-y-Coed, North Wales. *Grevillea* **1**: 57–60.

1370 —— (1872) *The Lichen-flora of Great Britain, Ireland and the Channel Islands.* Ed. 2. Shrewsbury: privately printed.

1371 —— (1873) Lichenological memorabilia.–No. 3. Hellbom's lichens of Lule Lapmark. *Grevillea* **1**: 121–126, 133–135.

1372 —— (1874) Lichenological memorabilia, no. 5. On *Lecidea dilleniana* (Ach.), and *Opegrapha grumulosa*, Duf. *Grevillea* **2**: 171–173.

1373 —— (1875) Lichenological memorabilia, No. 6. Lichenological researches in north and south Wales in 1874. *Grevillea* **3**: 113–116.

1374 —— (1875) On "*Parmelia millaniana*, Stirton". *Grevillea* **3**: 117–118.

1375 —— (1875) Lichenological memorabilia.–No. 7. Additions to the lichen-flora of Great Britain, &c. *Grevillea* **3**: 167–168.

1376 —— (1875) Lichenological memorabilia, no. 8. On *Lecidea trochodes*, (Tayl.), Leight. *Grevillea* **4**: 25–26.

1377 —— (1876) *Lichen pilularis*, Dav. *Grevillea* **4**: 126.

1378 —— (1876) *Lecanora angulosa*, (Schreb.) Ach. *Grevillea* **4**: 128.

1379 —— (1876) Lichenological memorabilia. No. 10. On the lichens of Fishguard, Pembrokeshire. *Grevillea* **5**: 81–86.

1380 —— (1876) New British lichens. *Trans. Linn. Soc. Lond.*, *ser.* 2, *Bot.* **1**: 145–147.

1381 —— (1877) New British lichens. *J. Linn. Soc.* (*Bot.*) **16**: 2.

1382 ——(1878) New British lichens. *Trans. Linn. Soc. Lond.*, *ser.* 2, *Bot.* **1**: 237–240.

112

1383 —— (1879) *The Lichen-flora of Great Britain, Ireland, and the Channel Islands.* Ed. 3. Shrewsbury: privately printed.
—— —See also Nos. 113, 1203 and 2423.

1384 **Leishman, M.** (1845) Parish of Govan. In *The New Statistical Account of Scotland.* Vol. 6. *Lanark*: 663–718. Edinburgh and London: W. Blackwood & Sons.

1385 **Leitch, S.** (1967) Lichen survey. In *Foula Expedition 1966*: 10–12. Ullswater: British Girls' Exploring Society.
Lester, P.—See No. 345.

1386 **Letrouit-Galinou, M. A.** (1974) Le développement des apothécies du *Gyalecta carneolutea* (Turn.) Oliv. (Discolichen Gyalectacée). *Bull. Soc. mycol. Fr.* **90**: 23–39.
—— —See also No. 740.

1387 **Lettau, G.** (1932–37) Monographische Bearbeitung einiger Flechtenfamilien. *Beih. Repert. Spec. nov. Regni veg.* **69**: 1–250.

1388 —— (1940) Flechten aus Mitteleuropa. III. u. IV. *Beih. Repert. Spec. nov. Regni veg.* **119**: 127–202.

1389 —— (1942) Flechten aus Mitteleuropa. VII. *Beih. Repert. Spec. nov. Regni veg.* **119**: 265–348.

1390 —— (1955) Flechten aus Mitteleuropa. X. *Feddes Reprium Spec. nov. veg.* **57**: 1–94.

1391 **Leuckert, C., Guse, K. and Poelt, J.** (1969) Zur Chemotaxonomie der *Pertusaria hymenea*-Gruppe (Lichenes, Pertusariaceae). *Herzogia* **1**: 159–171.

1392 —— **and Mathey, A.** (1975) Beitrage zur Chemotaxonomie einiger xanthonhaltiger Arten der Flechtengattung *Buellia*. *Herzogia* **3**: 461–488.

1393 —— **Poelt, J., Schultz, I. and Schwarz, B.** (1975) Chemotaxonomie und stammesgeschichtliche Differenzierung des Formenkreises von *Parmelia prolixa* in Europa (Lichenes, Parmeliaceae). *Decheniana* **127**: 1–36.
—— —See also No. 1005.

1394 **Lewis, D. H.** (1963) Field meeting at Preston Montford. *Lichenologist* **2**: 190–194.
—— —See also No. 1958.

1395 **Lewis, F. J.** (1905) The plant remains in the Scottish peat mosses. Part I. The Scottish Southern Uplands. *Trans. R. Soc. Edinb.* **41**: 699–723.

1396 —— (1907) The plant remains in the Scottish peat mosses. Part III. The Scottish Highlands and the Shetland Islands. *Trans. R. Soc. Edinb.* **46**: 33–70.

1397 **Lewis, J. R.** (1957) Intertidal communities of the northern and western coasts of Scotland. *Trans. R. Soc. Edinb.* **63**: 185–220.

1398 —— (1964) *The Ecology of Rocky Shores.* London: English Universities Press.

1399 **Liddle, M. J. and Greig-Smith, P.** (1975) A survey of tracks and paths in a sand dune ecosystem II. Vegetation. *J. appl. Ecol.* **12**: 909–930.

1400 **Lightfoot, J.** (1773) *Journal of a botanical excursion in Wales.* Manuscript in BM.

1401 —— (1777) *Flora Scotica*. Vol. 2. London: B. White. [Often stated to have been published in 1778 but 1777 is now known to be correct (see Price, J. H., *J. Soc. Biblphy nat. Hist.* **5**: 57–67, 1968); lichens pp. 800–898, 964–966.]

1402 —— (1789) *Flora Scotica*. Ed. 2. Vol. 2. London: R. Faulder. [Lichens pp. 800–898, 964–966; also re-issued 1792.]

—— —See also No. 1960.

1403 **Lillie, D.** (1912) Caithness lichens. *Scot. bot. Rev.* **1**: 146–153.

1404 **Lindahl, P.-O.** (1955) Some interesting lichens from the West of Scotland (Argyll). *Bot. Notiser* **108**: 17–21.

1405 **Lindgren, E. J.** (1975) The herd of reindeer. In *Glen More Forest Park, Cairngorms*, Ed. 5 (D. A. Woodburn, ed.): 45–47. [Forestry Commission Guide.] Edinburgh: H.M.S.O.

1406 **Lindsay, D. C.** (1970) The biology and taxonomy of lichens. *Proc. Bgham nat. Hist. Soc.* **21**: 251–270.

1407 [——] [1970] *Warwickshire lichens*. Manuscript. Report to Warwickshire Fungus Survey; copy with BLS.

1408 —— (1975) *Lichen distribution survey in Leicestershire*. Manuscript in LSR.

1409 **Lindsay, W. L.** (1853) The dyeing properties of lichens. *North British Agriculturalist*, 16 February and 26 April. [Not seen; journal published in Edinburgh 1843–1939 but no copies before 1860 located.]

1410 —— (1853–54) Dyeing properties of lichens. *Phytologist* **4**: 867–872, 901–909, 998–1003, 1068–1070; **5**: 179–183.

1411 —— (1854) Experiments on the dyeing properties of lichens. *Edinb. New phil. J.* **57**: 228–249, and *Appendix*: 385–401.

1412 —— (1855) Economical applications of British lichens. *Phytologist, n.s.* **1**: 286–289.

1413 —— (1855) The dyeing properties of lichens. *Edinb. New phil. J., n.s.* **2**: 56–80.

1414 —— (1855) On the dyeing properties of lichens. *Proc. bot. Soc. Edinb.* **1855**: 52–54.

1415 —— (1856) *A Popular History of British Lichens*. London: L. Reeve.

1416 —— (1857) On the structure of *Lecidea lugubris* (Sommf.). *Q. Jl microsc. Sci.* **5**: 177–185.

1417 —— (1857) Notes on the flora of Perth. *Phytologist, n.s.* **2**: 264–272.

1418 —— (1857) Monograph of the genus *Abrothallus* (De Notaris and Tulasne emend.). *Q. Jl microsc. Sci.* **5**: 27–63.

1419 —— (1859) Memoir on the spermogones and pycnides of filamentous, fruticulose, and foliaceous lichens. *Trans. R. Soc. Edinb.* **22**: 101–303.

1420 —— (1866) Contributions to the lichen-flora of northern Europe. *J. Linn. Soc. (Bot.)* **9**: 365–416.

1421 —— (1867) Is lichen-growth detrimental to trees? *Sci. Gossip* **3**: 241–242.

1422 —— (1867) On *Arthonia melaspermella*, Nyl., nov. sp. *J. Linn. Soc. (Bot.)* **9**: 268–286.

1423 —— (1867) To what extent is lichen-growth a test of age? *The Farmer* **1867**: 528.

114

1424 —— (1868) On the present domestic use of lichen dyestuffs in the Scottish Highlands and Islands. *J. Bot., Lond.* **6**: 84–89.

1425 —— (1868) On the present uses of lichens as dye-stuffs. *Rep. Br. Ass. Advmt Sci.* **37** [Transactions of the sections]: 38–40.

1426 —— (1868) On the present use of lichens as dyestuffs. *J. Bot., Lond.* **6**: 101–109.

1427 —— (1868) On the lichen flora of druidical stones in Scotland. *Trans. bot. Soc. Edinb.* **9**: 154–162.

1428 —— (1868) On the arctic *Cladoniae. Trans. bot. Soc. Edinb.* **9**: 166–187.

1429 —— (1869) On chemical reaction as a specific character in lichens. *J. Linn. Soc. (Bot.)* **8**: 36–63.

1430 —— (1869) Enumeration of micro-lichen parasites on other lichens. *Q. Jl microsc. Sci., n.s.* **9**: 49–57, 135–146, 342–358.

1431 —— (1869) Observations on new lichenicolous micro-fungi. *Trans. R. Soc. Edinb.* **25**: 513–555.

1432 —— (1870) Experiments on colour-reaction as a specific character in lichens. *Trans. bot. Soc. Edinb.* **10**: 82–98.

1433 —— (1871) Observations on lichenicolous micro-parasites. *Q. Jl microsc. Sci., n.s.* **11**: 28–42.

1434 —— (1872) Memoir on the spermogones and pycnides of crustaceous lichens. *Trans. Linn. Soc. Lond.* **28**: 189–318.

1435 **Linnaeus, C.** (1753) *Species plantarum.* Vol. 2. Stockholm. [Facsimile edition (1959) London: Ray Society.] [Lichens pp. 1140–1156, 1169 and 1185.]

Linton, D. L.—See No. 1831.

1436 **Lister, A.** (1901) Notes on Mycetozoa. *J. Bot., Lond.* **39**: 81–90.

1437 **Lister, G.** (1880) [*Field notebooks.*] Manuscripts in BM.

1438 **Little, J. E.** (1934) Botany. In *The Natural History of the Hitchin Region* (R. L. Hine, ed.): 56–72. Hitchin and District Regional Survey Association.

1439 **Livens, H. M.** (1911) The mosses and lichens. In *Handbook and Guide to Portsmouth* (J. C. Nicol, ed.): 211–214. Portsmouth: British Association.

1440 **Llano, G. A.** (1950) *A Monograph of the Lichen family Umbilicariaceae in the Western Hemisphere* [Navexos P-831]. Washington, D.C.: Office of Naval Research.

—— —See also No. 1805.

1441 **Lloyd, A. O.** (1972) An approach to the testing of lichen inhibitors. In *Biodeterioration of Materials* (A. H. Walters and E. H. Hueck-van der Plas, eds.) **2**: 185–191. London: Applied Science Publishers.

1441a —— (1976) Progress in studies of deteriogenic lichens. In *Proceedings of the Third International Biodegradation Symposium* (J. M. Sharpley and A. M. Kaplan, eds.): 395–402. London: Applied Science Publishers.

Lockley, R. M.—See No. 353.

1442 **Loder, J. de V.** (1935) *Colonsay and Oronsay in the Isles of Argyll: their History, Flora, Fauna and Topography.* Edinburgh and London: Oliver and Boyd.

1443 **Lonsdale, J.** (1974) *Lichens, with particular reference to the lichens of Kew and to pollution.* Kew Diploma thesis, Royal Botanic Gardens, Kew.

1444 **Loudon, J. C.** (1829) *An Encyclopaedia of Plants.* London: Longman etc. [Lichens pp. 948–977.]

1445 **Loudon, [J. W.] [assisted by G. Don and D. Wooster]** (1855) *Loudon's Encyclopedia of Plants.* London: Longman etc. [Lichens pp. 948–977.] [Lichens unchanged in the 1880 edition edited by Mrs Loudon.]

1446 **Low, G.** (1879) *A Tour through the Islands of Orkney and Schetland* [sic] *containing hints relative to their Ancient, Modern and Natural History collected in 1774.* Kirkwall: William Peace & Son.

Lukis, F. C.—See No. 126.

1447 **Lulham, R. B. J.** (1917) Lichens or crottles. *Sch. Nat. Study* **12**: 38–43.

1448 —— **and Haworth, F. M.** (1935) Lichens or crottles. *Sch. Nat. Study* **30**: 107–113. [Also issued separately as *Sch. Nat. Study Un. Special Leaflet* no. 5 with changed pagination.]

1449 **Lunan, D. A.** (1970) Lichenometric analysis in the Lake District. *Brycgstowe* [*J. Univ. Bristol Geogrl Soc.*] **5**: 56–57.

1450 **Lye, K. A.** (1969) The distribution and ecology of *Sphaerophorus melanocarpus. Svensk bot. Tidskr.* **63**: 300–318.

1451 **Lyle, L.** (1920) The marine algae of Guernsey. *J. Bot., Lond.* **58**, *Suppl.* **2**: 1–53.

1452 **Lynge, B.** (1913) On the world's "Lichenes exsiccati". *Nyt Mag. Naturvid.* **51**: 91–122.

1453 —— (1915–22) Index specierum et varietatum "Lichenum exsiccatorum". *Nyt Mag. Naturvid.* **53–60**. [Issued with separate pagination from journal; Pars I 1916–19; Pars II 1920–22.]

1454 —— (1935) Physciaceae. *Rabenh. Krypt.-Fl.* **9**, 6(1): 37–188.

1455 **Lysons, D. and Lysons, S.** (1816) *Magna Britannia; being a concise topographical account of the several counties of Great Britain.* Vol. 4. London: T. Cadell. [Lichens p. cxiv.]

1456 —— —— (1822) *Magna Britannia; being a concise topographical account of the several counties of Great Britain.* Vol. 6. London: T. Cadell.

Lysons, S.—See Nos. 1455 and 1456.

1457 **Maas Geesteranus, R. A.** (1948) Revision of the lichens of the Netherlands I. Parmeliaceae. *Blumea* **6**: 1–199.

1458 —— (1952) Revision of the lichens of the Netherlands II. Physciaceae. *Blumea* **7**: 206–287.

1459 **Mabey, R.** (1974) *The Pollution Handbook. The ACE/Sunday Times Clean Air and Water Surveys.* Harmondsworth: Penguin Education.

1460 **M'Andrew, J.** (1884) *Cladonia pyxidata* var. *leptophylla* Flk. in Scotland. *Scott. Nat.* **7**: 184.

1461 —— (1891) List of lichens gathered in Dumfriesshire, Kirkcudbrightshire &c. *Trans. J. Proc. Dumfries. Galloway nat. Hist. Antiq. Soc., ser. 2*, **7**: 28–36.

1462 —— (1893) A contribution to the cryptogamic botany of the Moffat district.

Trans. J. Proc. Dumfries. Galloway nat. Hist. Antiq. Soc., ser. 2, **8**: 30–33.

1463 —— (1894) Botanical notes for 1892. *Trans. J. Proc. Dumfries. Galloway nat. Hist. Antiq. Soc., ser.* 2, **9**: 29–31.

1464 —— (1895) Botanical notes for 1893. *Trans. J. Proc. Dumfries. Galloway nat. Hist. Antiq. Soc., ser.* 2, **10**: 10–12.

1465 —— (1896) Botanical notes for 1894. *Trans. J. Proc. Dumfries. Galloway nat. Hist. Antiq. Soc., ser.* 2, **11**: 10–11.

1466 —— (1897) Botanical notes for 1895. *Trans. J. Proc. Dumfries. Galloway nat. Hist. Antiq. Soc., ser.* 2, **12**: 22–23.

1467 —— (1904) A short talk on lichens, chiefly *Cladoniae. Trans. Edinb. Fld Nat. microsc. Soc.* **5**: 86–94.

1468 —— (1905) Botanical notes for 1899. *Trans. J. Proc. Dumfries. Galloway nat. Hist. Antiq. Soc., ser.* 2, **17**: 106–108.

1469 [——] (1905) Addenda and corrigenda to Mr M'Andrews lists of mosses, hepaticae and lichens of the district. *Trans. J. Proc. Dumfries. Galloway nat. Hist. Antiq. Soc., ser.* 2, **17**: 121–125.

1470 —— (1905) The botanical rarities of a sub-alpine parish. *Trans. Proc. bot. Soc. Edinb.* **22**: 166–169.

1471 [——] (1913) A new lichen. *Trans. Proc. bot. Soc. Edinb.* **26**: 184.

McCallum Webster, M.—See No. 2655.

McCarthy, P.—See No. 539.

1472 **McClintock, D.** (1964) Natural history of the garden of Buckingham Palace. Bryophytes and fungi. *Trans. Proc. S. Lond. ent. nat. Hist. Soc.* **1963**(2): 36–38.

1473 **M'Conachie, G.** (1900) On the ferns, mosses and lichens of Rerrick. *Trans. Proc. bot. Soc. Edinb.* **21**: 168–173.

1474 **McCrea, M. A.** (1924) Supplement to Marquand's "Flora of Guernsey and the Lesser Channel Islands." *Rep. Soc. guernés.* **2**: 167–182.

M'Crie, T.—See No. 182.

1475 **MacCulloch, J.** (1820) On peat. *Edinb. phil. J.* **2**: 40–59.

MacErlean, J. C.—See No. 1029.

1476 **McFarlane, A.** [*c.* 1900] *The Visitor's Guide to Malham.* Skipton: Edmondson. [Lichens p. 24.]

1477 **McGettigan, S.** (1972) *Some observations on the bark of* Corylus avellana *and its epiphytic lichen flora in the Burren, Co. Clare.* Honours thesis, University College, Galway. [Not seen.]

1478 **MacGillivray, J.** (1842) Account of the Island of St. Kilda, chiefly with respect to its natural history; from notes made during a visit in July 1840. *Edinb. New phil. J.* **32**: 47–70.

1479 **MacGillivray, W.** (1830) Account of the series of islands usually denominated the Outer Hebrides. Sect. [V]—Of the vegetation. *Edinb. J. nat. geogr. Sci.* **2**: 91–95.

1480 —— (1855) *The Natural History of Dee Side and Braemar.* London: privately printed. [Lichens pp. 375–383.]

1481 **M'Gregor, G.** (1845) United parish of Lismore and Appin. In *The New*

Statistical Account of Scotland. Vol. 7. *Argyle*: 223–256. Edinburgh and London: W. Blackwood & Sons. [Botany by J. Fleming and ? D. Carmichael.]

Macgregor, M.—See No. 442.

McIvor, W. G.—See No. 2151.

Mackenzie, E.—See Lamb, I. M.

1482 **McLean, R. C.** (1915) The ecology of the maritime lichens at Blakeney Point, Norfolk, *J. Ecol.* **3**: 129–148.

1483 —— **and Hyde, H. A.** (1924) The vegetation of Steep Holm. *J. Bot., Lond.* **62**: 167–175.

1484 **Macmillan, H.** (1855) On lichens collected on the Breadalbane Mountains and woods. *Proc. bot. Soc. Edinb.* **1855**: 23–24.

1485 —— (1856) On the rare lichens of Ben Lawers. *Edinb. New phil. J., n.s.* **3**: 257–268.

1486 —— (1856) On the rare lichens of Ben Lawers. *Edinb. New phil. J., n.s.* **4**: 27–38.

1487 —— (1857) Notice of cryptogamic plants found near New Abbey. *Ann. Mag. nat. Hist., ser.* 2, **20**: 303.

1488 —— (1858) Notice of cryptogamic plants found near New Abbey. *Trans. bot. Soc. Edinb.* **5**: 211–212.

1489 —— (1861) *Footnotes from the Page of Nature or First Forms of Vegetation.* Cambridge: Macmillan. [Lichens unchanged apart from pagination in Ed. 2 (1874) entitled only *First Forms of Vegetation.*]

1490 —— (1869) *Holidays on High Lands; or rambles and incidents in search of alpine plants.* London: Macmillan.

1491 —— (1876) The rare lichens of Glencroe. *Trans. bot. Soc. Edinb.* **12**: 289–299.

1492 —— (1889) The lichens of Inveraray. *Trans. Crypt. Soc. Scotland* **14**: 16–24.

1493 —— (1898) Larig Ghru. *J. Cairngorm Club* **2**: 297–310.

MacNair, R.—See No. 348.

1494 **M'Naughton, A.** (1845) Parish of Kilbride. In *The New Statistical Account of Scotland.* Vol. 5. *Bute*: 1–39. Edinburgh and London: W. Blackwood & Sons.

1495 **McNeill, M.** (1910) *Colonsay, one of the Hebrides.* Edinburgh: David & Douglas.

McRae, D. C.—See No. 226.

1496 **M'Rae, J.** (1845) Parish of Glensheil. In *The New Statistical Account of Scotland.* Vol. 14. *Ross and Cromarty*: 181–210. Edinburgh and London: W. Blackwood & Sons.

1497 **McVean, D. N.** (1955) *Nephromium arcticum* (L.) Fr., a lichen new to Britain. *Trans. Br. mycol. Soc.* **38**: 213–216.

1498 —— (1956) An arctic lichen *Siphula ceratites*, in Scotland. *Nature, Lond.* **177**: 797–798.

1499 —— (1957) Notes on *Siphula ceratites* in north-west Scotland. *New Phytol.* **56**: 150–153.

1500 —— (1961) Flora and vegetation of the islands of St Kilda and North Rona in 1958. *J. Ecol.* **49**: 39–54.

118

1501 —— (1962) *Cladonia elongata* (Jacq.) Hoffm. in the Cairngorms. *Lichenologist* **2**: 94–96.

1502 —— (1964) Dwarf shrub heaths. In *The Vegetation of Scotland* (J. H. Burnett, ed.): 481–498. Edinburgh and London: Oliver and Boyd.

1503 —— (1964) Grass heaths. In *The Vegetation of Scotland* (J. H. Burnett, ed.): 499–513. Edinburgh and London: Oliver and Boyd.

1504 —— (1964) Herb and fern meadows. In *The Vegetation of Scotland* (J. H. Burnett, ed.): 514–523. Edinburgh and London: Oliver and Boyd.

1505 —— (1964) Moss heaths. In *The Vegetation of Scotland* (J. H. Burnett, ed.): 524–535. Edinburgh and London: Oliver and Boyd.

1506 —— (1964) Regional pattern of the vegetation. In *The Vegetation of Scotland* (J. H. Burnett, ed.): 568–578. Edinburgh and London: Oliver and Boyd.

1507 —— **and Ratcliffe, D. A.** (1962) *Plant Communities of the Scottish Highlands.* [Monographs of the Nature Conservancy No. 1.] London: H.M.S.O.
—— —See also No. 1851.

1508 **Magnusson, A. H.** (1925) Studies in the *rivulosa*-group of the genus *Lecidea*. *Göteborgs K. Vetensk.-o. VitterhSamh. Handl., ser.* 4, **29**(4): 1–50.

1509 —— (1926) New or misunderstood European lichens. *Acta Horti gothoburg.* **2**: 71–82.

1510 —— (1929) A monograph of the genus *Acarospora. K. svenska VetenskAkad. Handl., ser.* 3, **7**(4): 1–400.

1511 —— (1930) New or otherwise interesting *Lecanora* species. *Acta Horti gothoburg.* **6**: 1–20.

1512 —— (1933) A monograph of the lichen genus *Ionaspis. Acta Horti gothoburg.* **8**: 1–47.

1513 —— (1935) Acarosporaceae und Thelocarpaceae. *Rabenh. Krypt.-Fl.* **9**, 5(1): 1–318.

1514 —— (1939) Studies in species of *Lecanora*, mainly the *Aspicilia gibbosa* group. *K. svenska VetenskAkad. Handl., ser.* 3, **17**(5): 1–182.

1515 —— (1940) Studies in species of *Pseudocyphellaria*—the *crocata* group. *Acta Horti gothoburg.* **14**: 1–36.

1516 —— (1944) Studies in the *ferruginea*-group of the genus *Caloplaca. Göteborgs K. Vetensk.-o. VitterhSamh. Handl., ser.* 6, B, **3**(1): 1–71.

1517 —— (1945) Contributions to the taxonomy of the *Lecidea goniophila* group. *Acta Horti gothoburg.* **16**: 125–134.

1518 —— (1947) Studies in non-saxicolous species of *Rinodina* mainly from Europe and Siberia. *Acta Horti gothoburg.* **17**: 191–338.

1519 **Maher, E.** (1968) *A preliminary study of the corticolous epiphytic communities of Irish woodlands.* Honours thesis, University College, Dublin. [Not seen.]

1520 **Maingay, A. C.** (1857) On the occurrence of *Pertusaria hutchinsiae* and other rare lichens, on the Breadalbane Mountains. *Ann. Mag. nat. Hist., ser.* 2, **20**: 303.

1521 —— (1858) On the occurrence of *Pertusaria hutchinsiae* and other rare lichens, on the Breadalbane Mountains. *Trans. bot. Soc. Edinb.* **5**: 212.

1522 **Mairet, E. M.** (1916) *A Book on Vegetable Dyes.* Hammersmith: Douglas Pepler. [And later editions to Ed. 5 (1931).]

1523 **Makarevich, M. F.** (1963) *Analiz līkhenoflori Ukraïns'kikh Karpat.* Kiev: Ukranian Academy of Sciences.

1524 **Malloch, A. J. C.** (1971) Vegetation of the maritime cliff-tops of the Lizard and Land's End peninsulas, West Cornwall. *New Phytol.* **70**: 1155–1197.

1525 —— (1972) A note on the ecology of *Gongylanthus ericetorum* in the Lizard peninsula, West Cornwall. *J. Bryol.* **7**: 81–85.

1526 **Manning, S. A.** (1938) Fauna and flora of Norfolk. Lichens. *Trans. Norfolk Norwich Nat. Soc.* **14**: 303–308.

1527 —— (1944) The natural history of Wheatfen Broad, Surlingham. Part VI. Lichens. *Trans. Norfolk Norwich Nat. Soc.* **15**: 420–422.

1528 —— (1960) New, rare and critical Norfolk lichens—I. *Trans. Norfolk Norwich Nat. Soc.* **19**: 65–68.

1529 —— (1960) New, rare and critical Norfolk lichens—II. *Trans. Norfolk Norwich Nat. Soc.* **19**: 168–172.

1530 —— (1968) Lichens from Lewis and Harris, Outer Hebrides (V.C. 110). *Trans. Proc. bot. Soc. Edinb.* **40**: 435–444.

1531 —— (1968) Records of Suffolk *Cladoniae* (lichens). *Trans. Suffolk Nat. Soc.* **14**: 11–15.

1532 —— (1974) *The Naturalist in South-East England. Kent, Surrey and Sussex.* Newton Abbot: David and Charles.

1533 —— (1976) *Nature in East Anglia.* Tadworth: World's Work.

Manson, R. T.—See No. 1197.

Manton, M.—See No. 2469.

1534 **Margerison, S.** (1907–9) The vegetation of some disused quarries. *Bradford Sci. J.* **1**: 359–364: **2**: 75–84, 141–147, 161–168, 206–214, 235–245.

1535 **Marquand, E. D.** (1891) The cryptogamic flora of Kelvedon and its neighbourhood, together with a few coast species; compiled from the herbarium and notes made by the late E. G. Varenne, M.R.C.S. *Essex Nat.* **5**: 1–30.

1536 —— (1893) The mosses, Hepaticae and lichens of Guernsey. *Rep. Trans. Guernsey Soc. nat. Sci.* **2**: 188–204.

1537 —— (1901) *Flora of Guernsey and the lesser Channel Islands.* London: Dulau. [Lichens pp. 278–302.] [See also No. 1474.]

1538 —— (1904) Report of the botanical section. Sark. *Rep. Trans. Guernsey Soc. nat. Sci.* **4**: 194–195.

1539 —— (1905) Report of section for marine zoology. *Rep. Trans. Guernsey Soc. nat. Sci.* **4**: 314–315.

1540 —— (1909) Botanical notes. *Rep. Trans. Guernsey Soc. nat. Sci.* **5**: 441–451.

1541 —— (1910) Report of the botanical section. *Rep. Trans. Guernsey Soc. nat. Sci.* **6**: 15–17.

1542 **Marrat, F. P.** (1860) On the hepatics and lichens of Liverpool and its vicinity. *Proc. lit. phil. Soc. Lpool* **14**, *Suppl.*: 1–14.

1543 —— (1863–64) A lichen new to the district. *Lecanora ventosa* (Ach.). In *Naturalists' Scrap Book, Liverpool District* **2**: 191.

1544 **Marriott, St J.** (1925) *British Woodlands as Illustrated by Lessness Abbey Woods.* London and Woolwich: Routledge & Pioneer Press. [Lichens pp. 47–48.]

Marsh, P.—See No. 1586.

1545 **Marshall, J. K.** (1967) Biological Flora of the British Isles: *Corynephorus canescens* (L.) Beauv. *J. Ecol.* **55**: 207–220.

Marshall, M. E.—See No. 2153.

Martin, A. F.—See No. 2463.

1546 **Martin, M.** (1716) *A Description of the Western Islands of Scotland.* Ed. 2 "very much corrected". London: A. Bell etc.

1547 **Martin, W. K. and Brokenshire, F. A.** (1948) Fortieth report on the botany of Devon. *Rep. Trans. Devon. Ass. Advmt Sci.* **80**: 39–49

1548 —— —— (1949) Forty-first annual report on the botany of Devon. *Rep. Trans. Devon. Ass. Advmt Sci.* **81**: 57–65.

1549 **Martindale, J. A.** (1872) Lichenographical notes. *J. Bot., Lond.* **10**: 17–21.

1550 [——] (1886) Lichens in Westmoreland. *Gdnrs' Chron.* **25**: 272.

1551 —— (1886) Additions to the lichen flora of Westmoreland. *Naturalist, Hull* **1886**: 49.

1552 —— (1886) Descriptions of new British lichens. *Naturalist, Hull* **1886**: 101.

1553 —— (1886) New British lichens. *Naturalist, Hull* **1886**: 279.

1554 —— (1886–90) The lichens of Westmoreland. *Naturalist, Hull* **1886**: 317–324; **1887**: 47–54; **1888**: 25–32; **1890**: 157–164.

1555 —— (1886) New Westmoreland lichens. *Naturalist, Hull* **1886**: 374.

1556 —— (1887) Sedbergh district lichens. *Naturalist, Hull* **1887**: 285–287.

1557 —— (1887) Notes on British lichens. *Naturalist, Hull* **1887**: 295–299.

1558 —— (1887) Notes on British lichens: *Lecanora murorum* and its more immediate allies. *Naturalist, Hull* **1887**: 355–364.

1559 —— (1888) Our district. *Westmorland nat. Hist. Record* **1**: 1–19.

1560 —— (1889) *The Study of Lichens with special reference to the Lake District.* Ambleside: G. Middleton.

1561 —— (1889) A new *Lecidea. Westmorland nat. Hist. Record* **1**: 108.

1562 —— (1889) *Gyrophora spadochroa. Westmorland nat. Hist. Record* **1**: 161–162.

1563 —— (1890) The study of lichens, with especial reference to those of the Lake District. *Trans. Cumb. Westm. Assoc. Adv. Lit. Sci.* **15**: 1–24.

1564 —— (1895) *Umbilicaria pustulata* near Keswick. *Naturalist, Hull* **1895**: 310.

1565 —— [*c.* 1906] Botany. In *The Victoria History of the Counties of England, Westmorland* (W. Page, ed.). Unpublished manuscript in BM. [Lichens pp. 97–107.]

1566 **Mason, F. A.** (1932) Yorkshire Naturalists' Union excursions in 1932: Sedbergh. *Naturalist, Hull* **1932**: 339–344.

1567 —— **and Pearsall, W. H.** (1928) Yorkshire naturalists at Spurn. *Naturalist, Hull* **1928**: 315–318.

1568 —— —— (1929) Yorkshire naturalists at Pickering, Yorks. *Naturalist, Hull* **1929**: 277–282, 313–316.

1569 —— —— (1930) The Colden Valley, Hebden Bridge. *Naturalist, Hull* **1930**: 269–271.

1570 —— —— (1930) Natural history of Horton-in-Ribblesdale. *Naturalist, Hull* **1930**: 333–338.

1571 —— —— (1931) Yorkshire naturalists at Leyburn. *Naturalist, Hull* **1931**: 276–281.

1572 —— **and Stainforth, T.** (1927) Allerthorpe Common. *Naturalist, Hull* **1927**: 303–309.

—— —See also Nos. 1785, 1786, 1787, 1788, 1789, 1790 and 1791.

1573 **Mason, P. A.** (1975) 86th autumn meeting at Glentress Forest, Peebles. *News bot. Soc. Edinb.* **15**: 26–27.

1574 [——] (1976) Cryptogamic meeting report, 1974. Glentress Forest, Peebles, 5 October. *Trans. Proc. bot. Soc. Edinb.* **42**: 523–527. [Lichens pp. 525–527 by B. J. Coppins.]

1575 **Massee, G.** (1892) Lichens. *Grevillea* **21**: 28–29.

1576 —— (1892) New or rare lichens. *Grevillea* **21**: 60–61.

1577 —— (1892) A new marine lichen. *J. Bot., Lond.* **30**: 193–194.

1578 —— (1912) *British Fungi, with a chapter on Lichens.* London: G. Routledge & Sons. [Lichens pp. 516–525.]

1579 —— **and Crossland, C.** (1905) The fungus flora of Yorkshire. *Trans. Yorks. Nat. Un., ser. Bot.*, **4**: 1–396.

—— —See also No. 56.

Mathews, W.—See No. 132.

Mathey, A.—See Nos. 1093 and 1392.

1580 **Maton, G.** (1843) *The Natural History of a Part of the County of Wilts., comprehended within the distance of ten miles around the City of Salisbury.* London.

1581 **Maughan, R.** (1811) A list of the rarer plants observed in the neighbourhood of Edinburgh. *Mem. Wernerian nat. Hist. Soc.* **1**: 215–248. [Lichens pp. 246–247.]

1582 **Mayfield, A.** [19—] [*Correspondence*]. Letters in NWH. [Not abstracted.]

1583 —— (1917) The lichens of a boulder-clay area. *J. Ipswich Distr. Fld Club* **5**: 34–40.

1584 —— (1930) The hepatics, mosses and lichens of Suffolk. *J. Ipswich Distr. nat. Hist. Soc.* **1**: 89–140.

1585 —— (1932) Notes on the plant life of Mendlesham with special reference to the hepatics. *J. Ipswich Distr. nat. Hist. Soc.* **1**: 192–196.

1586 **Mayhead, G. J., Broad, K., and Marsh, P.** (1974) *Tree Growth on the South Wales Coalfield.* [Forestry Commission Research and Development Paper No. 108.] London: Forestry Commission.

1587 **Mellanby, K.** (1967) *Pesticides and Pollution.* [New Naturalist No. 50.] London: Collins.

1588 —— (1972) *The Biology of Pollution.* [Studies in Biology No. 38.] London: E. Arnold.

1589 **Menzies, J.** (1926) Note on lichens at Ardchullarie. *Trans. Proc. Perthsh. Soc. nat. Sci.* **8**: 156.

1590 —— (1927) Lichens: with notes on local species. *Trans. Proc. Perthsh. Soc. nat. Sci.* **8**: 159–173.

122

1591 —— (1930) Nature notes from Buckie Braes. *Trans. Proc. Perthsh. Soc. nat. Sci.* **8**: 266–268.

1592 **Merret[t], C.** (1666) *Pinax rerum naturalium britannicarum*. London. [Lichens pp. 71–72; re-issued in 1667 and 1704.]

1593 **Merrifield, [M.P.],** ed. (1860) *A Sketch of the Natural History of Brighton and its Vicinity*. Brighton: W. Pearce.

Merrifield, R. C. J.—See No. 845.

1594 **Metcalfe, G.** (1950) The ecology of the Cairngorms. Part II. The mountain *Callunetum*. *J. Ecol.* **38**: 46–74.

1595 **Miall, L. C.** (1863) *The Physical Geography and Natural Productions of Bradford and the Neighbourhood*. Bradford.

1596 —— (1865) On the botany of Malham. Part IV.—Lichenes. *Naturalist, Hull, o.s.* **2**: 376–379.

1597 —— **and Carrington, B.** (1862) *The Flora of the West Riding*. London: W. Pamplin. [Lichens pp. 62–71 by B. Carrington.]

—— —See also No. 2620.

1598 **Micheli, P. A.** (1729) *Nova plantarum genera ivxta Tovrnefortii methodvm disposita*. Florence. [Lichens pp. 73–106.] [Micheli's personal copy and some of the original drawings in BL; reprinted (1976) Richmond: Richmond Publishing.]

Miles, D.—See No. 207.

1599 **Miles, J.** (1973) Natural recolonization of experimentally bared soil in *Callunetum* in north-east Scotland. *J. Ecol.* **61**: 399–412.

1600 **Miles, M. L.** (1903) A ramble on the moor at Blair-Atholl. *Trans. Proc. Perthsh. Soc. nat. Sci.* **3**: 217–223.

1601 **Mill, R. R.** (1967) *Flora of Helensburgh and District*. Helensburgh: Macneur & Bryden. [Lichens pp. 31–32.]

Millar, A. H.—See No. 281.

1602 **Millbank, J. W.** (1972) Nitrogen metabolism in lichens. IV. The nitrogenase activity of the *Nostoc* phycobiont in *Peltigera canina*. *New Phytol.* **71**: 1–10.

1603 —— (1974) Nitrogen metabolism in lichens V. The forms of nitrogen released by the blue-green phycobiont in *Peltigera* spp. *New Phytol.* **73**: 1171–1181.

1604 —— **and Kershaw, K. A.** (1969) Nitrogen metabolism in lichens. I. Nitrogen fixation in the cephalodia of *Peltigera aphthosa*. *New Phytol.* **68**: 721–729.

1605 —— —— (1970) Nitrogen metabolism in lichens III. Nitrogen fixation by internal cephalodia in *Lobaria pulmonaria*. *New Phytol.* **69**: 595–597.

—— —See also Nos. 1024 and 1025.

Miller, C. R.—See No. 813.

1606 **Millet, L. and Millet, M.** (1853) Wild flowers and ferns of the Isles of Scilly observed in June and July. *Trans. nat. Hist. antiq. Soc. Penzance* **2**: 75–78.

Millet, M.—See No. 1606.

1607 **Millett, J. N. R., Rodd, E. H. and Couch, R. C. E.** (1852) Report [of the Council]. *Trans. nat. Hist. antiq. Soc. Penzance* **2**: 8–12.

—— —See also No. 1977.

Milne, L. M.—See No. 813.

1608 **Milsom, F. E.** (1927) Yorkshire bryologists at Meltham. *Naturalist, Hull* **1927**: 146.

1609 —— (1928) Yorkshire bryologists at Holme Moss. *Naturalist, Hull* **1928**: 210.

Mitchell, B. D.—See No. 2325.

1610 **Mitchell, M. E.** (1961) L'elément eu-océanique dans la flore lichénique du sud-ouest de l'Irelande. *Revta Biol.* **2**: 177–256.

1611 —— **and Henssen, A.** (1966) New or noteworthy lichens from Ireland. *Ir. Nat. J.* **15**: 143–145.

—— —See also No. 1738.

1612 **Mogridge, T. H.** (1836) *A descriptive sketch of Sidmouth.* Sidmouth: J. Harvey. [Lichens p. 83.]

Molesworth, S.—See No. 2214.

Monkhouse, P.—See No. 1833.

1613 **Montgomery, J.** (1834) *List of plants collected at or near Renfrewshire.* Manuscript in K.

1614 **Moore, C. C.** (1974) Lichens as indicators of air pollution in Dublin. *Technology Ireland* **6**(5): 19–20.

1615 —— (1976) Factors affecting the distribution of saxicolous lichens within a four kilometre distance of Dublin city centre. *Proc. R. Ir. Acad., B,* **76**: 263–283.

1616 **Moore, E.** (1848) Botany of Dartmoor. In *A Perambulation of the Ancient and Royal Forest of Dartmoor, and the Venville Precincts* (S. Rowe, ed.): 224–225. Plymouth: J. B. Rowe. [See also No. 2057.]

1617 **Moore, P. D.** (1971) Computer analysis of sand dune vegetation in Norfolk, England, and its implications for conservation. *Vegetatio* **23**: 323–338.

1618 —— **and Beckett, P. J.** (1971) Vegetation and development of Llyn, a Welsh mire. *Nature, Lond.* **231**: 363–365.

1619 **Moorhouse, M.** (1971) *An ecological study of saxicolous lichens in an urban area.* Cert. Ed. thesis, University of Leeds. [Copy with M. R. D. Seaward.]

Morey, F.—See No. 2593.

Morgan, G.—See No. 2351.

1620 [**Morgan, T. O.**] (1849) *Flora Cereticae Superioris. A catalogue of plants indigenous in the neighbourhood of Aberystwith.* Aberystwyth. [Lichen p. 18.]

1621 **Morgan-Huws, D. I.** (1971) *Lichens and air pollution.* Ph.D. thesis, University of Leicester.

1622 —— **and Haynes, F. N.** (1973) Distribution of some epiphytic lichens around an oil refinery at Fawley, Hampshire. In *Air Pollution and Lichens* (B. W. Ferry, M. S. Baddeley and D. L. Hawksworth, eds.): 89–108. London: Athlone Press of the University of London.

—— —See also No. 976.

1623 **Morgan-Jones, G.** (1964) *Taxonomic and biological studies in the Coelomycetes.* Ph.D. thesis, University of Nottingham.

1624 —— (1971) Conidium ontogeny in Coelomycetes. III. Meristem thallo-
conidia. *Can. J. Bot.* **49**: 1939–1940.

1625 —— (1972) Studies on lichen asci. II. Further examples of the bitunicate
type. *Lichenologist* **5**: 275–282.

1626 —— **and Swinscow, T. D. V.** (1965) Pyrenocarpous lichens: 7. On the genus
Microglaena Körb. *Lichenologist* **3**: 42–54.

—— —See also No. 1955.

1627 **Morison, R.** (1699) *Plantarum historiae universalis Oxoniensis. Pars tertia.*
Oxford. [Lichens pp. 622–627.]

—— —See also No. 2422.

1628 **Morley, J. and Grove, W. B.** (1883) Twenty-fourth annual report of the
Birmingham Natural History and Microscopical Society, read at the
annual meeting, held February 6th 1883. *Rep. Trans. Bgham nat. Hist.
& microsc. Soc.* **1882**: xxvi-xxxvii.

1629 —— —— (1884) Twenty-fifth annual report of the Birmingham Natural
History and Microscopical Society, read at the annual meeting, held
February 5th, 1884. *Rep. Trans. Bgham nat. Hist. & microsc. Soc.*
1883: xix-xxxi. [Lichens mainly by W. H. Wilkinson.]

1630 **Morton, J.** (1712) *The Natural History of Northampton-shire.* London:
Knaplock & Wilkin. [Lichens pp. 361–364.]

Morris, M. G.—See No. 2038.

Moses, P.—See No. 347.

1631 **Mosley, C.** (1927) *Xanthoria parietina* Th. Fr. *Naturalist, Hull* **1927**: 245.

Mosley, O.—See No. 336.

1632 **Moss, C. E.** (1899) Local records in natural history: Lichens. *Halifax Nat.* **3**:
128.

1633 —— (1913) *Vegetation of the Peak District.* Cambridge: Cambridge University Press.

1634 **Mott, F. T., Carter, T., Cooper, E. F., Finch, J. E. M. and Cooper, C. W.**
(1886) *The Flora of Leicestershire, including the cryptogams.* London
and Edinburgh: Leicester Literary and Philosophical Society.

Mottram, R. H.—See No. 1661.

1635 **Motyka, J.** (1936, 1938) *Lichenum genus* Usnea *studium monographicum, pars
systematica.* Leopoli: J. Motyka. [Pp. 1–304 published 1936; pp. i-iv,
305–651 published 1938.]

1636 —— (1960) O niektórych mniej znanych i nowych gatunkach rodzaju
Alectoria Ach. *Fragm. flor. geobot.* **6**: 441–452.

1637 —— (1961) O niektórych nowych i mniej znanych europejskich gatunkach
rodzaju *Ramalina* Ach.—De speciebus quibusdam generis *Ramalina*
Ach. europaeis novis et minus cognitis. *Fragm. flor. geobot.* **6**: 637–
644.

1638 —— (1961) Przeglad gatunkow rodzaju *Ramalina* Ach. srodkowej i
zachodniej Europy. Conspectus Ramalinarum europae mediae et
occidentalis. *Fragm. flor. geobot.* **6**: 645–682.

1639 —— (1961) *Ramalina scopulorum* (Retz.) Ach. i pokrewne gatunki w Europie.
Fragm. flor. geobot. **6**: 683–708.

1640 **Moyse, J. and Nelson-Smith, A.** (1963) Zonation of animals and plants on rocky shores around Dale, Pembrokeshire. *Fld Stud.* **1** (5): 1–31.

1641 **Mudd, W.** (1854) An account of the lichens of Cleveland, with their localities. *Phytologist* **5**: 71–76, 97–102.

1642 —— (1861) *Herbarium Lichenum britannicorum.* 3 fasc., nos. 1–301. Great Ayton. [Intact set in BM.]

1643 —— (1861) *A Manual of British Lichens.* Darlington: privately printed.

1644 —— (1865) *A Monograph of British* Cladoniae, *Illustrated with Dried Specimens of Eighty Species and Varieties.* Cambridge: privately printed.

—— —See also No. 1675.

Muggleton, J.—See No. 236.

Murdoch, J. B.—See No. 666.

1645 **Murray, G.** (1877) On the nature of spermatia. *J. Bot., Lond.* **15**: 299–300.

1646 **Nádvorník, J.** (1942) Systematische Übersicht der mitteleuropäischen Arten der Flechtenfamilie Caliciaceae. *Studia bot. čsl.* **5**: 6–40.

1647 **Naylor, G. L.** (1930) Notes on the distribution of *Lichina confinis* and *L. pygmaea* in the Plymouth district. *J. mar. biol. Ass. U.K.* **16**: 909–918.

Needham, J.—See Nos. 517 and 518.

1648 **Neill, P.** (1805) Tour through some of the Orkney Islands, in 1804. *Scots Mag.* **67**: 179–184.

1649 —— (1806) *A tour through some of the Islands of Orkney and Shetland.* Edinburgh. [Lichens p. 190.]

1650 [——] (1830) Lichen. In *The Edinburgh Encyclopedia* (D. Brewster, ed.) **12**: 729–741. Edinburgh: W. Blackwood.

1651 **Nelson-Smith, A.** (1967) Marine biology of Milford Haven: The distribution of littoral plants and animals. *Fld Stud.* **2**: 435–477.

—— —See also No. 1640.

1652 **Nethersole-Thompson, D. and Watson, A.** (1974) *The Cairngorms.* London: Collins.

1653 [**Neville-Smith, C. H.**] (1976) [*Buellia canescens* in Orkney]. *Bull. Orkney Fld Club* **1**: 4.

1654 **New, J. K.** (1961) Biological Flora of the British Isles: *Spergula arvensis* L. *J. Ecol.* **49**: 205–215.

Newbigging, T.—See No. 2254.

1655 **Newbould, P. J.** (1960) The ecology of Cranesmoor, a New Forest valley bog. *J. Ecol.* **48**: 361–383.

Newcombe, C. F.—See No. 598.

1656 **Nichol, W.** (1860) Recent additions to the cryptogamic flora of Edinburgh. *Trans. bot. Soc. Edinb.* **6**: 109–110.

1657 —— (1860) List of cryptogamic plants collected near Moffat. *Trans. bot. Soc. Edinb.* **6**: 137–141.

1658 **Nichols, J.** (1795) *The History and Antiquities of the County of Leicester.* Vol. 1. London: J. Nichols. [Lichens by R. Pulteney pp. clxxxvii–clxxxix, and G. Crabbe pp. cxcix–cc.] [Reprinted (1971) Leicester: S. R. Publishers.]

1659 **Nicholson, B. E. and Brightman, F. H.** (1966) *The Oxford Book of Flowerless Plants*. London: Oxford University Press.

1660 **Nicholson, I. A. and Robertson, R. A.** (1958) Some observations on the ecology of an upland grazing in north-east Scotland with special reference to *Callunetum. J. Ecol.* **46**: 239–270.
—— —See also No. 1236.

1661 **Nicholson, W. A. and Ellis, E. A.** (1935) The botany of Norfolk. In *A Scientific Survey of Norwich and District* (R. H. Mottram, ed.): 24–34. London: British Association.

Nickless, G.—See No. 345.

Nicol, J. C.—See No. 1439.

1662 **Nicolson, J. and Burn, R.** (1777) *The History and Antiquities of the Counties of Westmorland and Cumberland*. Vol. 2. London: W. Strahan and T. Cadell. [Lichens pp. 589, 593.]

Nicolle, E. T.—See No. 104.

1663 **Non-Lichenologist** (1859) Cumbrian lichen. *Phytologist, n.s.* **3**: 223.

1664 **Noon, R. A. and Hawksworth, D. L.** (1973) The lichen flora of Lundy. *Rep. Lundy Fld Soc.* **23**: 52–58.

1665 **Nordin, I.** (1964) *Abrothallus suecicus*, a common lichenicolous fungus. *Svensk bot. Tidskr.* **58**: 225–232.

1666 —— (1972) Caloplaca, *sect.* Gasparrinia *i Nordeuropa*. Uppsala: Skriv Service.

1667 **Norfolk Naturalists Trust** (1976) *Nature in Norfolk: a Heritage in Trust*. Norwich: Jarrold.

1668 **Norkett, A. H.** (1938) Lichens collected on Limpsfield Common, 1937. *Lond. Nat.* **1937**: 58.

1669 **Norman, A. M.** (1884) Address to the members of the Tyneside Naturalists' Field Club. *Nat. Hist. Trans. Northumb.* **8**: 67–134. [Lichen p. 74.]

1670 **North, F. J., Campbell, B. and Scott, R.** (1949) *Snowdonia, the National Park of North Wales*. [New Naturalist no. 13.] London: Collins.

Norton, T. A.—See No. 2474.

1671 **Nourish, R.** (1972) *A chemotaxonomic study of British lichens of the subgenus* Cladina *by physico-chemical techniques*. M.Sc. thesis, University of Salford.

1672 —— **and Oliver, R. W. A.** (1974) Chemotaxonomic studies on British lichens. I. *Cladonia* subgenus *Cladina. Lichenologist* **6**: 73–95.

1673 —— —— (1976) Chemotaxonomic studies on the *Cladonia chlorophaea-pyxidata* complex and some allied species in Britain. In *Lichenology: Progress and Problems* (D. H. Brown, D. L. Hawksworth and R. H. Bailey, eds.): 185–214. [Systematics Association Special Volume No. 8.] London, New York and San Francisco: Academic Press.

Nowell, J.—See No. 2256.

1674 **Nylander, W.** (1858–60) *Synopsis Methodica Lichenum*. Paris.

1675 —— (1863) Observationes quaedam circa Herbarium Lichenum Britannicorum by William Mudd, fascic. I-III, 1861. *Flora, Jena* **46**: 77–79.

1676 —— (1864) Pyrenocarpei quidam europaei novi. *Flora, Jena* **47**: 353–358.

1677 —— (1865) Lecideae quaedam europaeae novae. *Flora, Jena* **48**: 3–7.

1678 —— (1865) Lecideae adhuc quaedam europaeae novae. *Flora, Jena* **48**: 144–148.

1679 —— (1865) Novitatiae quaedam Lichenum europaeorum variarum tribuum. *Flora, Jena* **48**: 209–213.

1680 —— (1865) Circa Thelocarpa europaea notula. *Flora, Jena* **48**: 260–262.

1681 —— (1865) Enumeratio synoptica Sticteorum. *Flora, Jena* **48**: 296–299.

1682 —— (1865) Adhuc novitiae quaedam lichenum europae variarum tribuum. *Flora, Jena* **48**: 353–358.

1683 —— (1865) Addenda nova ad lichenographiam europaeam. *Flora, Jena* **48**: 601–606.

1684 —— (1866) Addenda nova ad lichenographiam europaeam. Continuatio. *Flora, Jena* **49**: 84–87.

1685 —— (1866) Addenda ad lichenographiam europaeam. Continuatio altera. *Flora, Jena* **49**: 369–374.

1686 —— (1866) Addenda ad lichenographiam europaeam. Continuatio tertia. *Flora, Jena* **49**: 417–421.

1687 —— (1867) Addenda nova ad lichenographiam europaeam. Continuatio quinta. *Flora, Jena* **50**: 326–330.

1688 —— (1868) Addenda ad lichenographiam europaeam. Continuatio septima. *Flora, Jena* **51**: 161–165.

1689 —— (1868) Addenda ad lichenographiam europaeam. Continuatio octava. *Flora, Jena* **51**: 342–348.

1690 —— (1868) Addenda nova ad lichenographiam europaeam. Continuatio nona. *Flora, Jena* **51**: 473–478.

1691 —— (1869) Addenda nova ad lichenographiam europaeam. Continuatio decima. *Flora, Jena* **52**: 81–85.

1692 —— (1869) Addenda nova ad lichenographiam europaeam. Continuatio duodecima. *Flora, Jena* **52**: 409–413.

1693 —— (1870) Addenda nova ad lichenographiam europaeam. Continuatio tertiadecima. *Flora, Jena* **53**: 33–38.

1694 —— (1872) Addenda nova ad lichenographiam europaeam. Continuatio quarta decima. *Flora, Jena* **55**: 353–365.

1695 —— (1873) Addenda nova ad lichenographiam europaeam. Continuatio quinta decima. *Flora, Jena* **56**: 17–23.

1696 —— (1873) Addenda nova ad lichenographiam europaeam. Continuatio sexta decima. *Flora, Jena* **56**: 289–300.

1697 —— (1874) Addenda nova ad lichenographiam europaeam. Continuatio septima decima. *Flora, Jena* **57**: 6–16.

1698 —— (1874) Addenda nova ad lichenographiam europaeam. Continuatio octava decima. *Flora, Jena* **57**: 305–318.

1699 —— (1875) Addenda nova ad lichenographiam europaeam. Continuatio [nona] decima. *Flora, Jena* **58**: 6–15.

1700 —— (1875) Addenda nova ad lichenographiam europaeam. Continuatio vicesima. *Flora, Jena* **58**: 102–106.

1701 —— (1875) Addenda nova ad lichenographiam europaeam. Continuatio vicesima prima. *Flora, Jena* **58**: 297–303.

128

1702 —— (1875) Addenda nova ad lichenographiam europaeam. Continuatio tertia et vicesima. *Flora, Jena* **58**: 440–448.

1703 —— (1876) Addenda nova ad lichenographiam europaeam. Continuatio quarta et vicesima. *Flora, Jena* **59**: 231–239.

1704 —— (1876) Addenda nova ad lichenographiam europaeam. Continuatio quinta et vicesima. *Flora, Jena* **59**: 305–311.

1705 —— (1876) Addenda nova ad lichenographiam europaeam. Continuatio sexta et vicesima. *Flora, Jena* **59**: 571–578.

1706 —— (1877) Addenda nova ad lichenographiam europaeam. Continuatio septima et vicesima. *Flora, Jena* **60**: 220–233.

1707 —— (1877) Addenda nova ad lichenographiam europaeam. Continuatio octava et vicesima. *Flora, Jena* **60**: 457–463.

1708 —— (1879) Addenda nova ad lichenographiam europaeam. Continuatio secunda et tricesima. *Flora, Jena* **62**: 353–364.

1709 —— (1880) Addenda nova ad lichenographiam europaeam. Continuatio tertia et tricesima. *Flora, Jena* **63**: 10–15.

1710 —— (1880) Addenda nova ad lichenographiam europaeam. Continuatio quarta et tricesima. *Flora, Jena* **63**: 387–394.

1711 —— (1881) Addenda nova ad lichenographiam europaeam. Continuatio quinta et tricesima. *Flora, Jena* **64**: 2–8.

1712 —— (1881) Addenda nova ad lichenographiam europaeam. Continuatio sexta et tricesima. *Flora, Jena* **64**: 177–189.

1713 —— (1881) Addenda nova ad lichenographiam europaeam. Continuatio septima et tricesima. *Flora, Jena* **64**: 449–459.

1714 —— (1881) Addenda nova ad lichenographiam europaeam. Continuatio octava et tricesima. *Flora, Jena* **64**: 529–541.

1715 —— (1882) Addenda nova ad lichenographiam europaeam. Continuatio nona et tricesima. *Flora, Jena* **65**: 451–458.

1716 —— (1883) Addenda nova ad lichenographiam europaeam. Continuatio quadragesima. *Flora, Jena* **66**: 97–109.

1717 —— (1883) Addenda nova ad lichenographiam europaeam. Continuatio quadragesima prima. *Flora, Jena* **66**: 531–538.

1718 —— (1886) Addenda nova ad lichenographiam europaeam. Continuatio quadragesima quinta. *Flora, Jena* **69**: 97–102.

1719 —— (1886) Addenda nova ad lichenographiam europaeam. Continuatio quadragesima sexta. *Flora, Jena* **69**: 461–466.

1720 **Ogilvie, W. M.** (1849) Occurrence of *Stereocaulon tomentosum* in fruit. *Phytologist* **3**: 555.

1721 **O'Hare, G. [P.]** (1973) Lichen techniques of pollution assessment. *Area* **5**: 223–229.

1722 —— (1973) Lichens and sulphur dioxide pollution in West Central Scotland. *Northern Univ. Geogrl J.* **10**: 12–19.

1723 —— (1974) *Air pollution and lichens in the Western Central Lowlands of Scotland.* Ph.D. thesis, University of Glasgow.

1724 —— (1975) Lichens and bark acidification as indicators of air pollution in West Central Scotland. *J. Biogeogr.* **1**: 135–146.

1725 —— **and Williams P.** (1975) Some effects of sulphur dioxide flow on lichens. *Lichenologist* **7**: 116–120.

1726 **Oliver, F. W.** (1913) Some remarks on Blakeney Point, Norfolk. *J. Ecol.* **1**: 4–15.

1727 —— (1929) Report of the Blakeney Point Research Station, 1927–1929. *Trans. Norfolk Norwich Nat. Soc.* **12**: 630–653. [Lichens pp. 645–647 by W. Watson and P. W. Richards.]

Oliver, R. W. A.—See Nos. 1672 and 1673.

1728 **Olivier, H. l'Abbé** (1907) Lichens d'Europe. [Fasc. 1.] *Mém. Soc. natn. Sci. nat. math. Cherbourg* **36**: 77–274.

1729 —— (1909) Lichens d'Europe. [Fasc. 2.] *Mém. Soc. natn. Sci. nat. math. Cherbourg* **37**: 29–200.

1730 —— (1911) Étude synoptique et géographique des Lécidés de la flore d'Europe. *Bull. Géogr. bot.* **21**: 157–209.

1731 —— (1905–7) Les principaux parasites de nos lichens Français. *Bull. Géogr. bot.* **15**: 206–220, 273–284; **16**: 42–48, 187–200, 253–264; **17**: 123–128, 162–176, 232–240.

1732 —— (1912) Les *Pertusaria* de la flore d'Europe. *Bull. Géogr. bot.* **22**: 193–217.

1733 —— (1915) Les *Lecidea* de la flore d'Europe. Étude synoptique et géographique. *Bull. Géogr. bot.* **25**: 93–183.

1734 —— (1917) Les *Arthonia* de la flore d'Europe. *Bull. Géogr. bot.* **27**: 181–218.

1735 —— (1918–19) Les lichens pyrénocarpés de la flore d'Europe. *Bull. Géogr. bot.* **28**: 146–152, 168–183; **29**: 6–16, 35–48, 97–107.

1736 —— (1921) Prodromus lichenum europeorum. *Mems R. Acad. Cienc. Artes Barcelona, ser.* 3, **16**(14): 441–518.

Orton, J. H.—See No. 2238.

1737 **Östhagen, H.** (1972) The chemical strains in *Cladonia luteoalba* Wils. et Wheld. and their distribution. *Norw. J. Bot.* **19**: 37–41.

1738 **O'Sullivan, A. M. and Mitchell, M. E.** (1974) An unusual lichen habitat. *Ir. Nat. J.* **18**: 23–24.

1739 **Osvald, H.** (1949) Notes on the vegetation of British and Irish mosses. *Acta phytogeogr. suec.* **26**: 1–62.

1740 **Ozenda, P. and Clauzade, G.** (1970) *Les Lichens. Étude biologique et flore illustrée.* Paris: Masson & Cie.

Page, W.—See Nos. 135, 136, 176, 368, 556, 610, 611, 707, 1046, 1047, 1048, 1049, 1050, 1051, 1565, 1826, 1852 and 2168.

1741 **Paget, C. J. and Paget, J.** (1834) *Sketch of the Natural History of Yarmouth and its Neighbourhood containing Catalogues of the Species of Animals, Birds, Reptiles, Fish, Insects, and Plants, at present known.* Yarmouth: F. Skill & Quay. [Lichens pp. 79–83.]

Paget, J.—See No. 1741.

1742 **Palmer, M. G.**, ed. (1946) *The Fauna and Flora of the Ilfracombe District of North Devon.* Ilfracombe: Ilfracombe Field Club. [Lichens pp. 238–239.]

1743 **Parfitt, E.** (1883) The lichen flora of Devonshire. *Rep. Trans. Devon. Ass. Advmt Sci.* **15**: 290–345.

—— ——See also No. 674.

1744 **Park, J. J.** (1818) *The Topography and Natural History of Hampstead.* London: Nichols, Son & Bentley.

1744a **Park, R. B.** (1958) Reindeer in the Cairngorms. *Fld Nat.* **3**: 11–12.

1745 **Parkinson, J.** (1640) *Theatrum botanicum.* London. [Lichens pp. 1308–1315.]

1746 **Parsons, H. F.** (1875) The flora of East Somerset [abstract]. *Naturalist, Hull* **1**: 53–55.

1747 —— (1875) The flora of the eastern border of Somersetshire. *Proc. Somerset archaeol. nat. Hist. Soc.* **21**(2): 53–61.

1748 [——] (1877) Goole Scientific Society. *Naturalist, Hull* **2**: 155–156.

1749 —— (1877) Goole Scientific Society. *Naturalist, Hull* **2**: 171–172.

1750 —— (1877) Goole Scientific Society. *Naturalist, Hull* **2**: 187–188.

1751 —— (1877) Report of the botanical section: 1877. *Trans. Yorks. Nat. Un., Bot.* **1**: 1–8.

1752 —— (1878) Goole Scientific Society. *Naturalist, Hull* **4**: 13–14.

1753 —— (1878) Alpine plants on lowland heaths. *Naturalist, Hull* **3**: 114–116.

1754 —— (1878) Report of the botanical section: 1878. *Trans. Yorks. Nat. Un., Bot.,* **1**: 9–28, 43–50.

1755 —— (1912) The flora of the commons near Croydon. *Proc. Trans. Croydon nat. Hist. scient. Soc.* **7**(3): 57–58.

1756 [**Partaker, A**] (1884) A lichen supper. *Trans. Penzance nat. Hist. antiq. Soc.* **1**: 417–419.

Patil, G. P.—See Nos. 1233 and 1234.

Paton, A. W.—See No. 281.

1757 **Patton, D.** (1915) The flora of the Culbin Sands. *Trans. Proc. bot. Soc. Edinb.* **26**: 345–374.

—— ——See also No. 2265.

1758 **Paulson, R.** (1911) Some common lichens. *Sch. Nat. Study* **21**: 71–74. [Also issued separately as *Publ. Sch. Nat. Study Un.* no. 26 with changed pagination.]

1759 —— (1914) *Lecanora isidioides* Nyl. in the New Forest. *J. Bot., Lond.* **52**: 184–185.

1760 [——] (1914) [*Lecanora isidioides* Nyl. and *Parmelia revoluta* var. *concentrica* Cromb. exhibited.] *Proc. Linn. Soc. Lond.* **1913–14**: 70.

1761 —— (1915) A wandering lichen. *Essex Nat.* **18**: 25.

1762 [——] (1915) A wandering lichen. *J. Bot., Lond.* **53**: 308.

1763 —— (1917) The Varenne collection of lichens: a report on its present condition. *Essex Nat.* **18**: 133–134.

1764 —— (1917) *Chaenotheca melanophaea* (Ach.) Zwackh, var. nov. *flavocitrina. J. Bot., Lond.* **55**: 195–196.

1765 —— (1918) Notes on the ecology of lichens with especial reference to Epping Forest. *Essex Nat.* **18**: 276–286.

1766 —— (1919) Lichens found near Painswick. *Trans. Br. mycol. Soc.* **6**: 303–304.

1767 —— (1919) The lichen flora of Hertfordshire. *Trans. Herts. nat. Hist. Soc. Fld Club* **17**: 83–96.

1768 —— (1921) Ten years' progress in lichenology in the British Isles. *Essex Nat.* **19**: 273–287.

1769 —— (1921) Lichens additional to the Hertfordshire list of cryptogams. *Trans. Herts. nat. Hist. Soc. Fld Club* **17**: 290.

1770 —— (1927) Lichens of the Hereford foray. *Trans. Br. mycol. Soc.* **12**: 87–90.

1771 —— (1929) Lichens of the Oxford foray. *Trans. Br. mycol. Soc.* **14**: 183–185.

1772 —— (1930) Lichens of the Bristol foray. *Trans. Br. mycol. Soc.* **15**: 11–12.

1772a —— **and Hastings, S.** (1914) A wandering lichen. *Knowledge* **37**: 319–323.

1773 —— —— (1920) The relation between the alga and fungus of a lichen. *J. Linn. Soc., Bot.* **44**: 497–506.

1774 —— **and Thompson, P. G.** (1911) Report on the lichens of Epping Forest (first paper). *Essex Nat.* **16**: 136–145.

1775 —— —— (1913) Report on the lichens of Epping Forest (second paper). *Essex Nat.* **17**: 90–105.

1776 —— —— (1920) Supplemental report on the lichens of Epping Forest. *Essex Nat.* **19**: 27–30.

1777 **Paxton, F. V.** (1881) Ordinary meeting, June 8th, 1880. *Rep. Chichester nat. Hist. microsc. Soc.* [Meetings individually paged.]

1778 **Payne, E.** (1921) Liverpool Botanical Society. *Lancs. Chesh. Nat.* **13**: 215–218.

1779 **Peake, J. F. and James, P. W.** (1967) Lichens and Mollusca. *Lichenologist* **3**: 425–428.

1780 **Pearsall, W. H.** (1936) The botany of the Lake District. In *A Scientific Survey of Blackpool and District* (A. Grime, ed.): 134–138. London: British Association.

1781 —— (1938) The soil complex in relation to plant communities III. Moorlands and bogs. *J. Ecol.* **26**: 298–315.

1782 —— (1950) *Mountains and Moorlands* [New Naturalist No. 11.]. London: Collins.

1783 —— (1956) Two blanket-bogs in Sutherland. *J. Ecol.* **44**: 493–516.

1784 —— [**revised W. Pennington**] (1968) *Mountains and Moorlands* [New Naturalist No. 11.] Ed. 2. London: Collins.

1785 —— **and Mason, F. A.** (1921) Yorkshire naturalists at Dent. *Naturalist, Hull* **1921**: 273–281.

1786 —— —— (1922) Yorkshire naturalists at Thornton Dale. *Naturalist, Hull* **1922**: 289–296.

1787 —— —— (1923) Yorkshire naturalists at Helmsley. *Naturalist, Hull* **1923**: 246–255.

1788 —— —— (1923) Yorkshire naturalists at Penistone. *Naturalist, Hull* **1923**: 340–344.

1789 —— —— (1924) Yorkshire naturalists at Ravenscar. *Naturalist, Hull* **1924**: 272–278.

1790 —— —— (1925) Middleton-in-Teesdale and its natural history. *Naturalist, Hull* **1925**: 249–252.

1791 —— —— (1928) Yorkshire naturalists at Cawthorne. *Naturalist, Hull* **1928**: 379–383.

132

1792 —— **and Pennington, W.** (1973) *The Lake District: a landscape history.*
[New Naturalist no. 53.] London: Collins.
—— —See also Nos. 1567, 1568, 1569, 1570 and 1571.

1793 **Pearson, M. C. and Rogers, J. A.** (1962) Biological Flora of the British Isles:
Hippophaë rhamnoides L. *J. Ecol.* **50**: 501–513.
—— —See also No. 553.

1794 **Pearson, R. G. and White, E.** (1964) The phenology of some surface-active
arthropods of moorland and country in North Wales. *J. anim. Ecol.*
33: 245–258.

Peat, A.—See No. 264.

1795 **Peck, A. E.** (1917) Yorkshire mycologists at Buckden. *Naturalist, Hull* **1917**:
99–102.

Peck, H. J.—See No. 645.

1796 **Peck, W.** (1815) *A Topographical Account of the Isle of Axholme, being the
West Division of the Wapentake of Manley in the County of Lincoln.*
Vol. 1. Doncaster.

1797 **Pennant, T.** (1776) *A Tour in Scotland and Voyage to the Hebrides;
MDCCLXXII.* Part I. Chester. [Also Ed. 2(1776).]

1798 —— (1776) *A Tour in Scotland. MDCCLXXII.* Part II. London: B. White.
[also Ed. 2(1776).]

1799 —— (1796) *The History of the Parishes of Whiteford and Hollywell.* London:
B. and J. White. [Lichens pp. 152–155.]

1800 —— (1801) *A Tour from Downing to Alston-Moor.* London: E. Harding.

Pennington, W.—See Nos. 1784 and 1792.

1801 **Pentecost, A.** (1969) The lichen flora of the Lizard. *Lizard* **4**: 7–13.

1802 —— **and Fletcher, A.** (1974) Tufa: an interesting lichen substrate. *Lichenologist* **6**: 100–101.

1803 **Pentreath, J. B.** (1851) List of the lichens found in West Penwith. *Trans. nat.
Hist. antiq. Soc. Penzance* **2**: 24–29.

1804 **Percival, M. S.** (1960) Botany. In *The Cardiff Region* (J. F. Rees, ed.): 45–57.
Cardiff: University of Wales Press.

1805 **Perez-Llano, G. A.** (1944) Lichens. Their biological and economic significance.
Bot. Rev. **10**: 1–65.

1806 **Perring, F. H. and Randall, R. E.** (1972) An annotated flora of the Monach
Isles National Nature Reserve, Outer Hebrides. *Trans. Proc. bot. Soc.
Edinb.* **41**: 431–444.
—— —See also Nos. 266 and 2038.

1807 **Petch, T.** (1936) Notes on British Hypocreaceae. II. *J. Bot., Lond.* **74**: 185–
193.

1808 —— (1938) British Hypocreales. *Trans. Br. mycol. Soc.* **21**: 243–301.

1809 **Peterken, G. F.** (1969) Development of vegetation in Staverton ᵀark, Suffolk.
Fld Stud. **3**: 1–39. [Lichens pp. 37–39 by F. Rose.]

1810 —— (1969) *Wyre Forest (south-east) report.* Manuscript. Report to the
Nature Conservancy; copy with BLS:

1811 —— **and Rose, F.** (1970) *Felbrigg Wood, Norfolk.* Manuscript. Report to the
Nature Conservancy; copy with BLS.

———See also Nos. 795 and 2622.

1812 **Petiver, J.** (1695) *Museii Petiveriani. Centuria prima.* London. [Lichens pp. 11–13.]

1813 ——— (1699) *Museii Petiveriani. Centuria quarta & quinta.* London. [Lichens p. 40.]

1814 ——— (1702) *Gazophylacii Naturae et Artis, Decas prima.* London. [Lichen p. 16.]

1815 ——— (1764) *Opera historiam naturale spectantia or Gazophylacium.* Vol. 1. London: Millan.

1816 **Peveling, E. and Hill, D. J.** (1974) The localization of an insoluble intermediate in glucose production in the lichen *Peltigera polydactyla. New Phytol.* **73**: 767–769.

1817 **Phillips, H.,** ed. (1973) *Edlington Wood.* Doncaster: Doncaster Rural District Council. [Lichens p. 127 contributed by W. Bunting & D. W. Shimwell.]

1818 **Phillips, W.** (1873) Lichens in North Wales. *Grevillea* **2**: 27.

1819 ——— (1874) *Thelocarpon intermediellum.* Nyl. in Britain. *Grevillea* **2**: 125–126.

1820 ——— (1875) A new British *Sphinctrina. Gdnrs' Chron.* **4**: 165.

1821 ——— (1878) Lichens, their rate of growth. *Gdnrs' Chron.* **9**: 624.

1822 ——— (1878) Lichens: their rate of growth. *Gdnrs' Chron.* **10**: 75–76.

1823 ——— (1880) British lichens: hints how to study them. *Midl. Nat.* **3**: 125–128, 167–172, 196–199, 241–245.

1824 ——— (1881) British lichens: hints how to study them. *Rep. Trans. Bgham. nat. Hist. & microsc. Soc.* **1880**: 19–35.

1825 ——— (1887) *A Manual of the British Discomycetes.* London: Paul & Trench [Also Ed. 2 (1893).]

1826 ——— (1908) Lichenes (Lichens). In *The Victoria History of the Counties of England, Shropshire* (W. Page, ed.) **1**: 80–84. London: Constable.

1827 ——— **and Plowright, C. B.** (1876) New and rare British fungi. *Grevillea* **4**: 118–124.

1828 ——— ——— (1877) New and rare British fungi. *Grevillea* **6**: 22–29.

Pickup, J. D.—See No. 2567.

Pielou, E. C.—See Nos. 1233 and 1234.

1829 **Pigott, C. D.** (1955) Biological Flora of the British Isles: *Thymus* L. *J. Ecol.* **43**: 365–387.

1830 ——— (1956) The vegetation of Upper Teesdale in the North Pennines. *J. Ecol.* **44**: 545–586.

1831 ——— (1956) Vegetation. In *Sheffield and its Region* (D. L. Linton, ed.): 79–90. Sheffield: British Association.

1832 ——— (1958) Biological Flora of the British Isles: *Polemonium caeruleum* L. *J. Ecol.* **46**: 507–525.

1833 ——— (1960) Natural history. In *Peak District* (P. Monkhouse, ed.): 13–20. [National Park Guide No. 3.] London: H.M.S.O.

1834 ——— (1968) Biological Flora of the British Isles: *Cirsium acaulon* (L.) Scop. *J. Ecol.* **56**: 597–612.

134

—— —See also No. 1155.

1835 **Pilkington, J.** (1789) *View of the present state of Derbyshire.* 2 vols. Derby. [Also Ed. 2 (1803).] [Botany by E. Darwin.]

1836 **Plot, R.** (1686) *The Natural History of Staffordshire.* Oxford.
Plowright, C. B.—See Nos. 425, 1827 and 1828.

1837 **Plues, M.** (1864) *Rambles in search of Flowerless Plants.* London: Journal of Horticulture and Cottage Gardener Office. [Lichens pp. 187–238.]

1838 —— (1865) *Rambles in search of Flowerless Plants.* Ed. 2. London: Houlston & Wright.

1839 **Plukenet, L.** (1696) *Almagestum botanicum.* London. [Lichens pp. 216, 254–255.]

1840 **Poelt, J.** (1958) Die lobaten Arten der Flechtengattung *Lecanora* Ach. sensu ampl. in der Holarktis. *Mitt. bot. StSamml., Münch.* **2**: 411–589.

1841 —— (1961) Die mitteleuropäischen Arten der *Lecidea goniophila*-Gruppe. *Ber. bayer. bot. Ges.* **34**: 82–91.

1842 —— (1962) Bestimmungsschlüssel der höheren Flechten von Europa. *Mitt. bot. StSamml., Münch.* **4**: 301–571.

1843 —— (1969) *Bestimmungsschlüssel europäischer Flechten.* Lehre: J. Cramer.

1844 —— (1970) Das Konzept der Artenpaare bei den Flechten. *Votr. GesGeb. Bot.* [*Dtsch. bot. Ges.*], *n.f.* **4**: 187–198.

1845 —— **and Döbbeler, P.** (1975) Über moosparasitische Arten der Flechtengattung *Micarea* und *Vezdeae*. *Bot. Jb.* **96**: 328–352.

1846 —— **and Krüger, U.** (1970) Die Verbreitungsverhältnisse der Flechtengattung *Squamarina* in Europa. *Feddes Reprium* **81**: 187–201.

1847 —— **and Sulzer, M.** (1974) Die Erdflechte *Buellia epigaea*, eine Sammelart. *Nova Hedwigia* **25**: 173–194.

—— —See also Nos. 856, 1391 and 1393.
Polkinghorne, B. C.—See Nos. 229 and 842.

1848 **Polwhele, R.** (1797) *The History of Devonshire.* Vol. 1. London: Cadell, Johnson and Dilly. [Lichens p. 97.]

1849 —— (1816) *The History of Cornwall.* Vol. 4. London: Law and Whittaker. [Lichens p. 126.]

1850 **Poore, M. E. D.** (1954) *The principles of vegetation classification and the ecology of Woodwalton Fen.* Ph.D. thesis, University of Cambridge.

1851 —— **and McVean, D. N.** (1957) A new approach to Scottish mountain vegetation. *J. Ecol.* **45**: 401–439.

1852 **Potter, M. C.** (1905) Botany. In *The Victoria History of the Counties of England, Durham* (W. Page, ed.) **1**: 35–81. London: Constable. [Lichens pp. 70–71 by W. Johnson.]

1853 **Potter, T. R.** (1842) *The History and Antiquities of Charnwood Forest.* London: Hamilton & Adams. [Lichens pp. 59–62 by A. Bloxam and C. Babington.]

1854 **Poulton, E. M.** (1914) The structure and life-history of *Verrucaria margacea*, Wahl., an aquatic lichen. *Ann. Bot.* **28**: 241–249.

1855 **Powell, H. T. and Chamberlain, Y. M.** (1956) Plant life on Rockall. In *Rockhall* (J. Fisher, ed.): 171–176. London: G. Bles.

1856 **[Pownall, H.]** (1825) *Some particulars relating to the History of Epsom*. Epsom: W. Dorling. [Lichen p. 185.]

1857 **Praeger, R. Ll.** (1934) *The Botanist in Ireland*. Dublin: Hodges & Figgis.

1858 **Pratt, A.** (1856) *Chapters on the Common Things of the Sea-Coast*. London: Society for Promoting Christian Knowledge. [Lichens p. 112.]

Prentice, I. C.—See No. 10.

1859 **Preston, H.** (1901) Lincolnshire naturalists at Little Bytham. *Naturalist, Hull* **1901**: 57–62.

Prime, C. T.—See No. 684.

1860 **Prince, C. R.** (1972) *The role of the lichen synusia in certain dwarf heath communities*. Ph.D. thesis, University of Aberdeen.

1861 —— (1974) Growth rates and productivity of *Cladonia arbuscula* and *Cladonia impexa* on the Sands of Forvie, Scotland. *Can. J. Bot.* **52**: 431–433.

1862 —— (1974) A study of a lichen synusium on the sands of Forvie, Scotland. *Nova Hedwigia* **25**: 719–736.

1863 —— **and Sommerville, A. H.** (1974) The H. E. Wilson lichen collection. *Trans. Proc. bot. Soc. Edinb.* **42**: 173–179.

1864 **Proctor, M. C. F.** (1956) Biological Flora of the British Isles: *Helianthemum* L. *J. Ecol.* **44**: 675–692.

1865 —— (1960) Biological Flora of the British Isles: *Tuberaria guttata* (L.) Fourreau. *J. Ecol.* **48**: 243–253.

1866 —— (1962) The epiphytic bryophyte communities of the Dartmoor oak-woods. *Rep. Trans. Devon. Ass. Advmt Sci.* **94**: 531–554.

1867 —— (1969) Flora and vegetation. In *Exeter and its Region* (F. Barlow, ed.): 97–116. Exeter: University of Exeter.

—— —See also No. 2337.

1868 **Pryor, A. R. and Jackson, B. D.** (1887) *Flora of Hertfordshire*. London: Gurney & Jackson. [Lichens pp. 520–521, det. E. M. Holmes.]

1869 **Pulteney, R.** (17—) *Catalogue of English plants, with the names of the first describers, or discoverers, annex:d*. Manuscript in BM. [Lichens pp. 30–32.]

1870 —— (1747) *Catalogus stirpium circa Lactodorum nascentium*. Manuscript in LSR.

1871 —— [c. 1747] *A catalogue of some of the more rare plants found in the neighbourhood of Leicester, Loughborough and Charley Forest*. Manuscript in LINN.

1872 —— [c. 1747] *A catalogue of plants spontaneously growing about Loughborough and the adjacent villages*. Manuscript in BM.

1873 —— (1749) *A methodical distribution of plants according to Mr Rays method together with a compleater method of classifying mosses compounded by Dr Dillenius*. Manuscript in LINN.

1874 —— (1756) An account of some of the more rare English plants observed in Leicestershire. *Phil. Trans. R. Soc.* **49**: 803–866.

1875 —— (1790) *Historical and Biographical Sketches of the Progress of Botany in England*. 2 vols. London: Cadell.

1876 —— (1799) *Catalogues of the Birds, Shells and some more rare Plants of Dorsetshire.* London: Privately printed. [Not seen; reprinted from J. Hutchins, *History and Antiquities of the County of Dorset*, Ed. 2, vol. 3.]

1877 —— (1813) *Catalogues of the Birds, Shells and some of the more rare plants of Dorsetshire.* Ed. 2. London: Privately printed. ["From the new and enlarged edition of Mr. Hutchins's History of that County with additions; and a brief memoir of the author"; lichens p. 100.]

—— —See also No. 1658.

Punithalingam, E.—See No. 966.

1878 **Purton, T.** (1817) *A Botanical Description of British Plants, in the Midland Counties, particularly of those in the neighbourhood of Alcester.* Vol. 2. Stratford-upon-Avon: J. Ward. [Lichens pp. 575–606.]

1879 —— (1821) *An Appendix to the Midland Flora.* Vol. 3(1). London. [Lichens pp. 119–175.]

1880 —— (1833) Plants observed in the neighbourhood of Barmouth, North Wales. *Mag. nat. Hist.* **6**: 57–59.

1881 **Pyatt, F. B.** (1967) The inhibitory influence of *Peltigera canina* on the germination of graminaceous seeds and the subsequent growth of the seedlings. *Bryologist* **70**: 326–329.

1882 —— (1968) The effect of sulphur dioxide on the inhibitory influence of *Peltigera canina* on the germination and growth of grasses. *Bryologist* **71**: 97–101.

1883 —— (1968) The occurrence of a rotifer on the surfaces of apothecia of *Xanthoria parietina. Lichenologist* **4**: 74–75.

1884 —— (1969) *Atmospheric pollution in South Wales in relation to the growth and distribution of lichens.* Ph.D. thesis, University of Wales.

1885 —— (1969) Studies of the periodicity of spore discharge and germination in lichens. *Bryologist* **72**: 48–53.

1886 —— (1969) The ultrastructure of the ascospore wall of the lichen *Pertusaria pertusa. Trans. Br. mycol. Soc.* **52**: 167–169.

1887 —— (1970) Lichens as indicators of air pollution in a steel producing town in South Wales. *Envir. Poll.* **1**: 45–56.

1888 —— (1973) Plant sulphur content as an air pollution gauge in the vicinity of a steelworks. *Envir. Poll.* **5**: 103–115.

1889 —— (1973) Sporophores of *Laccaria laccata* on metal spoil. *Trans. Br. mycol. Soc.* **61**: 189–190.

1890 —— [1974] (1973) Lichen propagules. In *The Lichens* (V. Ahmadjian and M. E. Hale, eds.): 117–145. New York and London: Academic Press.

1891 —— (1975) *Clavaria argillacea* on a spoil tip in south west England. *Trans. Br. mycol. Soc.* **64**: 171.

1892 —— (1976) Lichen ecology of metal spoil tips: effects of metal ions on ascospore viability. *Bryologist* **79**: 172–179.

1893 —— **and Harvey, R.** (1973) *Tichothecium erraticum* Mass. on *Lecanora campestris* (Schaer.) Hue and *Caloplaca heppiana* (Müll. Arg.) Zahlbr. *Israel J. Bot.* **22**: 139–143.

1894 **Pyefinch, K. A.** (1943) The intertidal ecology of Bardsey Island, North Wales, with special reference to the recolonisation of rock surfaces, and the rock-pool environment. *J. anim. Ecol.* **12**: 82–108.

1895 **Rackham, O.** (1975) *Hayley Wood. Its History and Ecology.* Cambridge: Cambridge and Isle of Ely Naturalists' Trust. [Lichens pp. 165–168, 213–214.]

1896 **Ragg, J. M. and Futty, D. W.** (1967) *The Soils of the County round Haddington and Eyemouth.* [Memoirs of the Soil Survey of Great Britain. Scotland.] Edinburgh: H.M.S.O. [Vegetation by E. L. Birse and J. S. Robertson.]

1897 **Raistrick, A. and Gilbert, O. L.** (1963) Malham Tarn House: Its building materials, their weathering and colonization by plants. *Fld Stud.* **1** (5): 89–115.

1898 **Ralfs, J.** (1842) Localities of British algae in addition to those given in Harvey's Manual. *Phytologist* **1**: 193–196.

1899 —— (1883) The lichens of West Cornwall. *Trans. Penzance nat. Hist. antiq. Soc.* **1**: 205–219.

1900 [——] (1884) Lichens. *Trans. Penzance nat. Hist. antiq. Soc.* **1**: 422.
Randall, R. E.—See No. 1806.

1901 **Ranwell, D. S.** (1955) *Slack vegetation, dune system development and cyclical change at Newborough Warren, Anglesey.* Ph.D. thesis, University of London.

1902 —— (1960) Newborough Warren, Anglesey. III. Changes in the vegetation on parts of the dune system after the loss of rabbits by myxomatosis. *J. Ecol.* **48**: 385–395.

1903 —— (1961) Field meeting at Lyndhurst. *Lichenologist* **1**: 275–276.

1904 —— (1963) Field meeting at Wareham. *Lichenologist* **2**: 197–200.

1905 —— (1966) The lichen flora of Bryher, Isles of Scilly, and its geographical components. *Lichenologist* **3**: 224–232.

1906 —— (1968) Lichen mortality due to 'Torrey Canyon' oil and decontamination measures. *Lichenologist* **4**: 55–56.

1907 **Ratcliffe, D. A.** (1954) *An ecological study of the mountain vegetation in the Carneddau group.* Ph. D. thesis, University of Wales.

1908 —— (1958) The range and habitats of *Sphagnum lindbergii* Schp. in Scotland. *Trans. Br. bryol. Soc.* **3**: 386–391.

1909 —— (1959) The vegetation of the Carneddau, North Wales. *J. Ecol.* **47**: 371–413.

1910 —— (1959) The mountain plants of the Moffat Hills. *Trans. Proc. bot. Soc. Edinb.* **37**: 257–271.

1911 —— (1964) Mires and bogs. In *The Vegetation of Scotland* (J. H. Burnett, ed.): 426–478. Edinburgh and London: Oliver and Boyd.

1912 —— (1964) Montane mires and bogs. In *The Vegetation of Scotland* (J. H. Burnett, ed.): 536–558. Edinburgh and London: Oliver and Boyd.

1913 —— (1968) An ecological account of atlantic bryophytes in the British Isles. *New Phytol.* **67**: 365–439.

138

1914 —— **and Walker, D.** (1958) The Silver Flowe, Galloway, Scotland. *J. Ecol.* **46**: 407–445.

—— —See also No. 1507.

1915 **Rattray, J.** (1884) Animal and vegetable symbiosis or consortism. *Trans. Edinb. nat. Fld Club* **1**: 172–184.

1916 **Rattray, M. A.** (1886) The May Island; its archaeology; its algoid flora; its phanerogams and higher cryptogams. *Trans. bot. Soc. Edinb.* **16**: 115–121.

1917 **Ravenshaw, T. F.** (1874) *Botany of North Devon.* [Extracted from Stewart's *North Devon Hand-book.*] Ilfracombe: W. Stewart. [Lichens pp. 88–89.]

1918 —— (1877) Botany. In *The North Devon Hand-book*, Ed. 4 (G. Tugwell, ed.): 291–398. London: Simpkin Marshall. [Lichens pp. 376–377.]

Rawes, M.—See Nos. 653, 2573 and 2574.

1919 **Ray, J.** (1660) *Catalogus plantarum circa Cantabrigiam nascentium.* Cambridge. [Lichens pp. 86–87.]

1920 —— (1670) *Catalogus plantarum Angliae et insularum adjacentium.* London. [Lichens pp. 192–194.]

1921 —— (1677) *Catalogus plantarum Angliae et insularum adjacentium.* Ed. 2. London. [Lichens pp. 184–186.]

1922 —— (1686) *Historia plantarum.* Vol. 1. London.

1923 —— (1690) *Synopsis methodica stirpium Britannicarum.* London. [Lichens pp. 14–15.]

1924 —— (1693) *Historiae plantarum.* Vol. 2. London.

1925 —— (1696) *Synopsis methodica stirpium Britannicarum.* Ed. 2. London. [Lichens pp. 40–42.]

1926 —— (1704) *Historiae plantarum.* Vol. 3. London.

1927 —— (1724) *Synopsis methodica stirpium Britannicarum.* Ed. 3. Vol. 1. London. [Reprint (1973) London: Ray Society.] [Lichens pp. 64–77.] [Prepared largely by Dillenius, notes and some original drawings by Dillenius in OXF (Bodlein Library).]

—— —See also Nos. 604, 605 and 849.

1928 **Rayner, R. W.** [1976] (1975) Lichens. In *The Natural History of Pagham Harbour. Part II. Plants and Animals other than Birds and Mammals* (R. W. Rayner, ed.): 47–49. Bognor Regis: Bognor Regis Natural History Society.

Rea, C.—See Nos. 28 and 2211.

1929 **Reader, H. P.** [1901] Notes on a collection of lichens by the late Mr J. Hardy. *Hist. Berwicksh. Nat. Club* **17**: 261–264.

1930 [——] (1909) Notes on specimens contributed or sent for identification. *Rep. Lichen Exch. Club* **1908**: 6–8.

1931 —— (1910) Notes on specimens contributed or sent for identification. *Rep. Lichen. Exch. Club* **1909**: 6–8.

1932 [——] (1911) Notes on specimens contributed or sent for identification. *Rep. Lichen Exch. Club* **1910**: 6(*22*)–8a(*24a*).

1933 —— (1926) The flora of Hawkesyard. Part II—Cryptogams. *Trans. a. Rep. N. Staffs. Fld Club* **60**: 118–132.

—— —See also No. 72.

1934 **Redinger, K.** (1937) Familie Arthoniaceae. *Rabenh. Krypt.-Fl.* **9**, 2(1): 1–180.

1935 —— (1937) Familie Graphidaceae. *Rabenh. Krypt.-Fl.* **9**: 2(2): 181–404.

1936 —— (1938) Restitution und kritische Revision der Flechtengattungen *Enterographa* Fée und *Sclerophyton* Eschw. *Repert. Spec. nov. Regni veg.* **43**: 49–77.

Rees, J. F.—See No. 1804.

1937 **Rehm, A.** (1971) A chemical study of *Sphaerophorus globosus* and *S. fragilis*. *Bryologist* **74**: 199–202.

Reid, G.—See No. 645.

1938 **Relhan, R.** (1785) *Flora Cantabrigiensis*. Cambridge. [Lichens pp. 421–441.]

1939 —— (1786) *Florae Cantabrigiensi. Supplementum* [*Primum*]. Cambridge. [Lichens pp. 20–21.]

1940 —— (1788) *Florae Cantabrigiensi. Supplementum Alterum*. Cambridge. [Lichens pp. 20–21.]

1941 —— (1793) *Florae Cantabrigiensi. Supplementum Tertium*. Cambridge. [Lichens pp. 11–13.]

1942 —— (1802) *Flora Cantabrigiensis*. Ed. 2. Cambridge. [Lichens pp. 443–474.]

1943 —— (1820) *Flora Cantabrigiensis*. Ed. 3. Cambridge. [Lichens pp. 473–505.]

1944 **Rennie, R. C.** (1966) *The Third Statistical Account of Scotland. The County of Stirling*. Glasgow: Collins. [Lichens p. 25.]

1945 **Rhodes, P. M. G.** (1910) Notes on mosses, Hepaticae and lichens from the Channel Islands. *Rep. Trans. Guernsey Soc. nat. Sci.* **6**: 88–91.

1946 [——] (1914) Notes on specimens contributed or sent for identification. *Rep. Lichen Exch. Club* **1913**: 5(52)–8(55).

1947 —— (1931) The lichen-flora of Hartlebury Common. *Proc. Bgham nat. Hist. Soc.* **16**: 39–43.

—— —See also No. 677.

1948 **Richards, J. W.** (1968). A look at lichens. *Countryman* **70**: 281–285.

1949 **Richards, P. W. M.** (1928) Ecological notes on the bryophytes of Middlesex. *J. Ecol.* **16**: 269–300.

1950 —— (1929) Notes on the ecology of the bryophytes and lichens at Blakeney Point, Norfolk. *J. Ecol.* **17**: 127–140.

—— —See also Nos. 1211 and 1727.

1951 **Richardson, D. H. S.** (1967) The transplantation of lichen thalli to solve some taxonomic problems in *Xanthoria parietina* (L.) Th. Fr. *Lichenologist* **3**: 386–391.

1952 —— (1975) *The Vanishing Lichens. Their History, Biology and Importance*. Newton Abbot, London & Vancouver: David & Charles.

1953 —— **and Green, B. H.** (1965) A subfossil lichen. *Lichenologist* **3**: 89–90.

1954 —— **Hill, D. J. and Smith, D. C.** (1968) Lichen physiology. XI. The role of the alga in determining the pattern of carbohydrate movement between lichen symbionts. *New Phytol.* **67**: 469–486.

1955 —— **and Morgan-Jones, G.** (1964) Studies on lichen asci. I. The bitunicate type. *Lichenologist* **2**: 205–224.

1956 —— **and Smith, D. C.** (1966) The physiology of the symbiosis in *Xanthoria*

140

aureola (Ach.) Erichs. *Lichenologist* **3**: 202–206.

1957 —— —— (1968) Lichen physiology IX. Carbohydrate movement from the *Trebouxia* symbiont of *Xanthoria aureola* to the fungus. *New Phytol.* **67**: 61–68.

1958 —— —— **and Lewis, D. H.** (1967) Carbohydrate movement between the symbionts of lichens. *Nature, Lond.* **214**: 879–882.

1959 **Richardson, M. J.** (1968) Report of the Cryptogamic Section, 1967. *Trans. Proc. bot. Soc. Edinb.* **40**: 473–480. [Lichens p. 476 by E. P. Beattie.] —— —See also No. 2475.

Richardson, R.—See No. 2389a.

1960 **Riddelsdell, H. J.** (1905) Lightfoot's visit to Wales in 1773. *J. Bot., Lond.* **43**: 290–307.

1961 **Rideout, E. H.** (1923) Vegetation of the Liverpool district. In *Merseyside* (A. Holt, ed.): 257–268. London: Hodder & Stoughton.

1962 **Riedl, H.** (1964) Bemerkungen über *Dermatina*-Arten aus West- und Mitteleuropa Vorläufige Mitteilung. *Sydowia* **17**: 102–113.

1963 **Rilstone, F.** (1940) A new fungus from Devon. *J. Bot., Lond.* **78**: 192–193.

1964 **Rimington, F. C.** (1953) Lichens. In *The Natural History of the Scarborough District* (G. B. Walsh and F. C. Rimington, eds.) **1**: 167–183. Scarborough: Scarborough Field Naturalists' Society.

1965 **Ritchie, J. C.** (1954) Biological Flora of the British Isles: *Primula scotia* Hook. *J. Ecol.* **42**: 623–628.

1966 —— (1955) Biological Flora of the British Isles: *Vaccinium vitis-idaea* L. *J. Ecol.* **43**: 701–708.

1967 —— (1956) Biological Flora of the British Isles: *Vaccinium myrtillus* L. *J. Ecol.* **44**: 291–299.

1968 **Ritchings, C. R.** (1923) Field work in the Burnley district. *Lancs. Chesh. Nat.* **15**: 211–212.

1969 **Roberts, G. E.** (1857) *The Valley of Habberley*. Kidderminster. [Lichen p. 84.]

1970 **Roberts, K. A. and Edwards, P. J.** (1973) *Wildlife in the Suburbs—Perivale Wood Nature Reserve*. Ealing: The Selborne Society. [Lichens pp. 16–17.]

1971 **Robertson, D. A.** (1955) *The ecology of the sand dune vegetation of Ross Links, Northumberland with special reference to secondary succession in the blow-outs*. Ph.D. thesis, University of Durham.

1972 **Robertson, E. T. and Gimingham, C. H.** (1951) Contributions to the maritime ecology of St. Cyrus, Kincardineshire. *Trans. Proc. bot. Soc. Edinb.* **35**: 370–414.

1973 **Robertson, J.** (1908) Excursions to the islands of Bute and Great Cumbrae. *Ann. Anderson. Nat. Soc.* **3**: 71–79.

Robertson, J. S.—See No. 1896.

Robertson, L.—See No. 1996.

Robertson, R. A.—See No. 1660.

1974 **Robinson, H.** (1929) Emmott Hall. *NWest. Nat.* **4**: 105–108.

Robinson, H. D.—See No. 447.

1975 **Robinson, W.** (1823) *The History and Antiquities of Enfield.* Vol. 1. London: John Nichols & Son. [Lichens pp. 271–272.]

1975a **Robson, E.** [1794] *Plantae rariores agro Dunelmensi indigenae.* Darlington: privately printed. [See also manuscript with Darlington and Teesdale Natural History Society.]
—— —See also No. 1113.

1976 **Robson, S.** (1777) *The British Flora.* York: W. Blanchard. [Lichens pp. 295–307.]

1977 **Rodd, E. H., Millett, J. N. R. and Couch, R. C. E.** (1853) Report [of the Council]. *Trans. nat. Hist. antiq. Soc. Penzance* **2**: 61–65.
—— —See also No. 1607.

1978 **Roebuck, W. D.** (1875) Riccal Common. *Naturalist, Hull* **1**: 37–40.

1979 —— (1877) Yorkshire Naturalists' Union. *Naturalist, Hull* **3**: 47–48.

1980 —— (1878) The Yorkshire Naturalists' Union. *Naturalist, Hull* **3**: 174–176.

1981 —— (1878) The Yorkshire Naturalists' Union. *Naturalist, Hull* **3**: 188–192.

1982 —— (1878) Yorkshire Naturalists' Union. *Naturalist, Hull* **4**: 28–31.

1983 —— (1878) Yorkshire Naturalists' Union. *Naturalist, Hull* **4**: 46–48.

1984 —— (1879) Yorkshire Naturalists' Union. *Naturalist, Hull* **4**: 158–160.

1985 —— (1879) Yorkshire Naturalists' Union. *Naturalist, Hull* **4**: 174–176.

1986 —— (1879) The Leeds Naturalists' Club and Scientific Association. *Naturalist, Hull* **5**: 14–15.

1987 —— (1879) Yorkshire Naturalists' Union. *Naturalist, Hull* **5**: 15–16.

1988 —— (1879) Yorkshire Naturalists' Union. *Naturalist, Hull* **5**: 31–32.

1989 —— (1880) Yorkshire Naturalists' Union. *Naturalist, Hull* **6**: 31–32.

1990 —— (1883) Yorkshire Naturalists' Union. *Naturalist, Hull* **8**: 173–176.

1991 **Roger, J. G.** (1960) Report of the Alpine Section, 1959. Day excursion to Am Binnein, Perthshire (18th July 1959). *Trans. Proc. bot. Soc. Edinb.* **39**: 120–121.

1992 —— (1965) Report of the Alpine Section, 1964. Day excursion to Meall nan Tarmachan (3,421 ft.), Breadalbane, Perthshire. 27th June, 1964. *Trans. Proc. bot. Soc. Edinb.* **40**: 132–133.

1993 —— (1967) Report of the Alpine Section. Day excursion to Canness Glen, N.W. Angus, 26th June 1966. *Trans. Proc. bot. Soc. Edinb.* **40**: 353–355.
—— —See also Nos. 700 and 1118.

Rogers, J. A.—See No. 1793.

Rogers, R.—See No. 2501.

1994 **Rogers, T.** (1881) Manchester Cryptogamic Society. *Naturalist, Hull* **7**: 172.

1995 —— (1883) Manchester Cryptogamic Society. *Naturalist, Hull* **9**: 59–60.

1996 **Romans, J. C. C., Stevens, J. H., Robertson, L. and Jones, D.** (1966) Semi-fossil lichen fungi in Scottish hill soils. *Nature, Lond.* **209**: 96.

1997 **Roper, F. C. S.** (1875) *Flora of Eastbourne.* London. [Lichens pp. 149–150.]

1998 —— (1875) On additions to the fauna and flora of Eastbourne during 1875. *J. Trans. Eastbourne nat. Hist. Soc.* **1**: sine pagin.

1999 —— (1877) On additions to the fauna and flora of Eastbourne since 1875. *J. Trans. Eastbourne nat. Hist. Soc.* **1**: sine pagin.

2000 —— (1878) On additions to the fauna and flora of the Cuckmere District during the past year. *J. Trans. Eastbourne nat. Hist. Soc.* **1**: sine pagin.

2001 —— (1879) On additions to the fauna and flora of the Cuckmere District during the past year. *J. Trans. Eastbourne nat. Hist. Soc.* **1**: sine pagin.

2002 —— (1880) On the botany of Compton Place Wall. *J. Trans. Eastbourne nat. Hist. Soc.* **1**: sine pagin.

2003 **Rose, F.** (1953) A survey of the ecology of the British lowland bogs. *Proc. Linn. Soc. Lond.* **164**: 186–211.

2004 —— (1966) Reports of field meetings held in 1965. 2nd October—Oldbury Hill. *Bull. Kent Fld Club* **11**: 29–31. [Lichens by F. H. Brightman.]

2005 —— (1967) Reports of field meetings held in 1966. 23rd April—Biddenden area. *Bull. Kent Fld Club* **12**: i-iii.

2006 —— (1967) Reports of field meetings held in 1966. 23rd October—Hythe area. *Bull. Kent Fld Club* **12**: iii-iv.

2007 —— (1968) *Corticolous and lignicolous lichens of Staverton Park*. Manuscript. Report to the Nature Conservancy; copy with BLS.

2008 —— (1968) *Preliminary report on deciduous woodlands and old parklands containing fragments of old deciduous woodland, in Norfolk and Suffolk. I Suffolk. II Norfolk*. Manuscript. Report to the Nature Conservancy; copy with BLS.

2009 —— (1968) Reports of field meetings held in 1967. 14th October—Fairlight, East Sussex. *Bull. Kent Fld Club* **13**: 35–38.

2010 —— (1968) *Report on survey carried out in the Forest of Dean, April and May 1968*. Manuscript. Report to the Nature Conservancy; copy with BLS.

2011 —— (1968) *Survey report on woodland and forest relict areas of the Wealden region*. Manuscript. Report to the Nature Conservancy; copy with BLS.

2012 —— (1968) *Wood south-east of Alcester. Great Wood, Vowchurch. Moccas Park. Banses Wood, south-east of Cockyard*. Manuscript. Report to the Nature Conservancy; copy with BLS.

2013 —— (1968) *Wyre Forest (visited 28th and 29th May, 1968)*. Manuscript. Report to the Nature Conservancy; copy with BLS.

2014 —— (1969) Reports of field meetings. 20th April—Combwell Wood, Goudhurst. *Bull. Kent Fld Club* **14**: 15–18.

2015 —— (1969) [*Kent lichen check-list.*] Manuscript. Copy with BLS.

2016 —— [1969] [*Burnham Beeches. Longleat Woods and Park. Cranborne Chase. Severnake Forest.*] Manuscript. Report to the Nature Conservancy; copy with BLS.

2017 —— (1969) [*Hareshaw Linn, Bellingham. Ingleton Glens, West Riding. Hackfall, West Riding.*] Manuscript. Report to the Nature Conservancy; copy with BLS.

2018 —— (1969) *Report to the Nature Conservancy on a visit to Wyre Forest in April 1969 (Supplement to Report of May 1968)*. Manuscript. Report to the Nature Conservancy; copy with BLS.

2019 —— (1970) Lichens as pollution indicators. *Your Environ.* **1**: 185–189.

2020 —— (1971) *Report on surveys of woodland and parkland carried out in Corn-*

wall. 1–4 *October* 1971. Manuscript. Report to the Nature Conservancy; copy with BLS.

2021 —— (1971) *A survey of the woodlands of the Lake District, and an assessment of their conservation value based upon structure, age of trees, and lichen and bryophyte epiphyte flora.* Manuscript. Report to the Nature Conservancy; copy with BLS.

2022 —— (1971) *Report on sites visited in East Anglia, June 1970; March 1971; July 1971.* Manuscript. Report to the Nature Conservancy; copy with BLS.

2023 —— (1971) *Report on Sotterley Park, East Suffolk.* Manuscript. Report to the Nature Conservancy; copy with BLS.

2024 —— (1972) *Brampton Bryan Park, Herefordshire.* Manuscript. Report to the Nature Conservancy; copy with BLS.

2025 —— (1972) *Downton Castle Park & Gorge, Herefordshire.* Manuscript. Report to the Nature Conservancy; copy with BLS.

2026 —— (1972) [*Hull Wood, Glandford. Sheringham Park. Sturston Carr. Hilborough Park.*] Manuscript. Report to the Nature Conservancy; copy with BLS.

2027 —— (1972) [*Kentchurch Park, Homend Park. Eastnor Park. Holme Lacy Park & Wood, Herefordshire.*] Manuscript. Report to the Nature Conservancy; copy with BLS.

2028 —— (1972) *Report on lichenological and bryological survey of the woodlands etc in Morven and the Loch Sunart—Ardnamurchan area.* Manuscript. Report to the Nature Conservancy; copy with BLS.

2029 —— (1972) *Report on Wensleydale sites surveyed 5th & 6th August 1972.* Manuscript. Report to the Nature Conservancy; copy with BLS.

2030 —— (1972) *Survey report on woodland sites in North Yorkshire & Co. Durham.* Manuscript. Report to the Nature Conservancy; copy with BLS.

2031 —— (1972) *Wye Gorge—below Symonds Yat, Herefordshire.* Manuscript. Report to the Nature Conservancy; copy with BLS.

2032 —— (1973) A mobile lichen—in which grid square? *Bull. Br. Lichen. Soc.* **33**: 13.

2033 —— (1973) *Report on survey work on cryptogamic vegetation in Essex and Suffolk, 16th–19th March 1973.* Manuscript. Report to the Nature Conservancy; copy with BLS.

2034 —— (1973) *Report on woodland surveys in South Wales, June & September 1972.* Manuscript. Report to the Nature Conservancy; copy with BLS.

2035 —— (1973) *Report on surveys of woodlands in Kintyre—Knapdale and in Sunart with particular reference to their cryptogamic floras.* Manuscript. Report to the Nature Conservancy; copy with BLS.

2036 —— (1973) *Interim report on surveys done in Westmorland and Cumberland August 29th–September 1st, 1973.* Manuscript. Report to the Nature Conservancy; copy with BLS.

2037 —— (1973) Detailed mapping in south-east England. In *Air Pollution and Lichens* (B. W. Ferry, M. S. Baddeley and D. L. Hawksworth, eds.): 77–88. London: Athlone Press of the University of London.

144

2038 —— (1974) The epiphytes of oak. In *The British Oak; its History and Natural History* (M. G. Morris and F. H. Perring, eds.): 250–273. Faringdon: E. W. Classey.

2039 —— (1974) *Survey of woodland and parkland sites in the S. W. Midlands, April 12–18, 1973.* Manuscript. Report to the Nature Conservancy; copy with BLS.

2040 —— (1974) *Report on work done in East Anglia in botanical survey 16th–19th July, 1974.* Manuscript. Report to the Nature Conservancy; copy with BLS.

2041 —— (1974) *Report on survey of woodland and parkland sites in Cumbria, August, 1974.* Manuscript. Report to the Nature Conservancy; copy with BLS.

2042 —— (1975) The vegetation and flora of Tycanol Wood. *Nature Wales* **14**: 178–185.

2043 —— (1975) The lichens of the Gwaun Valley. *Nature Wales* **14**: 186–189.

2044 —— (1975) *Report on woodland sites surveyed in Cumbria September 16th–19th, 1975.* Manuscript. Report to the Nature Conservancy; copy with BLS.

2045 —— (1975) *Report on survey of Tyneham Ranges for bryophytes and lichens, 18th–20th April,* 1975. Manuscript. Report to the Nature Conservancy; copy with BLS.

2046 —— (1975) *Report on survey of chalk grasslands in the military lands about Tidworth March 13th 1975.* Manuscript. Report to the Nature Conservancy; copy with BLS.

2047 —— (1975) *The lichens and bryophytes of sites in Hereford, Salop, and Staffordshire, based on a survey made from April 21–24, 1975.* Manuscript. Report to the Nature Conservancy; copy with BLS.

2048 —— [197?5] *Scales Wood.* Manuscript. Report to the Nature Conservancy; copy with BLS.

2049 —— (1976) Lichenological indicators of age and environmental continuity in woodlands. In *Lichenology: Progress and Problems* (D. H. Brown, D. L. Hawksworth and R. H. Bailey, eds.): 279–307. [Systematics Association Special Volume No. 8.] London, New York and San Francisco: Academic Press.

2050 —— (1976) *Report on survey of cryptogam flora (bryophytes & lichens) of the Longleat Estate surveyed* 8.1969, 11.1973 & 27.11.1974. Manuscript. Report to the Nature Conservancy; copy with BLS.

2051 —— **and Bailey, R. H.** (1969) Lichens from the Forest of Dean—New or otherwise interesting records. *Jl N. Gloucs. Nat. Soc.* **20**: 77–79.

2052 —— **Hawksworth, D. L. and Coppins, B. J.** (1970) A lichenological excursion through the North of England. *Naturalist, Hull* **1970**: 49–55.

2053 —— **and James, P. W.** (1974) Regional studies on the British lichen flora I. The corticolous and lignicolous species of the New Forest, Hampshire. *Lichenologist* **6**: 1–72.

2054 —— **and Showell, J.** (1968) *Windsor Forest, 17/2/68.* Manuscript. Report to the Nature Conservancy; copy with BLS.

—— —See also Nos. 206, 321, 963, 967, 968, 969, 970, 1145, 1146, 1147,

1148, 1149, 1150, 1151, 1152, 1153, 1809 and 1811.

2055 **Rotheray, L.** (1900) *Flora of Skipton and District.* Skipton: Craven Naturalists' and Scientific Association. [Lichens pp. 70–85.]

2056 **[Rotheroe, M.]** (1971) The lichens of Warwickshire. *Newsl. Warwicks. Fungus Surv.* **5**: 6.

—— —See also No. 400.

Roux, C.—See Nos. 401 and 402.

2057 **Rowe, S.** (1896) *A Perambulation of the Ancient and Royal Forest of Dartmoor and the Venville Precincts.* Ed. 3. Exeter: J. Commin. [Lichens pp. 394–398 by F. Brent.]

—— —See also No. 1616.

2058 **Runemark, H.** (1956) Studies in *Rhizocarpon.* I. Taxonomy of the yellow species in Europe. *Op. bot. Soc. bot. Lund.* **2**(1): 1–152.

2059 —— (1956) Studies in *Rhizocarpon* II. Distribution and ecology of the yellow species in Europe. *Op. bot. Soc. bot. Lund.* **2**(2): 1–150.

Russell, B.—See No. 906.

Russell, F. S.—See No. 2250.

2060 **Russell, G.** (1962) Observations on the marine algae of the Isle of May. *Trans. Proc. bot. Soc. Edinb.* **39**: 271–289.

2061 **Sadler, J.** (1857) *Narrative of a ramble among the Wild Flowers on the Moffat Hills in August* 1857. Moffat: W. Muir. [Lichens pp. 63–64.]

2062 —— (1860) Muscological excursions to Ramsheugh and Glenfarg, Ochil Hills, Perthshire, and Habbie's Haw, Pentland Hills, in September last. *Trans. bot. Soc. Edinb.* **6**: 151–153.

2063 —— (1873) Notes on the flora of the Isle of May, Firth of Forth. *Trans. bot. Soc. Edinb.* **11**: 390–392.

2064 —— (1879) Notes on the alpine flora of Ben Nevis, Inverness-shire. *Trans. bot. Soc. Edinb.* **13**: 50–54.

—— —See also No. 187.

Sage, B. L.—See No. 1056.

2065 **Salisbury, E. J.** (1918) The oak-hornbeam woods of Hertfordshire. Parts III and IV. *J. Ecol.* **6**: 14–52.

2066 —— (1922) The soils of Blakeney Point: a study of soil reaction and succession in relation to the plant covering. *Ann. Bot.* **36**: 391–431.

2067 —— (1934) On the day temperatures of sand dunes in relation to the vegetation at Blakeney Point, Norfolk. *Trans. Norfolk Norwich Nat. Soc.* **13**: 333–355.

2068 —— (1952) *Downs and Dunes. Their plant life and its environment.* London: G. Bell and Sons.

—— —See also No. 735.

2069 **Salisbury, G.** (1953) A new species of *Arthopyrenia* with blue-green algal cells. *Naturalist, Hull* **1953**: 17–18.

2070 —— (1953) The genus *Thelocarpon* in Britain. *NWest. Nat.* **24**: 66–76.

2071 —— (1966) A monograph of the lichen genus *Thelocarpon* Nyl. *Lichenologist* **3**: 175–196.

146

2072 —— (1972) *Thelotrema* Ach. sect. *Thelotrema* 1. The *T. lepadinum* group. *Lichenologist* **5**: 262–274.

2073 —— (1975) *Thelotrema monosporum* Nyl. in Britain. *Lichenologist* **7**: 59–61.

2074 —— (1975) *Leucocarpopsis*, a new lichen genus. *Nova Hedwigia* **26**: 351–352.

Salt, J.—See No. 916.

2075 **Salt, L.** (1971) Taking a look at the insectiverous beds. *Bull. Bgham bot. Hort. Soc.*, April 1971: 5–6.

2076 **Salwey, T.** (18—) *Lichenes Britannici*. 4 fasc., nos. 1–397. Oswestry. [Not seen; see Sayre (1969).]

2076a —— (1823–51). *Correspondence.* 2 vols. of letters to Salwey in Ludlow Museum. [Not seen; not abstracted.]

2077 —— (1844) A list of lichens gathered in different parts of Wales, principally in the neighbourhood of Barmouth, with a few casual observations upon some of the species. *Ann. Mag. nat. Hist.* **13**: 25–32, 260–263.

2078 —— (1845) A list of the scarcer amongst the lichens which are found in the neighbourhood of Oswestry and Ludlow, with occasional observations upon some of them. *Ann. Mag. nat. Hist.* **16**: 90–99.

2079 —— (1845) Addendum to the list of Welsh lichens. *Ann. Mag. nat. Hist.* **16**: 99.

2080 —— (1846) A list of the scarcer amongst the lichens which are found in the neighbourhood of Oswestry and Ludlow, with occasional observations upon some of them. *Trans. bot. Soc. Edinb.* **2**: 203–213.

2081 —— (1849) Stirpes cryptogamae Sarnienses; or contributions towards the cryptogamic flora of Guernsey. *Ann. Mag. nat. Hist.*, ser. 2, **3**: 22–29.

2082 —— (1854) Observations on Penzance lichens. *Trans. nat. Hist. antiq. Soc. Penzance* **2**: 137–148.

2083 —— [1855] Botany. In *The History of Oswestry* (W. Cathrall, ed.): 213–221. Oswestry: G. Lewis.

2084 —— (1860) *Roccella tinctoria* and *R. phycopsis* in the Isle of Wight. *Phytologist, n.s.* **4**: 267–268.

2085 —— [1860–?62] *Lichenes centum ex herbario T. Salwey*. 4 fasc., nos. 1–402. [Intact set not seen.]

2086 —— (1863) On some new British lichens. *Trans. bot. Soc. Edinb.* **7**: 550–558.

2087 —— (1863) A list of certain plants to be met with in the neighbourhoods of Barmouth, Dolgelley, and Harlech, with a few habitats of plants in other parts of Merionethshire. In *The Tourist's and Visitor's Hand-Book and Guide to Harlech, Barmouth, Dolgelley, Towyn, Aberdovey, & their neighbourhoods* (D. Jones, ed.): 125–158. Barmouth: D. Jones. [Lichens pp. 146–153.]

—— —See also Nos. 252 and 1267.

Samuelsson, G.—See No. 559.

2088 **Sandstede, H.** (1931) Die Gattung *Cladonia*. *Rabenh. Krypt.-Fl.* **9**, 4(2): 1–531.

2088a —— (1932–39) Cladoniaceae A. Zahlbr. I–III. In *Die Pflanzenareale* (E. Hannig and H. Winkler, eds.) **3** (6): 63–71, **4** (7): 83–90, **4** (8): 93–100 [+ maps 51–80]. Jena: G. Fischer.

2089 **Sandwith, C. I.** (1943) Bristol botany in 1943. *Proc. Bristol Nat. Soc., ser.* 4, **9**: 471–473.

2090 **Sankey, J.** (1966) *Chalkland Ecology.* [The Scholarship Series in Biology.] London: Heinemann Educational Books.

2091 **Santesson, R.** (1939) Amphibious pyrenolichens I. *Ark. Bot.* **29A**(10): 1–67.

2092 —— (1952) Foliicolous lichens I. A revision of the taxonomy of the obligately foliicolous, lichenized fungi. *Symb. bot. upsal.* **12**(1): 1–590.

2093 **Sato, M.** (1963) Mixture ratio of the lichen genus *Thamnolia. Nova Hedwigia* **5**: 149–155.

2094 —— (1965) The mixture ratio of the lichen genus *Thamnolia* in New Zealand. *Bryologist* **68**: 320–324.

2095 —— (1968) The mixture ratio of the lichen genus *Thamnolia* in Tasmania and New Guinea. *J. Jap. Bot.* **43**: 328–334.

2096 **Saunders, J.** (1911) Lichenes, Fungi, and Mycetozoa. *Trans. Herts. nat. Hist. Soc. Fld Club* **14**: 229–232.
—— —See also No. 1065.

2097 **Saunders, P. J. W.** (1970) Air pollution in relation to lichens and fungi. *Lichenologist* **4**: 337–349.
Savage, E. J. (née Fry)—See Nos. 736–739.

2098 **Savage, S.** (1963) *Catalogue of the Smithian herbarium.* Vol. 7. Manuscript in LINN.

2099 **Savidge, J. P., Heywood, V. H. and Gordon, V.,** eds. (1963) *Travis's Flora of South Lancashire.* Liverpool: Liverpool Botanical Society.

2100 **Sawyer, E. N.** (1931) Note on the squamules of *Cladonia ochrochlora* Flk. var. *ceratodes* Flk. *Proc. Bristol Nat. Soc., ser.* 4, **7**: 252–258.

2101 **Sayre, G.** (1969) Cryptogamae Exsiccatae—An annotated bibliography of published exsiccatae of Algae, Lichenes, Hepaticae, and Musci. *Mem. N. Y. Bot. Gdn* **19**: 1–174.

2102 **Schaerer, L. E.** (1850) *Enumeratio critica Lichenum europaeorum.* Bern.

2103 **Schmidt, A.** (1970) Anatomisch-taxonomische Untersuchungen an europäischen Arten der Flechtenfamilie Caliciaceae. *Mitt. StInst. allg. Bot. Hamburg* **13**: 111–166.
Schultz, I.—See No. 1393.
Schwarz, B.—See No. 1393.

2104 **Scott, G. A. M.** (1963) The ecology of shingle beach plants. *J. Ecol.* **51**: 517–527.

2105 —— (1965) The shingle succession at Dungeness. *J. Ecol.* **53**: 21–31.

2106 **Scott, G. D.** (1960) Studies of the lichen symbiosis. 1. The relationship between nutrition and moisture content in the maintenance of the symbiotic state. *New Phytol.* **59**: 374–381.

2107 —— (1964) The lichen symbiosis. *Advmt. Sci., Lond.* **21**: 244–248.
—— —See also No. 258.
Scott, R.—See No. 1670.

2108 **Seaward, M. R. D.** (1961) Lichenology. *Trans. Lincs. Nat. Un.* **15**: 128–129.

2109 —— (1962) Lichenology. *Trans. Lincs. Nat. Un.* **15**: 200–201.

2110 —— (1963) Field meeting at Market Rasen. *Lichenologist* **2**: 194–197.

148

2111 —— (1963) Lichenology. *Trans. Lincs. Nat. Un.* **15**: 285–286.

2112 —— (1963) The British Lichen Society at Market Rasen. *Trans. Lincs. Nat. Un.* **15**: 287–289.

2113 —— (1964) Lichenology. *Trans. Lincs. Nat. Un.* **16**: 39–41.

2114 —— (1965) *The ecology of Twigmoor Warren, North Lincolnshire.* M.Sc. thesis, University of Nottingham.

2115 —— (1965) Lincolnshire psocids. *Trans. Lincs. Nat. Un.* **16**: 99–100.

2116 —— (1965) Lichenology. *Trans. Lincs. Nat. Un.* **16**: 110–111.

2117 —— (1966) Preliminary notes on the heathland lichen flora in the Scunthorpe area. *Lichenologist* **3**: 275–276.

2118 —— (1966) A check-list of Lincolnshire lichens. *Trans. Lincs. Nat. Un.* **16**: 153–159.

2119 —— (1966) Lichenology. *Trans. Lincs. Nat. Un.* **16**: 164–166.

2120 —— (1967) Lichenology. *Trans. Lincs. Nat. Un.* **16**: 233–234.

2121 —— (1968) F. A. Lees' botanical collections: Part 2. *Naturalist, Hull* **1968**: 133–135.

2122 —— (1968) Bryology and lichenology. *Trans. Lincs. Nat. Un.* **17**: 33.

2123 —— (1968) British Lichen Society 7th annual field meeting, Stamford. *Trans. Lincs. Nat. Un.* **17**: 33–34.

2124 —— (1969) Bryology and lichenology. *Trans. Lincs. Nat. Un.* **17**: 103–104.

2125 —— (1970) F. A. Lees' botanical collections: Part 3. *Naturalist, Hull* **1970**: 125–129.

2126 —— (1970) The lichen *Parmelia incurva* (Pers.) Fr. in Yorkshire. *Naturalist, Hull* **1970**: 130.

2127 —— (1970) *A revised check-list of Lincolnshire lichens.* Horsforth, Leeds: Trinity and All Saints' Colleges.

2128 —— (1970) Bryology and lichenology. *Trans. Lincs. Nat. Un.* **17**: 151–152.

2129 —— (1971) *A guide to the lichenological collection of Thomas Hebden (1849–1931).* Keighley: Cliffe Castle Museum.

2130 —— (1972) *Aspects of urban lichen ecology.* Ph.D. thesis, University of Bradford.

2131 —— (1972) Lichenology. *Trans. Lincs. Nat. Un.* **18**: 16–17.

2132 —— (1972) *A lichen check-list of the West Riding conurbation.* Horsforth, Leeds: Trinity and All Saints' Colleges.

2133 —— (1972) William Johnson's lichen collection. *Naturalist, Hull* **1972**: 13–14.

2134 —— (1973) The ecology of Scunthorpe heathlands with particular reference to Twigmoor Warren. *J. Scunthorpe Mus. Soc., ser.* 2, **2**: 1–28.

2135 —— (1973) Lichen ecology of the Scunthorpe heathlands I. Mineral accumulation. *Lichenologist* **5**: 423–433.

2136 —— (1973) Distribution maps of lichens in Britain. *Lichenologist* **5**: 464–466.

2137 —— (1973) F. A. Lees' botanical collections: Part 4. *Naturalist, Hull* **1973**: 35–36.

2138 —— (1973) Lichenology. *Trans. Lincs. Nat. Un.* **18**: 84–85.

2139 —— (1974) A note on *Phauloppia lucorum* C. L. Koch (Acari: Oribatidae) and lichens. *Lichenologist* **6**: 126–127.

2140 —— (1974) Some observations on heavy metal toxicity and tolerance in lichens. *Lichenologist* **6**: 158–164.

2141 —— (1974) Bryology and lichenology. *Trans. Lincs. Nat. Un.* **18**: 131–135.

2142 —— (1974) Contributions to the lichen flora of lowland Scotland – 1. Dumfriesshire. *Trans. Proc. bot. Soc. Edinb.* **42**: 143–151.

2143 —— (1975) Distribution maps of lichens in Britain. *Lichenologist* **7**: 180.

2144 —— (1975) Lichen flora of the West Yorkshire conurbation. *Proc. Leeds phil. lit. Soc., sci. sect.* **10**: 141–208.

2145 —— (1975) Contributions to the lichen flora of south-east Ireland—I. *Proc. R. Ir. Acad.* **B, 75**: 185–205.

2146 —— (1976) Performance of *Lecanora muralis* in an urban environment. In *Lichenology: Progress and Problems* (D. H. Brown, D. L. Hawksworth and R. H. Bailey, eds.): 323–357. [Systematics Association Special Volume No. 8.] London, New York and San Francisco: Academic Press.

2147 —— (1976) Further recent records of *Lobaria* from southeast Ireland. *Ir. Nat. J.* **18**: 336–337.

2148 —— (1976) The lichens of County Carlow, Ireland. *Revue bryol. lichén.* **42**: 665–676.

2149 —— (1976) Lichen mapping scheme: a progress report for Ireland. *Ir. Nat. J.* **18**: 335–336.

2150 —— (1976) Lichens in air-polluted environments: multivariate analysis of the factors involved. In *Proceedings of the Kuopio Meeting on Plant Damages Caused by Air Pollution* (L. Kärenlampi, ed.): 57–63. Kuopio: University of Kuopio and Kuopio Naturalists' Society.

2151 —— (1976) Yorkshire lichen material of W. G. McIvor in the herbarium of the National Botanic Gardens, Glasnevin, Dublin. *Naturalist, Hull* **1976**: 125–128

—— —See also Nos. 236, 437, 447, 652 and 979.

2152 **Segal, S.** (1969) *Ecological Notes on Wall Vegetation.* The Hague: W. Junk.

2153 **Service, R.** (1888) The natural history of Kirkcudbrightshire. In *Guide Book to the Stewartry of Kirkcudbright*, Ed. 5 (M. E. Marshall, ed.): 157–175. Castle-Douglas: Advertiser Steam Press. [Lichens pp. 161–162.]

Seviour, J. A.—See No. 1126.

2154 **Shackleton, A.** (1891) Disappearance of Yorkshire plants. *Naturalist, Hull* **1891**: 79–80.

2155 —— (1896) Symbiosis and mimicry in lichens. *Naturalist, Hull* **1896**: 16.

2156 —— **and Hebden T.** (1892) New British lichens. *Naturalist, Hull* **1892**: 17.

2157 —— —— (1893) Additions to the lichen flora of the West Riding of Yorkshire. *Naturalist, Hull* **1893**: 165–171.

Sharrock, J. T. R.—See No. 702.

2158 **Shaw, H. K. A.**, ed. (1961) Additions to the wild flora and fauna of the Royal Botanic Gardens, Kew: XXIV. *Kew Bull.* **15**: 169–191.

Shaw, P. J.—See No. 447.

Sheail, J.—See No. 2575.

2159 **Sheard, J. W.** (1964) The genus *Buellia* de Notaris in the British Isles (excluding section *Diploicia* (Massal.) Stiz.). *Lichenologist* **2**: 225–262.

2160 —— (1965) *Lecanora* (sect. *Aspicilia*) *leprosescens* Sandst. new to the British Isles. *Lichenologist* **3**: 93–94.

150

2161 —— (1966) *A revision of the lichen genus* Rinodina *in Europe and its taxonomic affinities*. Ph.D. thesis, University of London.

2162 —— (1967) A revision of the lichen genus *Rinodina* (Ach.) Gray in the British Isles. *Lichenologist* **3**: 328–367.

2163 —— (1973) *Rinodina interpolata* (Stirt.) Sheard, a new combination in the British and Scandinavian lichen floras. *Lichenologist* **5**: 461–463.

2164 —— **and Ferry, B. W.** (1964) The lichen flora of Bardsey. *Rep. Bardsey Obs.* **11**: 56–65.

2165 —— —— (1965) The lichen flora of Skokholm. *Rep. Skokholm Bird Obs.* **1964**: 23–35.

2166 —— —— (1967) The lichen flora of the Isle of May. *Trans. Proc. bot. Soc. Edinb.* **40**: 268–282.

2167 —— **and James, P. W.** (1976) Typification of the taxa belonging to the *Ramalina siliquosa* species aggregate. *Lichenologist* **8**: 35–46.

—— —See also No. 704.

2168 **Shenstone, J. C.** (1903) Botany. In *The Victoria History of the Counties of England, Essex* (H. A. Doubleday and W. Page, eds.) **1**: 31–67. Westminster: Constable. [Lichens pp. 53–57.]

2169 **Sheppard, T.** (1910) Naturalists at Scunthorpe. *Naturalist, Hull* **1910**: 391–398.

2170 **Shimwell, D. W.** [1969] *The Vegetation of the Derbyshire Dales*. Shrewsbury: The Nature Conservancy.

2171 —— (1971) *The Description and Classification of Vegetation*. London: Sidgwick & Jackson.

2172 —— **and Laurie, A. E.** (1972) Lead and zinc contamination of vegetation in the Southern Pennines. *Envir. Poll.* **3**: 291–301.

—— —See also Nos. 438 and 1817.

2173 **Shoosmith, F. H.** (1907) Lichens. *Nature Reader Monthly* **1907** (December): 1–12.

2174 **Showell, J. P.** (1967) A short guide to the study of lichens. *Nat. Sci. in Schools* **5**: 65–69.

2175 —— (1969) Lichens which grow on soil in lowland England and how to identify them. Part 1. *Nat. Sci. in Schools* **7**: 43–53.

2176 —— (1969) Lichens which grow on soil in lowland England and how to identify them. Part 2. *Nat. Sci. in Schools* **7**: 76–83.

2177 —— (1969) *Lichens*. [Publication No. 38.] London: School Natural Science Society. [Separate issue of Nos. 2174–2176.]

—— —See also No. 2054.

2178 **Sibbald, R.** (1684) *Nuncius Scoto-Britannus, sive admonitio de atlante Scotico*. Vol. 2(1). *Scotia illustrata sive prodromus historiae naturalis*. Edinburgh. [Lichens p. 35.]

2179 **Sibthorp, J.** (1794) *Flora Oxoniensis*. Oxford. [Lichens pp. 315–335.]

2180 **Side, A. G.** (1971) Reports of field meetings held in 1970. 26th April—Dungeness. *Bull. Kent Fld Club* **16**: 18.

2181 —— (1972) Reports of field meetings held in 1971. 9th May—'wall tour' from Maidstone. *Bull. Kent Fld Club* **17**: 18–19.

2182 —— (1973) Reports of field meetings held in 1972. 7th May—'wall tour' at Sevenoaks. *Bull. Kent Fld Club* **18**: 21–23.

2183 —— (1974) Report of annual exhibition. *Bull. Kent Fld Club* **19**: 33–34.

2184 —— **and Brightman, F. H.** (1965) Notes on the bryophytes and lichens of Ruxley Gravel Pit. *Trans. Kent Fld Club* **3**: 69–70.

2185 **Sim, J.** (1868) *On the Botany of Scotston Moor and its Neighbourhood. A paper read to the Aberdeen Natural History Society by John Sim, January 16, 1866.* Aberdeen: Aberdeen Natural History Society. [Lichens pp. 26–27.]

2186 **Simmons, I. G.** (1963) The blanket bog of Dartmoor. *Rep. Trans. Devon. Ass. Advmt Sci.* **95**: 180–196.

2187 **Simpson, F. R.** [1872] Address delivered to the Berwickshire Naturalists' Club, at Berwick, September 26th, 1872. *Hist. Berwicksh. Nat. Club* **6**: 288–316.

2188 **Sinker, C. A.** (1960) The vegetation of the Malham Tarn area. *Proc. Leeds phil. lit. Soc., sci. sect.* **8**: 139–175.

2189 —— (1962) The North Shropshire meres and mosses: A background for ecologists. *Fld Stud.* **1** (4): 101–138.

2190 **Sipman, H. J. M.** (1973) The *Cladonia pyxidata-fimbriata* complex in the Netherlands, with description of a new variety. *Acta bot. neerl.* **22**: 490–502.

Skepper, E.—See No. 981.

2191 **Skinner, J. F.** (1974) *Algae, lichens, and weathering on sandstone walls in Durham.* Duplicated report; copy with BLS.

2192 —— (1976) Lichen records and check-list for south-east Essex. *S. Essex Nat.* **1975**: 44–46.

2193 ——, ed. (1976) *Wildlife in South East Essex.* Southend-on-Sea: Southend-on-Sea Museums Service.

—— —See also Nos. 347 and 971.

2194 **Slack, A. and Dickson, J. H.** (1959) A further note on the limestone flora of Ben Sgulaird. *Glasg. Nat.* **18**: 106–108.

Sleigh, L. M.—See No. 813.

2195 **Slingsby, D. R.** (1969) *Aspects of nickel in lichens.* B.Sc. thesis, University of Bristol.

—— —See also No. 335.

2196 **Smith, A.** (1905) Cryptogams in the Grimsby district. *Naturalist, Hull* **1905**: 83–84.

2197 **Smith, A. L.** (19—) [*Manuscripts.*] 2 boxes in BM. [Notebooks and some letters; not abstracted.]

2198 —— (1906) British Coenogoniaceae. *J. Bot., Lond.* **44**: 266–268.

2199 —— (1907) Gall formation in *Ramalina. J. Bot., Lond.* **45**: 344–345.

2200 —— (1907) New localities of rare lichens. *J. Bot., Lond.* **45**: 345.

2201 —— (1910) Fungal parasites of lichens. *Trans. Br. mycol. Soc.* **3**: 174–178.

2202 —— (1910) New or rare microfungi. *Trans. Br. mycol. Soc.* **3**: 281–284.

2203 —— (1911) New lichens. *J. Bot., Lond.* **49**: 41–44.

2204 —— (1911) *A Monograph of the British Lichens.* Vol. 2. London: British

152

Museum (Natural History). [See No. 515 for Vol. 1.]

2205 —— (1918) *A Monograph of the British Lichens*. Vol. 1. Ed. 2. London: British Museum (Natural History).

2206 —— (1919) Lichens of the Baslow foray. *Trans. Br. mycol. Soc.* **6**: 252.

2207 —— (1921) *A Handbook of British Lichens*. London: British Museum (Natural History).

2208 —— (1921) *Lichens*. Cambridge: Cambridge University Press. [Reprint (1975) Richmond, Surrey: Richmond Publishing.]

2209 —— (1922) History of lichens in the British Isles. *SEast Nat.* **1922**: 19–35.

2210 —— (1926) *A Monograph of the British Lichens*. Vol. 2. Ed. 2. London: British Museum (Natural History).

2211 —— **and Rea, C.** (1904) Fungi new to Britain. *Trans. Br. mycol. Soc.* **2**: 59–67.

2212 **Smith, C.** (1845) Parish of Inverary. In *The New Statistical Account of Scotland*. Vol. 7. *Argyle*: 1–44. Edinburgh and London: W. Blackwood & Sons.

2213 **Smith, D. C.** (1961) The physiology of *Peltigera polydactyla* (Neck.) Hoffm. *Lichenologist* **1**: 209–226.

2214 —— **and Molesworth, S.** (1973) Lichen physiology. XIII. Effects of rewetting dry lichens. *New Phytol.* **72**: 525–533.

—— —See also Nos. 215, 827, 1019, 1954, 1956, 1957 and 1958.

2215 **Smith, G.** (1952) The Hereford foray. *Trans. Br. mycol. Soc.* **35**: 168–175.

2216 **Smith, J. A. and Gilchrist, J.** (1860) Notice of plants collected in the Isle of Skye. *Trans. bot. Soc. Edinb.* **6**: 44.

2217 **Smith, J. E.** (*c.* 1780–1828) *Collection of scientific and general correspondence*. Over 3,000 letters in LINN in 26 volumes. [Index by W. R. Dawson (1934) *The Smith Papers*. London.] [Not abstracted.]

2218 —— (1794) Remarks on the Abbé Wulfen's descriptions of lichens; published among his rare plants of Carniola, in Professor Jacquin's Collectanea, vol. II: 112. *Trans. Linn. Soc. Lond.* **2**: 10–14.

2219 [——] (1800) *Notes made at Yarmouth 1800*. Manuscript in LINN (Smith MSS no. 36).

2220 —— **and Sowerby, J.** (1790–1814, 1831–66). *English Botany*. 36 vols. and *Supplement* (W. J. Hooker, ed.) 5 vols. London: J. Sowerby. [Lichen plates distributed through the volumes 1792–1833; a complete set of these, bound in 3 vols., is in the BLS library; text by J. E. Smith, W. J. Hooker, W. Borrer, etc.]

2221 —— —— (1843–44) *English Botany*. Ed. 2 [arranged by C. Johnson]. Vols. 10–11. London: C. E. Sowerby.

—— —See also Nos. 36, 1059, 1060, 2098, 2223 and 2383.

2222 **Smith, J. E.** (1968) '*Torrey Canyon*', *Pollution and Marine Life*. Cambridge University Press.

2223 **Smith, [Lady]** (1832) *Memoir and Correspondence of the late Sir James Edward Smith, M.D.* 2 vols. London: Longman, etc.

2224 **Smith, R.** (1845) Parish of Lochwinnoch. In *The New Statistical Account of Scotland*. Vol. 7. *Renfrew*: 74–112. Edinburgh and London: W. Blackwood & Sons.

2225 **Smith, R.** (1898) Plant associations of the Tay basin. *Trans. Proc. Perthsh. Soc. nat. Sci.* **2**: 200–217.

2226 —— **and Smith, W. G.** (1905) Botanical survey of Scotland. III and IV. Forfar and Fife. *Scot. geogr. Mag.* **21**: 57–83.

Smith, S. T.—See No. 2351.

2227 **Smith, W.** (1916) Our lichen flora. *Trans. Aberd. wkg Men's nat. Hist. scient. Soc.* **3**: 252–253.

2228 **Smith, W. G.** (1878) A flint flake and its story. *Gdnrs' Chron.* **9**: 566–567.

2229 —— (1912) *Anthelia*: an arctic-alpine plant association. *Scot. bot. Rev.* **1**: 81–89.

2230 —— (1913) *Anthelia*: an arctic-alpine plant association. *Trans. Proc. bot. Soc. Edinb.* **26**: 36–44.

—— —See also No. 2226.

2230a **Smith, W. H.** (1885) *Walks in Weardale.* Ed. 2. Durham: Willan and Smith.

Smithson, S.—See No. 11a.

2231 **Smyth, E. S.** (1934) A contribution to the physiology and ecology of *Peltigera canina* and *P. polydactyla. Ann. Bot.* **48**: 781–818.

2232 **Snooke, W. D.** (1923) *Flora vectiana.* London.

Sommerville, A. H.—See No. 1863.

2233 **Soppitt, H. T.** (1881) Bradford Naturalists' Society. *Naturalist, Hull* **7**: 37–38.

2234 —— (1881) Bradford Naturalists' Society. *Naturalist, Hull* **7**: 68.

2235 —— **and West, W.** (1885) Plants of the Bradford district. *Naturalist, Hull* **10**: 178.

2236 —— —— (1886) Bradford Naturalists' Society – A year's botanical work. *Naturalist, Hull* **1886**: 60.

2237 **Southall, M.** [1825] *A Description of Malvern.* Ed. 2. Stourport: G. Nicholson. [Lichens pp. 215–216.]

2238 **Southward, A. J. and Orton, J. H.** (1954) The effects of wave-action on the distribution and numbers of the commoner plants and animals living on the Plymouth breakwater. *J. mar. biol. Ass. U.K.* **33**: 1–19.

Sowerby, J.—See Nos. 36, 467, 469, 471, 585, 2220, 2221, 2382 and 2392.

2239 **Sowter, F. A.** [*c.* 1941–72] [Card index and notebooks on Leicestershire lichens.] Manuscripts in LSR.

2240 —— (1945) *Thelidium aethioboloides* Zschacke. A lichen new to the British Isles. *NWest. Nat.* **20**: 73–74.

2241 —— (1945) Cumberland and Westmorland lichens. *NWest. Nat.* **20**: 74–75.

2242 —— (1947) *Lecidea botryoides* Zahl.—a lichen new to Britain. *NWest. Nat.* **22**: 283–284.

2243 —— (1950) *The Cryptogamic Flora of Leicestershire and Rutland. Lichenes.* Leicester: Leicester Literary and Philosophical Society. [Copy annotated by F. A. Sowter is now in LSR.]

2244 —— (1971) Mites (Acari) and lichens. *Lichenologist* **5**: 176.

2245 —— (1972) Leicestershire and Rutland cryptogamic notes, 2. *Trans. Leicester lit. phil. Soc.* **66**: 21–25.

2246 —— **and Hawksworth, D. L.** (1970) Leicestershire and Rutland cryptogamic notes, I. *Trans. Leicester lit. phil. Soc.* **64**: 89–100.

—— —See also No. 972.

2247 **Sparling, J. H.** (1962) *The autecology of* Schoenus nigricans *L.* Ph.D. thesis, University of London.

2248 **Speight, H. [J. Gray, pseud.]** (1892) *The Craven and North-West Yorkshire Highlands.* London: Elliot Stock. [Lichens p. 415.]

—— —See also Nos. 824 and 1339.

2249 **Spence, D. H. N.** (1974) Subarctic debris and scrub vegetation of Shetland. In *The Natural Environment of Shetland* (R. Goodier, ed.): 73–88. Edinburgh: Nature Conservancy Council.

2250 **Spooner, G. M. and Russell, F. S.,** eds. (1933) *Worth's Dartmoor.* Plymouth.

2251 **Spruce, R.** (1842) List of mosses &c. collected in Wharfedale, Yorkshire. *Phytologist* **1**: 197.

2252 **Stackhouse, J.** (1809) Tentamen marino-cryptogamicum. *Mém. Soc. Nat. Moscou* **2**: 50–97.

Stainforth, T.—See No. 1572.

2253 **Stanley, H.** (1907) *Sticta pulmonacea,* Ach., near Netherfield. *Hastings E. Suss. Nat.* **1**(2): 77.

2254 **Stansfield, A.** (1868) Observations on the botany of the Forest of Rossendale. In *History of the Forest of Rossendale* (T. Newbigging): 285–299. London: Simpkin Marshall.

2255 ——, ed. (1910) The flora of Todmorden. Order IV. – Lichenes. *Lancs. Nat.* **2**: 357–360.

2256 —— **and Nowell, J.** [1911] *Flora of Todmorden.* Manchester. [Lichens pp. 53–56.]

Steel, R. W.—See No. 2463.

Steele, R. C.—See Nos. 1317 and 2362.

2257 **Steers, D. A. and Jensen, H. A. P.** (1953) Winterton Ness. *Trans. Norfolk Norwich Nat. Soc.* **17**: 259–274.

Steers, J. A.—See Nos. 584 and 1154.

2258 **Stenhouse, J.** (1849) Examination of the proximate principles of some of the lichens. *Phil. Trans. R. Soc.* **138**: 63–89.

2259 —— (1870) Note on certain lichens. *Proc. R. Soc.* **18**: 222–227.

2260 **Stephenson, T. and Brokenshire, F. A.** (1946) Thirty-eighth report on the botany of Devon. *Rep. Trans. Devon. Ass. Advmt. Sci.* **78**: 51–64.

2261 —— —— (1947) Thirty-ninth report on the botany of Devon. *Rep. Trans. Devon. Ass. Advmt Sci.* **79**: 35–41.

2262 **Steven, H. M. and Carlisle, A.** (1959) *The Native Pine Woods of Scotland.* Edinburgh and London: Oliver and Boyd.

Stevens, J. H.—See No. 1996.

2263 **Stevenson, J.** (1879) *Mycologia Scotica.* Edinburgh: Cryptogamic Society of Scotland.

2264 **Stewart, E. J. A.** (1928) Lichens. In *Glasgow Sketches by Various Authors* (J. G. Kerr, ed.): 318–321. Glasgow: University Press.

2265 —— **and Patton, D.** (1924) Additional notes on the flora of the Culbin Sands. *Trans. Proc. bot. Soc. Edinb.* **29**: 27–40.

Stewart, W.—See No. 1917.

Stewart, W. D. P.—See No. 1026.

2266 **Stirton, J.** (1873) Additions to the lichen flora of Great Britain. *Grevillea* **2**: 71.

2257 —— (1874) On *Solorina bispora. Grevillea* **2**: 106–108.

2268 —— (1874) Lichen from Ben Lawers. *Grevillea* **3**: 24–25.

2269 —— (1874) New British lichens. *Grevillea* **3**: 33–37.

2270 —— (1874) New British lichen. *Grevillea* **3**: 79.

2271 —— (1874) 23rd September, 1873. *Rep. Trans. Glasgow Soc. Fld Nat.* **2**: 37–38.

2272 —— (1875) *Parmelia lillaniana* (a rejoinder). *Grevillea* **3**: 173–175.

2273 —— (1875) 25th August, 1874. Excursions. Ben Lawers. *Rep. Trans. Glasgow Soc. Fld Nat.* **3**: 70–71.

2274 —— (1876) On the cryptogamic botany of the West of Scotland. In *Notes on the Fauna and Flora of the West of Scotland*: xxv-xxxi. Glasgow: Blackie & Son.

2275 —— (1876) Lichenes. In *A Contribution Towards a Complete List of the Fauna and Flora of Clydesdale and the West of Scotland*: 99–108. Glasgow: The Glasgow Society of Field Naturalists.

2276 —— (1876) Lichens, British and foreign. *Rep. Trans. Glasgow Soc. Fld Nat.* **4**: 85–95.

2277 —— (1876) 6th July, 1875. Paper read. 'Cosmopolitan lichens.' *Rep. Trans. Glasgow Soc. Fld Nat.* **4**: 102.

2278 —— (1877) 30th May, 1876. Papers read. *Rep. Trans. Glasgow Soc. Fld Nat.* **5**: 176.

2279 —— (1877) Additions to the lichen flora of South Africa. *Rep. Trans. Glasgow Soc. Fld Nat.* **5**: 211–220.

2280 —— (1877) New or rare lichens. *Scott. Nat.* **4**: 27–29.

2281 —— (1878) On certain lichens belonging to the genus *Parmelia. Scott. Nat.* **4**: 298–299.

2282 —— (1878) A new Scottish lichen. *Scott. Nat.* **4**: 300.

2283 —— (1879) Descriptions of new Scottish lichens. *Scott. Nat.* **5**: 16–17.

2284 —— (1879) New and rare Scottish lichens. *Scott. Nat.* **5**: 217–221.

2285 —— (1881) On the genus *Usnea* and another (*Eumitria*) allied to it. *Scott. Nat.* **6**: 99–109.

2286 —— (1882–83) Notes on the genus *Usnea* with descriptions of new species. *Scott. Nat.* **6**: 292–297; **7** [*n.s.* **1**]: 74–79.

2287 —— (1883) On lichens (1) from Newfoundland, collected by Mr A. Gray, with a list of the species; (2) from New Zealand; (3) from the south of Scotland. *Trans. bot. Soc. Edinb.* **14**: 355–362.

2288 [——] (1885) [Kelso lichens]. *Scott. Nat.* **8** [*n.s.* **2**]: 35.

2289 —— (1885) Notes on British *Cladoniae. Scott. Nat.* **8** [*n.s.* **2**]: 118–122.

2290 —— (1887) A curious lichen from Ben Lawers. *Scott. Nat.* **9** [*n.s.* **3**]: 37–39.

2291 —— (1888) Lichens. *Scott. Nat.* **9** [*n.s.* **3**]: 307–309.

2292 —— (1899) Lichens and mosses from Carsaig, Argyle. *Ann. Scot. nat. Hist.* **1899**: 41–45.

—— —See also Nos. 48, 50 and 51.

2293 **Stizenberger, E.** (1863) Kritische Bemerkungen über die Lecideaceen mit nadelförmigen Sporen. *Nova Acta Acad. Caesar. Leop. Carol.* **30**(3): 1–76.

2294 —— (1865) Ueber die steinbewohnenden *Opegrapha*–Arten. *Nova Acta Acad. Caesar. Leop. Carol.* **32**(4): 1–36.

2295 —— (1867) *Lecidea sabuletorum* Flörke und die ihr verwandten Flechten-Arten. *Nova Acta Acad. Caesar. Leop. Carol.* **34**(2): 1–84.

2296 —— (1892) Die Alectorienarten und ihre geographische Verbreitung. *Annln naturh. Mus. Wien* **7**: 117–134.

Stott, P. A.—See Nos. 167 and 1160.

2297 **Stow, S. C.** (1900) A list of flowering plants, Filices, Equisetaceae, mosses, hepatics, and lichens, noted or taken at Woodhall Spa, Lincolnshire, July, 1899. *Naturalist, Hull* **1900**: 241–245.

Stubbs, F. B.—See No. 347.

2298 **Sugden, D. E.** (1965) *Aspects of the glaciation of the Cairngorm Mountains.* D.Phil. thesis, University of Oxford.

Sulzer, M.—See No. 1847.

2299 **Summerfield, R. J.** (1974) Biological Flora of the British Isles: *Narthecium ossifragum* (L.) Huds. *J. Ecol.* **62**: 325–339.

2300 **Summerhayes, E. S., Cole, L. W. and Williams, P. H.** (1924) Studies on the ecology of English heaths I. The vegetation of the unfelled portions of Oxshott Heath and Esher Common, Surrey. *J. Ecol.* **12**: 287–306.

2301 —— —— —— (1926) Studies on the ecology of English heaths II. Early stages in the recolonisation of felled pinewood at Oxshott Heath and Esher Common, Surrey. *J. Ecol.* **14**: 203–243.

Swinnerton, H. H.—See No. 658.

2302 **Swinscow, T. D. V.** (1958) An arctic-alpine lichen new to England. *Lichenologist* **1**: 29–30.

2303 —— (1959) Field meeting at Chagford. *Lichenologist* **1**: 115–118.

2304 —— (1960) Pyrenocarpous lichens: 1. *Lichenologist* **1**: 169–178.

2305 —— (1960) *Cavernularia hultenii* Degelius in Scotland. *Lichenologist* **1**: 179–183.

2306 —— (1961) Pyrenocarpous lichens: 2. *Gongylia* Körb. in the British Isles with first British record of *G. incarnata. Lichenologist* **1**: 242–250.

2307 —— (1962) Pyrenocarpous lichens: 3. The genus *Porina* in the British Isles. *Lichenologist* **2**: 6–56.

2308 —— (1962) An unusual parasymbiont of marine lichens. *Nature, Lond.* **194**: 500–501.

2309 —— (1963) Pyrenocarpous lichens: 4. Guide to the British species of *Staurothele. Lichenologist* **2**: 152–171.

2310 —— (1963) Pyrenocarpous lichens: 5. Fruiting *Normandina pulchella* (Borr.) Nyl. A further *Porina* species. *Lichenologist* **2**: 167–171.

2311 —— (1964) The classification of lichens. *Advmt Sci., Lond.* **21**: 241–244.

2312 —— (1964) Pyrenocarpous lichens: 6. The genus *Thrombium* in the British Isles. Species of *Belonia* Körb. in Britain. *Lichenologist* **2**: 276–283.

2313 —— (1965) Pyrenocarpous lichens: 8. The marine species of *Arthopyrenia* in the British Isles. *Lichenologist* **3**: 55–64.

2314 —— (1965) Pyrenocarpous lichens: 9. Notes on various species. *Lichenologist* **3**: 72–83.

2315 —— (1966) Pyrenocarpous lichens: 10. *Polyblastia quartzina* Lynge new to British Isles. *Microthelia marmorata* (Kremp.) Hepp in Körb. *Lichenologist* **3**: 233–235.

2316 —— (1967) Pyrenocarpous lichens: 11. A new species of *Arthopyrenia*. *Lichenologist* **3**: 415–417.

2317 —— (1967) Pyrenocarpous lichens: 12. The genus *Geisleria* Nitschke. *Lichenologist* **3**: 418–422.

2318 —— (1968) Pyrenocarpous lichens: 13. Fresh-water species of *Verrucaria* in the British Isles. *Lichenologist* **4**: 34–54.

2319 —— (1968) Almost immortal. *Scots Mag., n.s.* **88**: 559–561.

2320 —— (1968) Tenth anniversary. *Bull. Br. Lichen Soc.* **22**: 1–2.

2321 —— (1970) Pyrenocarpous lichens: 14. *Arthopyrenia* Massal. sect. *Acrocordia* (Massal.) Müll. Arg. in the British Isles. *Lichenologist* **4**: 218–233.

2322 —— (1971) Pyrenocarpous lichens: 15. Key to *Polyblastia* Massal. in the British Isles. *Lichenologist* **5**: 92–113.

2323 —— **and Krog, H.** (1975) The genus *Dermatocarpon* in East Africa with an overlooked species in Britain. *Lichenologist* **7**: 148–154.
—— —See also No. 1626.

2324 **Syers, J. K.** (1964) *A study of soil formation on Carboniferous limestone with particular reference to lichens as pedogenic agents.* Ph.D. thesis, University of Durham.

2325 —— **Birnie, A. C. and Mitchell, B. D.** (1967) The calcium oxalate content of some lichens growing on limestone. *Lichenologist* **3**: 409–414.

Sykes, J. M.—See No. 1067.

2326 **Symington, R.** (1845) Parish of Muirkirk. In *The New Statistical Account of Scotland*. Vol. 5. *Ayr*: 147–158. Edinburgh and London: W. Blackwood & Sons.

Szweykowskiego, J.—See No. 2363a.

2327 **Tallis, J. H.** (1957) *A study of the biology and ecology of* Rhacomitrium lanuginosum *Brid.* Ph.D. thesis, University of Wales.

2328 —— (1958) Studies in the biology and ecology of *Rhacomitrium lanuginosum* Brid. *J. Ecol.* **46**: 271–288.

2329 —— (1958) The British species of the genus *Cladonia*. *Lichenologist* **1**: 3–20.

2330 —— (1959) The British species of the genus *Usnea*. *Lichenologist* **1**: 49–83.

2331 —— (1959) *Usnea pendulina* Mot. new to Britain. *Lichenologist* **1**: 86–87.

2332 —— (1964) Lichens and atmospheric pollution. *Advmt Sci., Lond.* **21**: 250–252.

2333 —— (1969) The blanket bog vegetation of the Berwyn Mountains, North Wales. *J. Ecol.* **57**: 765–787.
—— —See also No. 216.

2334 **Tansley, A. G.,** ed. (1911) *Types of British Vegetation.* Cambridge: Cambridge University Press.

2335 —— (1939) *The British Islands and their Vegetation.* Cambridge: Cambridge

University Press. [Re-issued (1949) in 2 vols.]

2336 —— (1949) *Britain's Green Mantle*. London: Allen & Unwin.

2337 —— (1968) *Britain's Green Mantle*. Ed. 2 [Revised M. C. F. Proctor.]. London: Allen & Unwin.

2337a **Tapper, R.** (1976) Dispersal and changes in the local distribution of *Evernia prunastri* and *Ramalina farinacea*. *New Phytol*. **77**: 725–734.

2338 **Tavares, C. N.** (1965) The genus *Pannaria* in Portugal. *Port. Acta biol*. **8**: 1–16.

Taylor, A. M.—See No. 168.

Taylor, I. F.—See No. 2569.

2339 **Taylor, K.** (1971) Biological Flora of the British Isles: *Rubus chamaemorus* L. *J. Ecol*. **59**: 293–306.

2340 **Teesdale, R.** (1794) Plantae Eboracenses; or, a catalogue of the more rare plants which grow wild in the neighbourhood of Castle Howard, in the North Riding of Yorkshire, disposed according to the Linnean system. *Trans. Linn. Soc. Lond*. **2**: 103–125.

2341 —— (1800) A supplement to the Plantae Eboracenses printed in the second volume of these Transactions. *Trans. Linn. Soc. Lond*. **5**: 36–95.

2342 **Tellam, R. V.** (1885) The lichens of East Cornwall. *Trans. Penzance nat. Hist. antiq. Soc*. **2**: 73–81.

2343 [——] (1888) Lichens. *Trans. Penzance nat. Hist. antiq. Soc*. **2**: 379.

2344 **Templeton, J.** (17—) [*Collection of* 300 *drawings of Irish cryptogams*.] Manuscript in BM. [With text; about 57 lichens; not abstracted.]

2345 **Theakston, S. W.** (1840) *Theakston's Guide to Scarborough*. Scarborough: S. W. Theakston. [Lichen p. 99.] [And later editions to at least Ed. 5.]

2346 **Thomas, A.,** ed. (1972) B.M.S. day forays 1971. *Bull. Br. mycol. Soc*. **6**: 9–10.

2347 **Thomas, C. F.** (1939) *A version in lichens*. B.Sc. thesis, University of Bristol.

2348 **Thomas, M.** (1900) On the alpine flora of Clova. Part 1. *Trans. Proc. Perthsh. Soc. nat. Sci*. **3**: 60–69.

2349 **Thomas, R. N.** (1930) Flora of paper-mill lime waste dumps near Glasgow. *J. Ecol*. **18**: 333–351.

2350 **Thompson, C.** (1845) Parish of Wick. In *The New Statistical Account of Scotland*. Vol. 15. *Caithness*: 117–178. Edinburgh and London: W. Blackwood & Sons.

Thompson, P. G.—See Nos. 1774, 1775 and 1776.

2351 **Thompson, T. E., Smith, S. T., Jenkins, M., Benson-Evans, K., Fisk, D., Morgan, G., Delhanty, J. E. and Wade, A. E.** (1966) Contributions to the biology of the Inner Farne. *Trans. nat. Hist. Soc. Northumb., n.s.* **15**: 197–225. [Lichens by A. E. Wade.]

2352 **Thompson, V.** (1807) *Catalogue of Plants Growing in the vicinity of Berwick upon Tweed*. London: J. White. [Lichens pp. 109–111.]

2353 **Thomson, J. W.** (1955) *Peltigera pulverulenta* (Tayl.) Nyl. takes precedence over *Peltigera scabrosa* T. Fr. and becomes of considerable phytogeographic interest. *Bryologist* **58**: 45–49.

2354 **Thomson R.** (1900) *The Natural History of a Highland Parish* (*Ardclach, Nairnshire*). Nairn: G. Bain. [Lichens p. 267.]

2355 **Thomson, R. D.** (1844) On parietin, a yellow colouring matter, and on the inorganic food of lichens. *Edinb. New phil. J.* **37**: 187–198.

2356 —— (1844) On parietin, a yellow colouring matter, and on the inorganic food of lichens. *Proc. phil. Soc. Glasg.* **1**: 182–191.

Thornton, I. W. B.—See No. 325.

2357 **Thorp, T. K.** (1967) *An investigation into the epiphytic flora on* Corylus avellana L. *in relation to the occurrence of die-back, in the Burren, Co. Clare.* B.A. thesis, Trinity College, Dublin. [Not seen.]

2358 **Thurstan, V.** [1930] *The Use of Vegetable Dyes for Beginners.* Leicester: Dryad Press. [And later editions to at least Ed. 10 (1967).] [Lichens pp. 26–28.]

2359 **Thwaites, G. H. K.** (1849) Note on *Cystocoleus*, a new genus of minute plants. *Ann. Mag. nat. Hist., ser.* 2, **3**: 241–242.

2360 **Tibell, L.** (1971) The genus *Cyphelium* in Europe. *Svensk bot. Tidskr.* **65**: 138–164.

2361 —— (1975) The Caliciales of boreal North America. *Symb. bot. upsal.* **21**(2): 1–128.

2362 **Tittensor, R. M. and Steele, R. C.** (1971) Plant communities of the Loch Lomond oakwoods. *J. Ecol.* **59**: 561–582.

2363 **Tittley, I.** (1971) Reports of field meetings held in 1970. 11th October—Birchington-on-Sea. *Bull. Kent Fld Club* **16**: 34–36.

2363a **Tobolewski, Z.** [**and Kupczyk, B.**] (1971–76) Porosty (Lichenes). In *Atlas Rozmieszczenia Roślin Zarodnikowych w Polsce* (J. Szweykowskiego and T. Wojterskiego, eds.). Vol. 3 (1–3). Poznań: Polska Akademia Nauk.

2364 **Tonkin, J. C. and Tonkin, R. W.** (1887) *Guide to the Isles of Scilly.* Ed. 2. Penzance. [Lichens p. 105.]

Tonkin, R. W.—See No. 2364.

2365 **Townsend, C. C.** (1959) List of lichens around Cheltenham. *Jl N. Gloucs. Nat. Soc.* **10**: 2.

2366 **Trail, J. W. H.** (1887) On the influence of cryptogams on mankind. *Scott. Nat.* **9** [*n.s.* **3**]: 66–77.

2367 **Travis, W.** (1815) Natural productions. In *The Scarborough Guide*: 111–118. Scarborough: Thomas Coultas.

2368 **Travis, W. G.** (1910) *Lecanora atrynea* Nyl. in South Lancashire. *Lancs. Nat.* **3**: 71.

2369 —— (1913) A contribution to the flora of Arran. *Trans. Proc. bot. Soc. Edinb.* **26**: 120–129.

2370 —— (1914) The plant-associations of some South Lancashire peat-mosses. *Lancs. Chesh. Nat.* **7**: 171–176.

2371 —— (1914) Bryological notes in the Ingleton district. *Lancs. Chesh. Nat.* **7**: 323–326.

2372 —— (1915) Cheshire lichens. *J. Bot., Lond.* **53**: 219.

2373 —— (1917) Anglesea lichens. *J. Bot., Lond.* **55**: 54–55.

2374 —— (1921) The lichens of the Wirral. *Lancs. Chesh. Nat.* **14**: 177–190.

2375 —— (1925) Additions to the lichen-flora of Wirral. *Lancs. Chesh. Nat.* **17**: 152–154.

2376 —— (1947) A new British lichen: *Polyblastia wheldoni* sp. nov. *NWest. Nat.* **22**: 240–241.

—— —See also Nos. 2609, 2610, 2611, 2612 and 2613.

Trimen, H.—See No. 456.

Trotet, G.—See No. 1093.

2377 **Tugwell, G.**, ed. [1857] *The North Devon Hand Book.* London: Simpkin Marshall. [Lichens pp. 225–226.] [And later editions to at least Ed. 4 (1877) with changed pagination (see Nos. 44 and 1918).]

2378 **Turk, F. A. and Turk, S. M.** (1976) *A Handbook to the Natural History of the Lizard Peninsula.* Exeter: University of Exeter Department of Extra-Mural Studies. [Lichens p. 48.]

2379 **Turk, S. M.** (1971) Marine life of the Fowey foreshore. *J. Camborne-Redruth nat. Hist. Soc.* **2**(1): 6–20.

2380 —— (1972) A survey of the 'Island', St. Ives. *J. Camborne-Redruth nat. Hist. Soc.* **2**(2): 4–33.

2381 —— (1974) Marine life of the Marazion area. *J. Camborne-Redruth nat. Hist. Soc.* **3**(1): 17–38.

—— —See also No. 2378.

2382 **Turner, D.** [1799] *Journal of the first ten days of a tour made by Dawson Turner Esq. in company with James Sowerby Esq., author of British Fungi, to Lands End in the months of June and July* 1799. Manuscript in NWH.

2383 —— (1800–06) *Catalogue of British lichens.* Manuscript in LINN (Smith MSS no. 37.1). [Copied by J. E. Smith from a manuscript of Turner's.]

2384 —— (1800–10) *Botanical Memoranda.* 2 vols. Manuscripts in K. [Not abstracted.]

2385 —— (1800–1829) [*Correspondence*]. Letters 1809–29 in BM; other letters in K (5 vols.) and Trinity College, Cambridge (82 vols.). [Not abstracted.]

2386 —— (1804) Descriptions of four new British lichens. *Trans. Linn. Soc. Lond.* **7**: 86–95.

2387 —— (1804) Remarks upon the Dillenian herbarium. *Trans. Linn. Soc. Lond.* **7**: 101–115.

2388 —— (1807) Description of a new species of lichen. *Trans. Linn. Soc. Lond.* **8**: 260–261.

2389 —— (1808) Descriptions of eight new British lichens. *Trans. Linn. Soc. Lond.* **9**: 135–150.

2389a —— (1835) *Extracts from the Literary and Scientific Correspondence of Richard Richardson.* Yarmouth: privately printed.

2390 —— **and Borrer, W.** (1839) *Specimen of a Lichenographia Britannica.* Yarmouth: "for private circulation". [Pp. 1–208 printed by *c.* 1812 and widely circulated (see Tibell, 1975); pp. 209–240, the title page, Preface and Index printed and labelled "Supplement" in 1839.]

2391 —— **and Dillwyn, L. W.** (1805) *The Botanist's Guide through England and Wales.* 2 vols. London: Phillips & Fardon.

2392 —— **and Sowerby, J.** (1800) Catalogue of some of the more rare plants observed in a tour through the western counties of England, made in

June 1799. *Trans. Linn. Soc. Lond.* **5**: 234–241.
—— —See also No. 614.
2393 **Turner, W.** (1568) *Herball, the thirde parte.* Collen. [Lichen p. 56.]
2394 **Turrill, W. B.** (1948) *British Plant Life.* [New Naturalist No. 10.] London: Collins.
2395 **Turton, W. and Kingston, J. F.** [1830] The natural history of the district. In *The Teignmouth, Dawlish and Torquay Guide* (N.T. Carrington): sine pagin. Teignmouth: E. Croydon. [See also No. 763.]

2396 **Underwood, A. J.** (1973) Studies on zonation of intertidal prosobranch molluscs in the Plymouth region. *J. anim. Ecol.* **42**: 353–372.

2397 **Vainio, E. A.** (1887–97) Monographia Cladoniarum universalis. I-III. *Acta Soc. Fauna Flora fenn.* **4**: 1–510; **10**: 1–499; **14**: 1–268.
2398 ——(1921–34) Lichenographia Fennica I-IV. *Acta Soc. Fauna Flora fenn.* **49**(2): 1–274; **53**(1): 1–341; **57**(1): 1–138; **57**(2): 1–531.
Varenne, E. G.—See Nos. 1535 and 1763.
Venables, A.—See No. 252.
2399 **Verseghy, K.** (1962) Die Gattung *Ochrolechia. Beih. Nova Hedwigia* **1**: 1–146.
2400 **Vevers, H. G.** (1936) The land vegetation of Ailsa Craig. *J. Ecol.* **24**: 424–445.
2401 **Vězda, A.** (1959) K taxonomii, rozšířeni a ekologii lišejníku *Belonia russula* Kbr. *Přírodov. Čas. slezský* **20**: 241–253.
2402 —— (1965) Flechtensystematische Studien I. Die Gattung *Petractis* Fr. *Preslia* **37**: 127–143.
2403 —— (1966) Flechtensystematische Studien IV. Die Gattung *Gyalidea* Lett. *Folia geobot. phytotax. bohemoslavaca* **1**: 311–340.
2404 —— (1966) *Lichenes selecti exsiccati.* Fasc. XX. Prague: Instituto Botanico Academiae Scientiarum Čechoslovacae.
2405 —— (1967) Flechtensystematische Studien VI. Die Gattung *Sagiolechia* Massal. *Folia geobot. phytotax. bohemoslavaca* **2**: 383–396.
2406 —— (1967) *Lichenes selecti exsiccati.* Fasc. XXII. Prague: Instituto Botanico Academiae Scientiarum Čechoslovacae.
2407 —— (1967) Lichenes selecti exsiccati. Fasc. XXIII. Prague: Instituto Botanico Academiae Scientiarum Čechoslovacae.
2408 —— (1967) *Lichenes selecti exsiccati.* Fasc. XXVI. Prague: Instituto Botanico Academiae Scientiarum Čechoslovacae.
2409 —— (1968) Taxonomische Revision der Gattung *Thelopsis* Nyl. (Lichenisierte Fungi). *Folia geobot. phytotax. bohemoslovaca* **3**: 363–406.
2410 —— (1968) *Lichenes selecti exsiccati.* Fasc. XXVII. Prague: Instituto Botanico Academiae Scientiarum Čechoslovacae.
2411 —— (1969) Neue Taxa und Kombinationen in der Familie Gyalectaceae (Lichenisierte Fungi). *Folia geobot. phytotax. bohemoslavaca* **4**: 443–446.
2412 —— (1969) *Lichenes selecti exsiccati.* Fasc. XXX. Prague: Instituto Botanico Academiae Scientiarum Čechoslovacae.

162

2413 —— (1969) *Lichenes selecti exsiccati.* Fasc. XXXII. Prague: Instituto Botanico Academiae Scientiarum Čechoslovacae.

2414 —— (1971) *Lichenes selecti exsiccati.* Fasc. XXXIX. Prague: Instituto Botanico Academiae Scientiarum Čechoslovacae.

2415 —— (1971) *Lichenes selecti exsiccati.* Fasc. XL. Prague: Instituto Botanico Academiae Scientiarum Čechoslovacae.

2416 —— (1971) *Lichenes selecti exsiccati.* Fasc. XLI. Prague: Instituto Botanico Academiae Scientiarum Čechoslovacae.

2417 —— (1972) Flechtensystematische Studien VII. *Gyalideopsis,* eine neue Flechtengattung. *Folia geobot. phytotax. bohemoslovaca* **7**: 203–215.

2418 —— (1972) *Lichenes selecti exsiccati.* Fasc. XLII. Prague: Instituto Botanico Academiae Scientiarum Čechoslovacae.

2419 —— (1972) *Lichenes selecti exsiccati.* Fasc. XLIII. Prague: Instituto Botanico Academiae Scientiarum Čechoslovacae.

2420 —— (1973) *Lichenes selecti exsiccati.* Fasc. XLVII. Prague: Instituto Botanico Academiae Scientiarum Čechoslovacae.

2420a **Vick, C. M. and Bevan, R.** (1976) Lichens and tar spot fungus (*Rhytisma acerinum*) as indicators of sulphur dioxide pollution on Merseyside. *Envir. Poll.* **11**: 203–216.

2421 **Vickery, A. R.** (1975) The use of lichens in well-dressing. *Lichenologist* **7**: 178–179.

2422 **Vines, S. H. and Druce, G. C.** (1914) *An Account of the Morisonian Herbarium.* Oxford: Clarendon Press.
—— —See also No. 613.

2423 **Vize, J. E.** (1882) The parish of Forden. *Montgomeryshire Collections* **15**: 155–182. [Lichens pp. 177–178 by W. A. Leighton.]

2424 **Vouaux, L. l'Abbé** (1912–1914) Synopsis des champignons parasites de lichens. *Bull. Soc. mycol. Fr.* **28**: 177–256; **29**: 33–128, 399–494; **30**: 135–198, 281–329.

2425 **Wade, A. E.** [19—] *Register of the specimens in A. R. Horwood's herbarium, NMW accession no.* 23.94. Manuscript in NMW. [Not indexed.]

2426 —— (1919) The flora of Aylestone and Narborough bogs. *Trans. Leicester lit. phil. Soc.* **20**: 20–46.

2427 —— (1950) Botanical notes, 1947–48. *Trans. Cardiff Nat. Soc.* **79**: 52–54.

2428 —— (1952) The bryophytes and lichens of Clyne Common. *Proc. Swansea scient. Fld. Nat. Soc.* **2**: 338–342.

2429 —— (1953) *Parmelia laciniatula* (Flag.) Zahlbr. in Perthshire. *Ann. Mag. nat. Hist., ser.* 12, **6**: 879.

2430 —— (1953) Botanical notes, 1951–52. *Trans. Cardiff Nat. Soc.* **81**: 100–101.

2431 —— (1954) Lichens of Pembrokeshire. *NWest. Nat.* **25**: 242–254.

2432 —— (1956) Lichenological notes. I. *Trans. Br. mycol. Soc.* **39**: 416–422.

2433 —— (1956) Botanical notes, 1953–54. *Trans. Cardiff Nat. Soc.* **83**: 25–26.

2434 —— (1957–59) *A supplement to Watson's Census Catalogue of British Lichens.* Manuscripts and cards at BM.

2435 —— (1958) The British species of *Collema. Lichenologist* **1**: 21–29.

2436 —— (1958) *Cetraria islandica* (L.) Ach. var. *tenuifolia* (Retz) Wain.–C.

crispa (Ach.) Nyl. *Lichenologist* **1**: 40.

2437 —— (1958) Lichens of Carmarthenshire, South Wales. *Revue bryol. lichén.* **27**: 82–103.

2438 —— (1958) Glamorgan botanical notes. 1956. *Trans. Cardiff Nat. Soc.* **85**: 25–26.

2439 —— (1959) *Lepraria chlorina* Ach. in Britain. *Lichenologist* **1**: 86.

2440 —— (1959) The British species of *Alectoria*. *Lichenologist* **1**: 89–97.

2441 —— (1959) Glamorgan botanical notes. 1957. *Trans. Cardiff Nat. Soc.* **86**: 22–23.

2442 —— (1960) The British *Anaptychiae* and *Physciae*. *Lichenologist* **1**: 126–144.

2443 —— (1960) The lichens of Dale, Pembrokeshire. *Nature Wales* **6**: 49–55.

2444 —— (1960) Glamorgan botanical notes. 1958. *Trans. Cardiff Nat. Soc.* **87**: 27.

2445 —— (1961) The genus *Ramalina* in the British Isles. *Lichenologist* **1**: 226–241.

2446 —— (1961) Glamorgan botanical notes. 1959. *Trans. Cardiff Nat. Soc.* **88**: 21.

2447 —— (1962) Glamorgan botanical notes. 1960. *Trans. Cardiff Nat. Soc.* **89**: 31.

2448 —— (1965) The genus *Caloplaca* Th. Fr. in the British Isles. *Lichenologist* **3**: 1–28.

2449 —— (1967) Glamorgan botanical notes. 1963–66. *Trans. Cardiff Nat. Soc.* **93**: 47–48.

2450 —— **and Watson, W.** (1936) Lichens of Glamorgan. In *Glamorgan County History* (W. M. Tattersall, ed.) **1**: 183–190. Cardiff: W. Lewis.
—— —See also No. 2351.

2451 **Walker, A. K.** (1970) A discomycete parasitic on *Thamnolia vermicularis*. *Trans. Proc. bot. Soc. Edinb.* **41**: 59–60.
Walker, D.—See No. 1914.

2452 **Walker, M.** (1963) An amateur looks at north Cornish lichens. *J. Camborne-Redruth nat. Hist. Soc.* **1**(1): 1–3.

2453 **Wallace, J.** (1700) *An account of the Islands of Orkney.* London: J. Tonson. [Reprinted (1883) ed. J. Small.] [Lichens pp. 27–28.]

2454 **Wallace, N.** [1961–64] [*Notes made on field courses.*] Manuscript in BM.

2455 —— (1969) *The Lichens of Pilgrim Fort.* [Croydon Field Studies No. 1.] Croydon: London Borough of Croydon.

2456 —— [1970] *The Lichens of Pilgrim Fort.* Ed. 2. [Croydon Field Studies No. 1.] Croydon: London Borough of Croydon.

2457 —— (1972) Reports of field meetings held in 1971. 2nd October—Dungeness. *Bull. Kent Fld Club* **17**: 28–29.

2458 **Wallace, T. J.** (1953) The plant ecology of Dawlish Warren. Part I. *Rep. Trans. Devon. Ass. Advmt Sci.* **85**: 86–94.

2459 —— [1963] *The Axmouth-Lyme Regis Undercliffs National Nature Reserve.* Lyme Regis: Allhallows School.

2460 **Wallis, J.** (1769) *The Natural History and Antiquities of Northumberland: And of so much of the county of Durham as lies between the rivers Tyne and Tweed: and commonly called, North Bishoprick.* Vol. 1. London: privately printed.

Walpole, M.—See No. 451.

Walpole, P. R.—See No. 973.

Walsh, G. B.—See No. 1964.

Walters, A. H.—See No. 1441.

2461 Walters, W., ed. [1892] *Twiss and Sons' Illustrated Guide to Ilfracombe and North Devon*. Ilfracombe: Twiss and Sons. [Lichens p. 174.]

2462 [Walton, G. A., ed.] (1976) *An Tiaracht. Interim Report to the Royal Irish Academy on Expeditions to Inishtearaght 23rd to 29th July* 1973 *and 9th to 13th May* 1975. Manuscript. Copy in BM. [Lichens pp. 66–67, 100–101; latter section by D. L. Hawksworth.]

2463 Warburg, E. F. (1954) Vegetation and flora. In *The Oxford Region* (A. F. Martin and R. W. Steel, eds.): 56–62. Oxford: Oxford University Press.

2464 Ward, J. (1841) A notice of *Lecanora rubra*, found near Richmond, Yorkshire. *Proc. bot. Soc. Edinb.* **1840–41**: 50.

Ward, L. K.—See No. 2575.

Ward, S.—See No. 1279.

2465 Ward, S. D. (1968) *A study of the distribution and vegetational composition of* Calluna-Arctostaphylos *heaths in north-east Scotland and of related Scandinavian communities*. Ph.D. thesis, University of Aberdeen.

2466 —— (1970) The phytosociology of *Calluna-Arctostaphylos* heaths in Scotland and Scandinavia. I. Dinnet Moor, Aberdeenshire. *J. Ecol.* **58**: 847–863.

2467 —— (1971) The phytosociology of *Calluna-Arctostaphylos* heaths in Scotland and Scandinavia. II. The north-east Scottish heaths. *J. Ecol.* **59**: 679–696.

2468 —— (1971) The phytosociology of *Calluna-Arctostaphylos* heaths in Scotland and Scandinavia. III. A critical examination of the *Arctostaphyleto-Callunetum*, *J. Ecol.* **59**: 697–712.

2469 —— Jones, A. D. and Manton, M. (1972) The vegetation of Dartmoor. *Fld Stud.* **3**: 505–533.

Warden, A. J.—See No. 697.

2470 Warhurst, E. (1918) Liverpool Botanical Society. *Lancs. Chesh. Nat.* **10**: 342–343.

2471 Warner, R. (1795) *The History of the Isle of Wight*. Southampton: T. Cadell and W. Davies. [Lichens pp. 251–252.]

Waters, W. E.—See Nos. 1233 and 1234.

Waterston, A. R.—See No. 362.

Wathern, P.—See No. 801.

2472 Watling, R. [1967] *The Fungus and Lichen Flora of the Halifax Parish* (1775–1965). Halifax: Halifax Scientific Society. [Lichens pp. 37–45.]

2473 —— (1976) Scottish fungi. *Scottish Wildlife* **12**(2): 9–12.

2474 —— Irvine, L. M. and Norton, T. A. (1970) The marine algae of St. Kilda. *Trans. Proc. bot. Soc. Edinb.* **41**: 31–42.

2475 —— and Richardson, M. J. (1971) The agarics of St. Kilda. *Trans. Proc. bot. Soc. Edinb.* **41**: 165–187.

Watson, A.—See No. 1652.

2476 Watson, E. V. (1960) Further observations on the bryophyte flora of the Isle of May. II. Rate of succession in selected communities involving bryophytes. *Trans. Proc. bot. Soc. Edinb.* **39**: 85–106.

—— —See also No. 362.

2477 **Watson, J.** (1775) *The History and Antiquities of the Parish of Halifax, in Yorkshire.* London: T. Lowndes. [Lichens pp. 762–763, probably by J. H. Bolton.]

2478 **Watson, J. W.** (1854) Airy Holme Wood, as a locality for land shells. *Naturalist, o.s.* **4**: 228–230.

2479 **Watson, W.** (1909) The distribution of bryophytes in the woodlands of Somerset. *New Phytol.* **8**: 90–96.

2480 —— (1917) New, rare or critical lichens. *J. Bot., Lond.* **55**: 107–111, 204–210, 310–316.

2481 —— (1918) Cryptogamic vegetation of the sand-dunes of the west coast of England. *J. Ecol.* **6**: 126–143.

2482 —— (1918) The bryophytes and lichens of calcareous soil. *J. Ecol.* **6**: 189–198.

2483 —— (1919) The bryophytes and lichens of fresh water. *J. Ecol.* **7**: 71–83.

2484 —— (1920) Lichens of Llanberis and District. *J. Bot., Lond.* **58**: 108–110.

2485 —— (1922) The determination of lichens in the field. *J. Bot., Lond.* **60**, *Suppl.* **1**: 1–28.

2486 —— (1922) List of lichens, etc. from Chesil Beach. *J. Ecol.* **10**: 255–256.

2487 —— (1924) The lichens of the Bristol foray. *Trans. Br. mycol. Soc.* **8**: 133–135.

2488 —— (1925) Lichenological notes.—I. *J. Bot., Lond.* **63**: 130–132.

2489 —— (1925) The bryophytes and lichens of arctic-alpine vegetation. *J. Ecol.* **13**: 1–26.

2490 —— (1927) Lichenological notes.—II. *J. Bot., Lond.* **65**: 109–113.

2491 —— (1927) The lichens of Somerset, Part I. *Proc. Somerset archaeol. nat. Hist. Soc.* **73** (3): Appendix II.

2492 —— (1928) Lichenological notes—III. *J. Bot., Lond.* **66**: 17–21.

2493 —— (1928) The lichens of Somerset, Part II. *Proc. Somerset archaeol. nat. Hist. Soc.* **74**(3): Appendix II.

2494 —— (1929) Lichenological notes.—IV. *J. Bot., Lond.* **67**: 74–79.

2495 —— (1929) The lichens of Somerset, Part III (completion). *Proc. Somerset archaeol. nat. Hist. Soc.* **75**(3): Appendix II.

2496 —— (1930) Lichenological notes.—V. *J. Bot., Lond.* **68**: 265–270.

2497 —— (1930) *The Lichens of Somerset.* Taunton: Somersetshire Archaeological & Natural History Society. [Separate issue of Nos. 2491, 2493 and 2495.]

2498 —— (1932) Lichenological notes.—VI. *J. Bot., Lond.* **70**: 67–72, 96–100.

2499 —— (1932) The bryophytes and lichens of moorland. *J. Ecol.* **20**: 284–313.

2500 —— (1933) Lichenological notes.—VII. *J. Bot., Lond.* **71**: 314–318, 327–338.

2501 —— (1933) A list of Northamptonshire lichens contained in the herbaria of Robert Rogers, now deposited in the Museum, Public Library, Kettering. *J. Northampt. nat. Hist. Soc.* **27**: 24.

2502 —— (1935) Lichenological notes.—VIII. *J. Bot., Lond.* **73**: 149–160.

2503 —— (1935) Lichens of Ingleton and district. *Naturalist, Hull* **1935**: 235–238.

2504 [——] (1935) [In discussion on the origin and relationship of the British flora.] *Proc. R. Soc., B,* **118**: 229–230.

2505 —— (1935) Notes on lichens, mainly from Orkney, in the herbarium of the Royal Botanic Garden, Edinburgh. *Trans. Proc. bot. Soc. Edinb.* **31**: 505–520.

2506 —— (1936) The bryophytes and lichens of British woods. Part I. Beechwoods. *J. Ecol.* **24**: 139–161.

2507 —— (1936) The bryophytes and lichens of British woods. Part II. Other woodland types. *J. Ecol.* **24**: 446–478.

2508 —— (1937) Lichens of the Totnes district. *Trans. Br. mycol. Soc.* **21**: 12–15.

2509 —— (1939) Lichenological notes.—IX. *J. Bot., Lond.* **77**: 22–25, 33–44.

2510 —— (1939) Notes on lichens in the herbarium of the Royal Botanic Garden, Edinburgh. II. *Trans. Proc. bot. Soc. Edinb.* **32**: 502–515.

2511 —— (1940) The botanical section—Recorder's notes. *Proc. Somerset archaeol. nat. Hist. Soc.* **86**: 142–145.

2512 —— (1941) Yorkshire associations, lichenological or otherwise. *Naturalist, Hull* **1941**: 29–40.

2513 —— (1941) The botanical section—Recorder's notes. *Proc. Somerset archaeol. nat. Hist. Soc.* **87**: 120–126.

2514 —— (1942) Notes on British *Aspicilias* referred to in Magnussons' monograph on the *Aspicilia gibbosa* group. *J. Bot., Lond.* **80**: 74–76.

2515 —— (1942) Lichenological notes.—X. *J. Bot., Lond.* **80**: 137–149.

2516 —— (1942) The botanical section—Recorder's notes. *Proc. Somerset archaeol. nat. Hist. Soc.* **88**: 109–112.

2517 —— (1942) Notes on lichens in the herbarium of the Royal Botanic Garden, Edinburgh. III. *Trans. Proc. bot. Soc. Edinb.* **33**: 183–208.

2518 —— (1944) Report of the Botanical Section for the year 1943. Recorder's notes. *Proc. Somerset archaeol. nat. Hist. Soc.* **89**, *Rep. bot. sect.* [Separately printed]: 3–5.

2519 —— (1945) Notes on lichens in the herbarium of the Royal Botanic Garden, Edinburgh. IV. *Trans. Proc. bot. Soc. Edinb.* **34**: 233–243.

2520 —— (1946) The botanical section—Recorder's notes. *Proc. Somerset archaeol. nat. Hist. Soc.* **92**: 98–101.

2521 —— (1946) The lichens of Yorkshire. *Trans. Yorks. Nat. Un.* **37**: 1–64.

2522 —— (1947) The botanical section—Recorder's notes. *Proc. Somerset archaeol. nat. Hist. Soc.* **93**: 131–134.

2523 —— (1948) List of British fungi parasitic on lichens or which have been included as lichens (or vice versa), with some notes on their characters and distribution. *Trans. Br. mycol. Soc.* **31**: 305–339.

2524 —— (1950) The botanical section—Recorder's notes. *Proc. Somerset archaeol. nat. Hist. Soc.* **94**: 152–155.

2525 —— (1950) The botanical section—Recorder's notes. *Proc. Somerset archaeol. nat. Hist. Soc.* **95**: 162–164.

2526 [——] [*c.* 1950] [*Usnea* vice-county records.] Manuscript in BM with Wade (1957–59).

2527 —— (1951) The species of *Usnea* in Great Britain and Ireland. *Trans. Br.*

mycol. Soc. **34**: 368–375.

2528 —— (1953) *Census Catalogue of British Lichens.* London: Cambridge University Press.

2529 —— [1954] *The lichens of Devonshire.* Manuscript. [Intended for publication by the Devonshire Association; now with D. L. Hawksworth.]
—— —See also Nos. 636, 1255, 1727 and 2450.

2530 **Watson, [Sir] W[illiam]** (1759) An historical memoir on a genus of plants called *Lichen* by Micheli, Haller, and Linnaeus; and comprehended by Dillenius under the terms *Usnea, Coralloides,* and *Lichenoides:* tending principally to illustrate their several uses. *Phil. Trans. R. Soc.* **50**: 652–688.

2531 **Watson, William** (1913) Lichens. *J. Scott. Mountaineering Club* **12**: 267–274.

2532 **Watt, A. S.** (1937) Studies in the ecology of Breckland. II. On the origin and development of blow-outs. *J. Ecol.* **25**: 91–112.

2533 —— (1938) Studies in the ecology of Breckland. III. The origin and development of the *Festuco-Agrostidetum* on eroded sand. *J. Ecol.* **26**: 4–37.

2534 —— (1940) Studies in the ecology of Breckland. IV. The grass-heath. *J. Ecol.* **28**: 42–70.

2535 —— (1947) Contributions to the ecology of bracken (*Pteridium aquilinum*). IV. The structure of the community. *New Phytol.* **46**: 97–121.

2536 —— (1955) Bracken versus heather, a study in plant sociology. *J. Ecol.* **43**: 490–506.

2537 —— (1957) The effect of excluding rabbits from grassland B (*Mesobrometum*) in Breckland. *J. Ecol.* **45**: 861–878.

2538 —— (1960) The effect of excluding rabbits from acidiphilous grassland in Breckland. *J. Ecol.* **48**: 601–604.

2539 **Watt, L.** (1901) 24th April, 1900. [Lichens from Dumbartonshire]. *Trans. nat. Hist. Soc. Glasg., n.s.* **6**: 174–175.

2540 **Wattam, W. E. L.** (1913) Yorkshire naturalists at Roche Abbey. *Naturalist, Hull* **1913**: 205–210.

2541 —— (1913) Yorkshire naturalists at Kirkby Stephen. *Naturalist, Hull* **1913**: 230–237.

2542 —— (1915) Natural history of Sawley and Eavestone, near Ripon. *Naturalist, Hull* **1915**: 231–237.

2543 —— (1924) Sandsend lichen records, Yorks. *Naturalist, Hull* **1924**: 137–140.

2544 —— (1926) East Riding lichen records. *Naturalist, Hull* **1926**: 49.

2545 —— (1926) Yorkshire naturalists at Hornsea. *Naturalist, Hull* **1926**: 277–283.

2546 —— (1926) Lichens in north-east Cheshire. *Naturalist, Hull* **1926**: 283.

2547 —— (1927) Lichens in north-east Cheshire. *Naturalist, Hull* **1927**: 298.

2548 —— (1929) Lichen notes from Rhosneigr (Anglesey). *Naturalist, Hull* **1929**: 183–185.

2549 —— (1929) Yorkshire peat and other lichen habitats. *Naturalist, Hull* **1929**: 335–339.

2550 —— (1931) Pwllheli lichen records. *NWest. Nat.* **6**: 32.

2551 —— (1934) Yorkshire naturalists at Semerwater—Lichens. *Naturalist, Hull* **1934**: 162.

168

2552 —— (1934) Yorkshire naturalists at Fairburn—Lichens. *Naturalist, Hull* **1934**: 211–212.

2553 —— (1935) Dent Dale lichen records. *Naturalist, Hull* **1935**: 89.

2554 —— (1935) Bishopdale lichen notes. *Naturalist, Hull* **1935**: 89.

2555 —— (1935) Yorkshire naturalists at Bentham and Burton-in-Lonsdale— Lichens. *Naturalist, Hull* **1935**: 209–210.

2556 —— (1935) Yorkshire naturalists at Langsett—Lichens. *Naturalist, Hull* **1935**: 234.

2557 —— (1936) Yorkshire naturalists at Drop Clough—Lichens. *Naturalist, Hull* **1936**: 163.

2558 —— (1936) Yorkshire naturalists at Hawes—Lichens. *Naturalist, Hull* **1936**: 184–186.

2559 —— (1937) Yorkshire Naturalists' Union at Keld—Lichens. *Naturalist, Hull* **1937**: 165–166, 170–171.

2560 —— (1938) Yorkshire naturalists in Upper Nidderdale—Lichens. *Naturalist, Hull* **1938**: 235.

2561 —— (1938) Yorkshire naturalists at Pickering—Lichens. *Naturalist, Hull* **1938**: 241–242.

2562 —— (1939) Yorkshire naturalists at Wentbridge—Lichens. *Naturalist, Hull* **1939**: 217.

2563 —— (1939) Yorkshire naturalists in Teesdale—Lichens. *Naturalist, Hull* **1939**: 266.

2564 —— (1939) *Baeomyces rufus* DC. in the Halifax district. *Naturalist, Hull* **1939**: 292.

2565 —— (1942) Yorkshire naturalists at Stocksmoor, Huddersfield. *Naturalist, Hull* **1942**: 175–176.

2566 —— (1953) The lichen flora of the Huddersfield district. *Naturalist, Hull* **1953**: 57–60.

2567 —— (1973) Lichens. In *Fairburn and its Nature Reserve* (R. F. Dickens and J. D. Pickup, eds.): 27–28. Clapham: Dalesman Books.

2568 **Watts, G. D., Hornby, R., Lambley, P. W. and Ismay, J.** (1976) An ecological review of the Yare Valley near Norwich. *Trans. Norfolk Norwich Nat. Soc.* **23**: 231–248.

2569 **Webley, D. M., Henderson, M. E. K., and Taylor, I. F.** (1963) The microbiology of rocks and weathered stones. *J. Soil Sci.* **14**: 102–112.

2570 **Welch, D.** (1966) Biological Flora of the British Isles: *Juncus squarrosus* L. *J. Ecol.* **54**: 535–548.

2571 —— (1967) Communities containing *Juncus squarrosus* in Upper Teesdale, England. *Vegetatio* **14**: 229–240.

2572 —— (1970) *Saxifraga hirculus* L. in north-east Scotland. *Trans. Proc. bot. Soc. Edinb.* **41**: 27–30.

2573 —— **and Rawes, M.** (1964) The early effects of excluding sheep from high-level grasslands in the north Pennines. *J. appl. Ecol.* **1**: 281–300.

2574 —— —— (1966) The intensity of sheep grazing on high-level blanket bog in Upper Teesdale. *Ir. J. Agric. Research* **5**: 185–196.

—— —See also No. 653.

Welch, R. C.—See Nos. 795 and 1317.

2575 **Wells, T. C. E., Sheail, J., Ball, D. F. and Ward, L. K.** (1976) Ecological studies on the Porton Ranges: relationships between vegetation, soils and land-use history. *J. Ecol.* **64**: 589–626.

West, G. S.—See No. 2587.

2576 **West, W.** (1879) Report of the botanical section: 1879. *Trans. Yorks. Nat. Un., Bot.* **1**: 64–110.

2577 —— (1880) Buckinghamshire lichens. *Naturalist, Hull* **6**: 69–70.

2578 —— (1881) Yorkshire Naturalists' Union—Cryptogamic report for 1880. *Naturalist, Hull* **6**: 166–168.

2579 —— (1881–82) A few days of field botany in Scotland. *Naturalist, Hull* **7**: 73–74, 94–98, 112–116.

2580 —— (1882) A stroll near Baildon, in February. *Naturalist, Hull* **7**: 125–126.

2581 —— (1883) A new British lichen. *J. Bot., Lond.* **21**: 281.

2582 —— (1883) The principal plants of Malham. *Naturalist, Hull* **9**: 25–27.

2583 —— (1900) Botany. In *Handbook to Bradford and the Neighbourhood* (R. M. Wheeler, ed.): 112–117. Bradford: British Association.

2584 —— (1905) *Physcia parietina. J. Bot., Lond.* **43**: 31–32.

2585 —— (1912) Notes on the flora of Shetland, with some ecological observations. *J. Bot., Lond.* **50**: 265–275, 297–306.

2586 —— (1915) Ecological notes; chiefly cryptogamic. *J. Linn. Soc. (Bot.)* **43**: 57–85.

2587 —— **and West, G. S.** (1900–01) The algal-flora of Yorkshire. *Trans. Yorks. Nat. Un.* **5**: 1–239. [Lichen p. 129.]
—— —See also Nos. 1340, 2235 and 2236.

2588 **Westcott, F.** (1843) Plants observed in the neighbourhood of Ludlow, Shropshire. *Phytologist* **1**: 567–570.

Westhoff, V.—See No. 206.

2589 **Wheeler, B. D. and Whitton, B. A.** (1971) Ecology of Hell Kettles 4. Terrestrial and subaquatic vegetation. *Vasculum* **56**: 25–37.

Wheeler, R. M.—See No. 2583.

2590 **Wheldon, H. J.** (1911) Some Argyll and Perthshire fungi. *Scott. Nat.* **77**: 34–38.

2591 —— (1914) The fungi of the sand-dune formation of the Lancashire coast. *Lancs. Chesh. Nat.* **7**: 5–10.

2592 —— (1914) Parasitic fungi on lichens. *Lancs. Chesh. Nat.* **7**: 242.

2593 **Wheldon, J. A.** (1909) Lichens. In *A Guide to the Natural History of the Isle of Wight* (F. Morey, ed.): 89–102. Newport, Isle of Wight: County Press.

2594 —— (1909) *Cladonia deformis*, lichen new to Lancashire. *Lancs. Nat.* **2**: 22–23.

2595 —— (1910) New Lancashire cryptogams. *Lancs. Nat.* **3**: 81–83.

2596 —— (1910) New Lancashire lichens. *Lancs. Nat.* **3**: 192–194.

2597 —— (1912) Winter work in the Ribble estuary. *Lancs. Nat.* **4**: 346.

2598 —— (1912) The sand-dunes in April. *Lancs. Nat.* **5**: 38–38A.

2599 —— (1912) A new Lancashire lichen. *Lancs. Nat.* **5**: 260.

170

2600 [——] (1912) Notes on specimens contributed or sent for identification. *Rep. Lichen Exch. Club* **1911**: 12(*36*)–19(*43*).

2601 [——] (1913) Notes on specimens contributed or sent for identification. *Rep. Lichen Exch. Club* **1912**: 3(*46*)–4(*47*).

2602 —— (1914) Stone mites in West Lancashire. *Lancs. Chesh. Nat.* **7**: 31–32.

2603 —— (1915) A new British *Acrocordia. Lancs. Chesh. Nat.* **8**: 196–197.

2604 —— (1915) Liverpool Botanical Society. *Lancs. Chesh. Nat.* **8**: 198–199.

2605 —— (1916) A Westmorland *Pilophorus. Lancs. Chesh. Nat.* **8**: 343.

2606 —— (1918) Further notes on the Manx flora. *Lancs. Chesh. Nat.* **11**: 127–130.

2607 —— (1920) Llanberis lichens. *J. Bot., Lond.* **58**: 11–15.

2608 —— (1923) A botanical visit to the Isle of Man. *Lancs. Chesh. Nat.* **15**: 109–110, 150–152, 213–215. [See also **15**: 67 (1922).]

2609 —— **and Travis, W. G.** (1913) Lichens of Arran (V.C.–100). *J. Bot., Lond.* **51**: 248–253.

2610 —— —— (1913) A new Lancashire lichen. *Lancs. Nat.* **6**: 324.

2611 —— —— (1915) The lichens of South Lancashire. *J. Linn. Soc. (Bot.)* **43**: 87–136.

2612 —— —— (1923) The south Lancashire peat mosses. In *Merseyside* (A. Holt, ed.): 268–270. London: Hodder & Stoughton.

2613 —— —— (1923) The Lancashire sand dunes. In *Merseyside* (A. Holt, ed.): 271–281. London: Hodder & Stoughton.

2614 —— **and Wilson, A.** (1904) West Lancashire lichens. *J. Bot., Lond.* **42**: 255–261.

2615 —— —— (1907) *The Flora of West Lancashire.* Eastbourne: privately printed. [Lichens pp. 438–496.]

2616 —— —— (1909) *Gyrophora spodochroa* Ach. *J. Bot., Lond.* **47**: 447–448.

2617 —— —— (1914) Alpine vegetation on Ben-y-Gloe, Perthshire. *J. Bot., Lond.* **52**: 227–235.

2618 —— —— (1915) The lichens of Perthshire. *J. Bot., Lond.* **53**, *Suppl.*: 1–73.

2619 —— —— (1925) West Lancashire flora: Notes, additions and extinctions. *Lancs. Chesh. Nat.* **17**: 117–125.

—— —See also Nos. 904 and 905.

2620 **Whitaker, T. D.** (1805) *The History and Antiquities of the Deanery of Craven in the County of York.* London: Nichols & Son. [Lichens in Appendix I, pp. 9–11, by S. Hailstone; list unchanged in Ed. 2 (1812), pp. 517–519, Ed. 3 (1878), pp. 629–631, and in reprint: Miall, L.C. (1878) *The Geology Natural History and Pre-Historic Antiquities of Craven in Yorkshire*, pp. 39–41. Leeds: Dodgson.]

2621 **White, D. J. B.** (1961) Some observations on the vegetation of Blakeney Point, Norfolk, following the disappearance of the rabbits in 1954. *J. Ecol.* **49**: 113–118.

2622 —— **White, M. F. and Peterken, G. F.** (1970) *Polypodium* on Blakeney Point, Norfolk. *Trans. Norfolk Norwich Nat. Soc.* **21**: 372–376.

—— —See also No. 1154.

White, E.—See No. 1794.

2623 [**White, F. B.**] (1863) List of plants found near the Rumbling Bridge. *Trans. bot. Soc. Edinb.* **7**: 572–573.

2624 —— (1870) Mamsoul. *Proc. Perthsh. Soc. nat. Sci.* **1869–70**: 23–29.

2625 —— (1879) Glen Tilt: its fauna and flora. The cryptogamic flora. *Scott. Nat.* **5**: 85–96.

2626 —— (1881) The cryptogamic flora of Mull. *Scott. Nat.* **6**: 155–162.

White, M. F.—See No. 2622.

White, W.—See Nos. 415 and 1338.

2627 **Whittaker, E.** (1960) *Ecological effects of moor burning.* Ph.D. thesis, University of Aberdeen.

Whitton, B. A.—See No. 2589.

2628 **Wilberforce, P.** (1970) Field meetings. 26th April—Acrise, Denton and Womenswold. *Bull. Kent Fld Club* **15**: 16.

2629 **Wilkinson, W. H.** (1884) The study of a lichen from Oban (*Ricasolia amplissima*). *Midland Nat.* **7**: 273–277.

2630 —— (1887) A ramble amongst lichens in the Island of Bute. *Midland Nat.* **10**: 81–85.

2631 [——] (1888) Botanical excursion to Fawsley. *J. Northampt. nat. Hist. Soc.* **5**: 148–149.

2632 —— (1893) Lichens of the Isle of Man. *Midland Nat.* **16**: 245–248, 272–276.

2633 —— (1900) Merionethshire lichens. *J. Bot., Lond.* **38**: 182–184.

2634 —— (1900) The lichens (Lichenes). In *The Victoria History of the Counties of England, Hampshire and the Isle of Wight* (H. A. Doubleday, ed.) **1**(2): 75–79. Westminster: Constable.

2635 —— (1904) Radnorshire lichens. *J. Bot., Lond.* **42**: 111–113.

2636 [——] (1912) Two new lichens from Sutton Park. *Rep. Bgham nat. Hist. phil. Soc.* **1911**: 8.

2637 [——] (1913) Lichens. In *A Handbook for Birmingham and the Neighbourhood* (G. A. Auden, ed.): 483–485. Birmingham: Cornish Brothers.

Williams, E.—See No. 1360.

2638 **Williams, G.** (1895) Sockaugh. *J. Cairngorm Club.* **1**: 284–293.

Williams, P.—See No. 1725.

Williams, P. H.—See Nos. 2300 and 2301.

2639 **Willis, A. J.** (1963) Braunton Burrows: The effects on the vegetation of the addition of mineral nutrients to the dune soils. *J. Ecol.* **51**: 353–374.

2640 —— **Folkes, B. F., Hope-Simpson, J. F. and Yemm, E. W.** (1959) Braunton Burrows: The system and its vegetation. Part II. *J. Ecol.* **47**: 249–288.

2641 —— **and Yemm, E. W.** (1961) Braunton Burrows: Mineral nutrient status of the dune soils. *J. Ecol.* **49**: 377–390.

2642 **Willis, J. C. N.** (1953) Lichens. *Trans. Suffolk Nat. Soc.* **8**: 149–152.

2643 —— (1964) The great Schwendener controversy. *Trans. Suffolk Nat. Soc.* **12**: 379–381.

Willis-Bund, J. W.—See No. 134.

2644 **Wilson, A.** (1911) New Lancashire lichens. *Lancs. Nat.* **4**: 13–14.

2645 —— (1914) Lancashire lichens. *Lancs. Chesh. Nat.* **7**: 1.

2646 —— (1919) West Yorkshire botanical notes. *Naturalist, Hull* **1919**: 369.

2647 —— (1922) West Yorkshire botanical notes. *Naturalist, Hull* **1922**: 397–398.

2648 —— (1924) West Yorkshire botanical notes. *Naturalist, Hull* **1924**: 48–50.

2648a —— (1928) The botany of Tal-y-Fan. *Proc. Llandudno Distr. Fld Club* **14**: 82–98.

2649 —— (1929) Notes on the flora of Carnedd Llewelyn. *NWest. Nat.* **4**: 53–56.

2649a —— (1930) Botanical excursion Pont-y-Pant to Bettws-y-Coed. *Proc. Llandudno Distr. Fld Club* **15**: 39–40.

2649b —— (1933) Botanical ramble at Machno Hill. *Proc. Llandudno Distr. Fld Club* **17**: 9.

2650 —— (1938) *The Flora of Westmorland.* Arbroath: T. Buncle. [Lichens pp. 356–400.]

2651 —— (1947) The flora of a portion of north-east Caernarvonshire [continued]. *NWest. Nat.* **22**: 191–211.

2652 —— **and Wheldon, J. A.** (1908) Inverness-shire cryptogams. *J. Bot., Lond.* **46**: 347–356.

2653 —— —— (1909) *Gyrophora spodochroa* Ach. *J. Bot., Lond.* **47**: 431.

2654 —— —— (1909) A new lichen, *Cladonia luteoalba*, from Lancashire. *Trans. Lpool bot. Soc.* **1**: 6–7.

—— —See also Nos. 2614, 2615, 2616, 2617, 2618 and 2619.

2655 **Wilson, H.** (1968) Lichens. In *A check list of the Flora of the Culbin State Forest* (M. McCallum Webster, ed.): 31. Elgin: M. McCullum Webster.

Wilson, H. E.—See No. 1863.

2656 **Wilson, J. H.** (1910) *Nature Study Rambles round St. Andrews.* St Andrews: St Andrews University Press.

2657 **Winch, N. J.** (1791–1837) *Collection of scientific correspondence.* Letters in LINN in 8 vols. [Not abstracted.]

2658 —— (18—) [Annotations to a copy of No. 2391 in LINN.]

2659 —— (1807) *The Botanist's Guide through the Counties of Northumberland and Durham.* Vol. 2. Newcastle. [Lichens pp. 29–66.] [Copy annotated by Winch in LINN.]

2660 —— (1819) *An Essay on the Geographical Distribution of Plants through the counties of Northumberland, Cumberland and Durham.* Newcastle upon Tyne: E. Walker. [Lichens unchanged in Ed. 2 (1825).]

2661 —— (1831) Flora of Northumberland and Durham. *Trans. nat. Hist. Soc. Northumb.* **1831**: 1–150.

2662 —— (1833) *Remarks on the flora of Cumberland.* Newcastle: Mercury Press.

2663 —— (1837) Addenda to the Flora of Northumberland and Durham. *Trans. nat. Hist. Soc. Northumb.* **1837**: 151–159.

—— —See also No. 1195.

2664 **Windsor, J.** (1858) Lichens growing near Settle. *Phytologist, n.s.* **2**: 464–466.

2665 —— (1873) *Flora Cravoniensis, or a Flora of the Vicinity of Settle in Craven, Yorkshire.* Manchester: Cave & Sever. [Lichens pp. 146–171.]

Winkler, H.—See Nos. 559 and 2088a.

2666 **Wirth, V.** (1970) Studien zu den silicolen *Opegrapha*-Arten *O. horistica, O. zonata* und *O. gyrocarpa. Herzogia* **1**: 469–475.

2667 **Wise, J. R. [de C.]** (1883) *The New Forest: its History and its Scenery.* London:

Smith & Elder. [Also later editions; reprinted (1971) London: H. Sotheran.]

Wishart, D.—See Nos. 444, 445 and 446.

2668 **Withering, W.** (1776) *A Botanical Arrangement of all the Vegetables Naturally Growing in Great Britain.* Vol. 2. Birmingham: T. Cadell, etc. [Lichens pp. 704–731.]

2669 —— (1792) *A Botanical Arrangement of British Plants.* Ed. 2. Vol. 3. Birmingham: C. G. J. and J. Robinson, etc. [Lichens pp. 164–223, 252.]

2670 —— (1796) *An Arrangement of British Plants.* Ed. 3. Vol. 4. Birmingham: Privately printed. [Lichens pp. 1–77.]

2671 —— (1801) *A Systematic Arrangement of British Plants.* Ed. 4. Vol. 4. London: T. Cadell, etc. [Lichens pp. 1–76.]

2672 —— (1812) *A Systematic Arrangement of British Plants.* Ed. 5. Vol. 4. Birmingham: R. Scholey. [Lichens pp. 1–83, 109.]

2673 —— (1818) *An Arrangement of British Plants.* Ed. 6. Vol. 4. London: R. Scholey. [Lichens pp. 1–90, 118.]

2674 —— (1830) *An Arrangement of British Plants.* Ed. 7. Vol. 4. London: C. J. G. and F. Rivington, etc. [Lichens pp. 1–67, 87.]

—— —See also No. 507.

Wojterskiego, T.—See No. 2363a.

Woodburn, D. A.—See No. 1405.

2675 **Woodhead, N.** (1933) The alpine plants of the Snowdon range. *Bull. alp. Gdn Soc..* **2**: 12–21, 45–51.

2676 **Woodruffe-Peacock, E. A.** (1896) Lincolnshire naturalists at Market Rasen. *Naturalist, Hull* **1896**: 11–16.

2677 —— (1897) Lincolnshire naturalists in the Gainsborough neighbourhood. *Naturalist, Hull* **1897**: 253–256.

2678 —— (1898) Lincolnshire Naturalists' Union at Grantham. *Naturalist, Hull* **1898**: 241–245.

2679 —— (1900) Lincolnshire naturalists at Somercotes and Saltfleetby. *Naturalist, Hull* **1900**: 75–79.

2680 —— (1900) Naturalists at Newark. *Naturalist, Hull* **1900**: 117–124.

2681 —— (1900) Naturalists at Lincoln. *Naturalist, Hull* **1900**: 247–252.

2682 —— (1902) Lincolnshire naturalists at Revesby. *Naturalist, Hull* **1902**: 145–148.

2683 —— (1902) Lincolnshire naturalists at Scunthorpe. *Naturalist, Hull* **1902**: 375–380.

Woolhouse, H. W. — See No. 1020.

Wooster, D. — See No. 1445.

2684 **Worth, R. H.** (1930) Address of the President. *Rep. Trans. Devon. Ass. Advmt Sci.* **62**: 49–115.

2685 —— (1937) The vegetation of Dartmoor. *Rep. Trans. Plymouth Inst.* **17**: 285–296.

—— —See also Nos. 399 and 2250.

2686 [**Wright, C. H.**] (1890) British hymenolichen. *J. R. microsc. Soc.* **1890**: 647.

174

2687 **Wunder, H.** (1974) Schwarzfrüchtige, saxicole Sippen der Gattung *Caloplaca* (Lichenes, Teloschistaceae) in Mitteleuropa, dem Mittelmeergebiet und Vorderasien. *Bibthca Lichenol., Lehre* **3**: 1–186.

2688 **Yarranton, G. A.** (1967) A quantitative study of the bryophyte and macro-lichen vegetation of the Dartmoor granite. *Lichenologist* **3**: 392–408.

Yemm, E. W. — See Nos. 2640 and 2641.

2689 **Yonge, C. M.** (1949) *The Sea Shore.* [New Naturalist no. 12] London: Collins.

————See also No. 208.

2690 **Yoshimura, I. and Hawksworth, D. L.** (1970) The typification and chemical races of *Lobaria pulmonaria* (L.) Hoffm. *J. Jap. Bot.* **45**: 33–41.

2691 —— **and Isoviita, P.** (1969) Scopoli's lichen specimens and the typification of *Lobaria scrobiculata* (Scop.) DC. *Annls bot. fenn.* **6**: 348–352.

Young, A. R. — See No. 348.

Young, G. — See No. 1124.

2692 **Zschacke, H.** (1914) Die mitteleuropäischen Verrucariaceen. II. *Hedwigia* **55**: 286–324.

2693 —— (1924) Die mitteleuropäischen Verrucariaceen. IV. *Hedwigia* **65**: 46–64.

2694 —— (1927) Die mitteleuropäischen Verrucariaceen. V. *Hedwigia* **67**: 45–85.

2695 —— (1933–34) Epigloeaceae, Verrucariaceae und Dermatocarpaceae. *Rabenh. Krypt.-Fl.* **9**, 1(1): 44–695.

Index

The vice-county system of botanical recording in the British Isles (Fig. 22) was originated by Hewett C. Watson (1804–1881) in his *Cybele Britannica* of 1852, and a comparable system was produced for Ireland by Robert Ll. Praeger in 1896. It has been employed very extensively by British botanists since that time, not least by lichenologists, especially through their use of W. Watson's *Census Catalogue* of 1953[2528]. Recording on a 10 km × 10 km grid square basis (see p. 37) is now widely used by British lichenologists, but in local floras this is usually employed *within* particular vice-counties[269, 920, etc.]; for this reason it has been utilized in the preparation of this Index. In determining to which vice-county particular localities belong, the gazetteers of Bartholomew (1966) and the Ordnance Survey (1953) were used in conjunction with the 1 : 625 000 (about ten miles to one inch) maps included in Dandy (1969). In particularly difficult cases, the Ordnance Survey's Seventh Series 1 : 63 360 (one mile to one inch) maps were consulted, backed up in many cases by the local knowledge of colleagues.

Where references to certain administrative areas could not be localized to particular vice-counties, then all possibly relevant vice-counties were scored (e.g. 'Devonshire' as 3–4, 'Sussex' as 13–14, 'Yorkshire' as 61–65). References to 'Cairngorm' and 'London' were scored as vice-counties 96 and 21 respectively. The Isles of Scilly, traditionally incorporated within West Cornwall (v.c. 1), have been separated out.

Each title examined was scored for vice-counties with the exception of those listed as not abstracted (p. 195). Within each vice-county listing, the major lichen flora(s) and/or check-list(s) are indicated in bold face (i.e. **1537**) as are any major supplements to them; in cases where there are several detailed floras, only the most recent are emphasized in this way.

Channel Islands: 22, 58, 103, 104, 126, 157, 254, 267, 306, 371, 460, 461, 462, 466, 470, 475, 480, 482, 488, 489, 491, 498, 501, 505, 506, 508, 510, 554, 574, 577, 582, 585, 591, 622, 630, 675, 711, 712, 854, 932, 949, 963, 965, 978, 998, 1004, 1047, 1053, 1054, 1059, 1060, 1079, 1092, 1097, 1108, 1130 1135, 1137, 1138, 1140, 1143, 1148, 1190, 1225, 1226, 1229, 1230, 1258, 1269, 1270, 1272, 1279, 1290, 1309, 1319, 1343, 1344, 1351, 1361, 1362, 1367, 1372, 1373, 1380, 1415, 1419, 1420, 1429, 1451, 1454, 1474, 1489, 1513, 1518, 1536, **1537**, 1538, 1539, 1540, 1541, 1545, 1610, 1635, 1639, 1643, 1650, 1654, 1678, 1679, 1688, 1690, 1694, 1695, 1696, 1697, 1702, 1703, 1709, 1712, 1714, 1728, 1729, 1730, 1732, 1733, 1734, 1837, 1838, 1865, 1905, 1930, 1931, 1932, 1935, 1936, 1945, 1946, 1954, 2053, 2059, 2081, 2082, 2084, 2086, 2091, 2159, 2161, 2200, 2220, 2221, 2307, 2313, 2321, 2397, 2411, 2434, 2435, 2442, 2445, 2448, 2510, 2523, 2526, 2528, 2668, 2669, 2670, 2671, 2672, 2673, 2674 and 2693.

Isles of Scilly: 433, 440, 461, 476, 543, 554, 574, 622, 630, 711, 712, 963, **1047**, 1059, 1060, 1137, 1138, 1140, 1143, 1419, 1510, 1606, 1643, 1666, 1728, **1905**, 1906, 1930, 2053, 2082, 2091, 2200, 2221, 2321, 2364, 2413, 2445, 2515, 2526, and 2528.

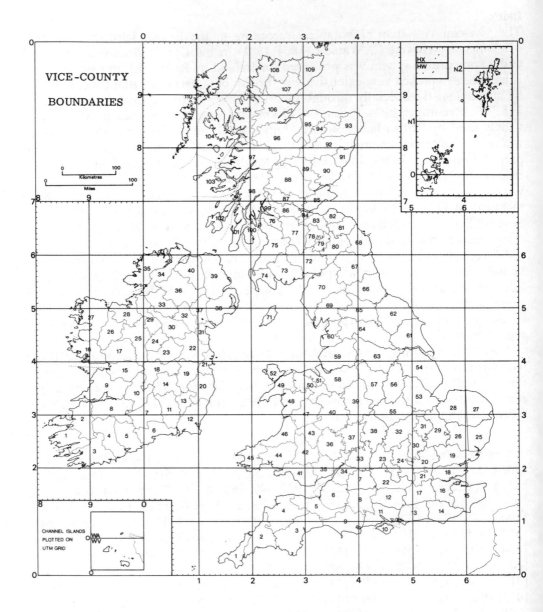

Fig. 22. Vice-county boundaries of Great Britain and Ireland (reproduced from a map kindly supplied by the Biological Records Centre of the Institute of Terrestrial Ecology).

West Cornwall (v.c. 1): 36, 41, 152, 239, 240, 306, 314, 329, 330, 350, 357, 358, 401, 433, 440, 452, 460, 461, 466, 467, 473, 476, 477, 480, 487, 488, 491, 496, 498, 505, 534, 535, 542, 543, 546, 578, 585, 711, 720, 796, 827, 828, 834, 854, 865, 909, 923, 931, 940, 1005, 1022, 1034, 1036, **1047**, 1059, 1060, 1076, 1086, 1087, 1088, 1130, 1132, 1137, 1138, 1140, 1209, 1223, 1229, 1230, 1269, 1270, 1271, 1273, 1346, 1350, 1364, 1367, 1397, 1398, 1419, 1448, 1510, 1516, 1524, 1525, 1607, 1610, 1635, 1639, 1643, 1647, 1666, 1671, 1698, 1702, 1707, 1728, 1729, 1733, 1801, 1803, 1837, 1838, 1849, 1898, **1899**, 1900, 1905, 1906, 1927, 1931, 1932, 1935, 1952, 1976, 1977, 2049, 2059, 2082, 2086, 2098, 2159, 2162, 2167, 2195, 2200, 2203, 2209, 2220, 2221, 2222, 2307, 2309, 2310, 2330, 2364, 2378, 2380, 2381, 2390, 2391, 2392, 2432, 2434, 2435, 2439, 2445, 2448, 2452, 2490, 2492, 2494, 2496, 2498, 2500, 2502, 2505, 2509, 2515, 2517, 2523, 2526, 2669, 2670, 2671, 2672, 2673 and 2674.

East Cornwall (v.c. 2): 22, 41, 152, 306, 314, 350, 357, 358, 393, 401, 428, 433, 460, 461, 466, 467, 507, 578, 585, 682, 711, 720, 796, 834, 909, 923, 931, 940, 949, 960, 963, 965, 970, 986, 1022, 1034, 1036, **1047**, 1059, 1060, 1075, 1086, 1087, 1088, 1122, 1132, 1137, 1140, 1209, 1229, 1230, 1270, 1271, 1272, 1273, 1274, 1277, 1346, 1364, 1397, 1415, 1448, 1514, 1516, 1537, 1610, 1635, 1637, 1643, 1647, 1666, 1728, 1837, 1838, 1889, 1891, 1892, 1898, 1927, 1930, 1932, 1935, 1946, 1952, 1976, 2020, 2038, 2049, 2053, 2086, 2098, 2159, 2160, 2209, 2220, 2221, 2307, 2309, 2312, 2327, 2330, 2335, **2342**, 2343, 2379, 2390, 2391, 2392, 2397, 2432, 2434, 2435, 2445, 2448, 2496, 2498, 2500, 2502, 2509, 2523, 2526, 2669, 2670, 2671, 2672, 2673 and 2674.

South Devon (v.c. 3): 1, 5, 99, 129, 212, 255, 259, 269, 306, 327, 334, 350, 357, 358, 371, 399, 419, 420, 430, 433, 448, 459, 461, 466, 471, 486, 489, 507, 546, 547, 571, 572, 573, 574, 575, 578, 582, 585, 599, 622, 630, 674, 681, 682, 711, 712, 720, 725, 726, 727, 728, 729, 763, 833, 834, 888, 889, 890, 891, 892, 893, 894, 895, 896, 897, 898, 899, 900, 901, 902, 909, 923, 924, **930**, 931, **932**, 934, 935, 936, 938, 939, 940, 942, 943, 947, 949, 951, 952, 956, 957, 959, 960, 963, 965, 966, 970, 971, 1008, 1009, 1012, 1013, 1014, 1015, 1016, 1019, 1034, 1035, 1036, 1048, 1059, 1060, 1087, 1088, 1092, 1122, 1123, 1130, 1136, 1137, 1138, 1140, 1196, 1209, 1210, 1225, 1226, 1229, 1230, 1234, 1269, 1270, 1271, 1272, 1273, 1274, 1277, 1279, 1293, 1294, 1303, 1304, 1307, 1310, 1318, 1327, 1344, 1346, 1350, 1356, 1364, 1372, 1386, 1393, 1397, 1415, 1418, 1419, 1433, 1434, 1437, 1456, 1486, 1490, 1510, 1558, 1577, 1610, 1612, 1616, 1635, 1643, 1647, 1664, 1671, 1704, 1728, 1729, 1732, **1743**, 1756, 1848, 1849, 1864, 1866, 1867, 1892, 1898, 1899, 1930, 1932, 1934, 1935, 1946, 1951, 1954, 1955, 1963, 2038, 2049, 2053, 2057, 2059, 2091, 2098, 2159, 2186, 2198, 2220, 2221, 2238, 2250, 2260, 2261, 2303, 2307, 2308, 2309, 2310, 2313, 2330, 2335, 2358, 2391, 2395, 2396, 2399, 2415, 2424, 2434, 2435, 2442, 2445, 2448, 2454, 2458, 2459, 2469, 2480, 2490, 2492, 2494, 2498, 2499, 2500, 2502, 2508, 2509, 2515, 2523, 2526, 2529, 2600, 2601, 2669, 2670, 2671, 2672, 2673, 2674, 2684, 2685 and 2688.

North Devon (v.c. 4): 5, 36, 37, 39, 44, 93, 94, 129, 255, 306, 326, 335, 350, 357, 358, 433, 441, 461, 547, 573, 578, 585, 630, 674, 711, 712, 725, 726, 833, 834, 895, 899, 900, 902, 923, **930**, 931, 932, 934, 935, 937, 938, 939, 940, 947, 949, 951, 952, 956, 963, 970, 1008, 1010, 1011, 1012, 1013, 1014, 1015, 1034, 1048, 1059, 1060, 1092, 1129, 1143, 1209, 1210, 1226, 1229, 1279, 1293, 1294, 1345, 1346, 1364, 1393, 1419, 1433, 1456, 1490, 1547, 1548, 1610, 1635, 1639, 1642, 1643, 1664, 1704, 1728,

1732, 1742, **1743**, 1848, 1866, 1867, 1898, 1917, 1918, 1934, 1952, 2049, 2098, 2159, 2198, 2220, 2221, 2250, 2261, 2303, 2307, 2309, 2318, 2330, 2335, 2377, 2391, 2392, 2424, 2432, 2434, 2435, 2442, 2445, 2448, 2461, 2469, 2480, 2481, 2490, 2498, 2500, 2502, 2509, 2523, 2526, 2529, 2600, 2639, 2640, 2641, 2669, 2670, 2671, 2672, 2673, 2674, 2684 and 2688.

South Somerset (v.c. 5): 75, 79, 80, 82, 83, 85, 86, 87, 88, 89, 90, 91, 92, 357, 358, 519, 546, 547, 577, 578, 909, 931, 932, 965, 1001, 1050, 1138, 1208, 1226, 1229, 1238, 1270, 1271, 1272, 1273, 1279, 1304, 1307, 1329, 1392, 1518, 1610, 1611, 1635, 1671, 1930, 1946, 2049, 2098, 2159, 2309, 2318, 2330, 2335, 2391, 2432, 2434, 2435, 2442, 2445, 2448, 2479, 2480, 2482, 2488, 2490, 2491, 2492, 2493, 2494, 2495, 2496, **2497**, 2498, 2499, 2500, 2502, 2507, 2509, 2510, 2511, 2513, 2515, 2516, 2518, 2520, 2522, 2523, 2524, 2525, 2526, 2600, 2669, 2670, 2671, 2672, 2673 and 2674.

North Somerset (v.c. 6): 11, 79, 80, 85, 86, 87, 88, 89, 90, 91, 92, 98, 221, 255, 306, 328, 329, 331, 335, 356, 357, 358, 374, 394, 461, 476, 482, 486, 487, 491, 501, 505, 547, 577, 578, 591, 613, 622, 668, 720, 791, 932, 1050, 1059, 1060, 1087, 1129, 1138, 1225, 1237, 1279, 1343, 1459, 1483, 1510, 1518, 1610, 1611, 1642, 1643, 1666, 1729, 1746, 1747, 1772, 1837, 1838, 1864, 1930, 1946, 1976, 2018, 2098, 2100, 2159, 2172, 2195, 2220, 2221, 2231, 2307, 2309, 2312, 2313, 2327, 2330, 2335, 2347, 2391, 2392, 2401, 2402, 2434, 2435, 2442, 2445, 2448, 2479, 2480, 2482, 2487, 2490, 2491, 2492, 2493, 2494, 2495, 2496, **2497**, 2498, 2499, 2500, 2502, 2509, 2511, 2513, 2515, 2516, 2520, 2522, 2523, 2526, 2669, 2670, 2671, 2672, 2673 and 2674.

North Wiltshire (v.c. 7): 357, 358, 578, 582, 591, 635, 720, 970, 1086, 1094, 1095, 1096, **1097**, 1098, 1099, 1100, 1101, 1102, 1103, 1104, 1105, 1106, 1107, 1108, 1109, 1138, 1234, 1247, 1256, 1259, 1306, 1307, 1310, 1320, 1518, 1643, 1837, 1838, 1931, 1976, 2049, 2051, 2167, 2330, 2361, 2434, 2448, 2494, 2496, 2500, 2502, 2672, 2673 and 2674.

South Wiltshire (v.c. 8): 357, 358, 466, 534, 578, 585, 613, **636**, 826, 948, 970, 1059, 1060, 1086, 1092, 1094, 1095, 1096, **1097**, 1099, 1100, 1101, 1102, 1108, 1138, 1234, 1256, 1259, 1307, 1310, 1580, 1643, 1834, 1837, 1838, 1976, 2016, 2046, 2049, 2050, 2053, 2159, 2220, 2330, 2382, 2391, 2392, 2490, 2498, 2500, 2526, 2575, 2668, 2669, 2670, 2671, 2672, 2673 and 2674.

Dorset (v.c. 9): 25, 42, 129, 219, 220, 267, **269**, 358, 382, 547, 549, 574, 578, 582, 681, 711, 834, 923, 931, 932, 948, 949, 969, 970, 1025, 1039, 1049, 1052, 1059, 1060, 1092, 1122, 1138, 1196, 1273, 1344, 1415, 1418, 1419, 1489, 1610, 1635, 1643, 1650, 1671, 1728, 1761, 1762, 1876, 1877, 1904, 1930, 1935, 2016, 2045, 2049, 2053, 2098, 2159, 2220, 2309, 2317, 2321, 2327, 2328, 2330, 2334, 2335, 2360, 2361, 2382, 2391, 2392, 2402, 2414, 2417, 2419, 2432, 2434, 2445, 2448, 2486, 2490, 2494, 2496, 2498, 2500, 2502, 2509, 2510, 2515, 2523, 2526, 2670, 2672, 2673, 2674 and 2686.

Isle of Wight (v.c. 10): 252, 306, 461, 462, 477, 487, 507, 554, 578, 740, 963, 965, 1059, 1060, 1122, 1138, **1251**, 1302, 1304, 1343, 1344, 1367, 1429, 1439, 1454, 1610, 1635, 1642, 1643, 1728, 1729, 1930, 1931, 1935, 1936, 2053, 2084, 2086, 2159, 2220, 2221, 2232, 2330, 2390, 2399, 2432, 2434, 2435, 2445, 2448, 2471, 2498, 2502, 2517, 2526, 2593, 2600, 2634, 2670, 2671, 2672, 2673 and 2674.

South Hampshire (v.c. 11): 1, 129, 198, 255, 262, 306, 357, 358, 371, 433, 435, 457, 459, 460, 461, 463, 464, 466, 475, 476, 480, 498, 507, 510, 574, 578, 582, 585, 591, 706, 720, 740, 865, 891, 923, 928, 931, 932, 940, 948, 949, 959, 961, 965, 970, 976, 978, 1022, 1032, 1053, 1054, 1059, 1060, 1063, 1122, 1129, 1138, 1139, 1148, 1195, 1269, 1271, 1272, 1274, 1304, 1343, 1344, 1345, 1351, 1355, 1356, 1362, 1415, 1439, 1510, 1518, 1551, 1610, 1621, 1622, 1635, 1643, 1655, 1671, 1679, 1689, 1692, 1696, 1697, 1728, 1732, 1733, 1759, 1760, 1903, 1927, 1931, 1932, 1934, 1935, 1936, 1946, 1948, 1955, 2019, 2037, 2038, 2049, 2051, **2053**, 2086, 2098, 2159, 2162, 2220, 2221, 2322, 2327, 2330, 2334, 2360, 2383, 2389, 2390, 2391, 2397, 2413, 2415, 2419, 2432, 2434, 2442, 2445, 2448, 2480, 2494, 2496, 2498, 2499, 2500, 2505, 2506, 2509, 2510, 2515, 2517, 2523, 2526, 2575, 2600, **2634**, 2667, 2668, 2669, 2670, 2671, 2672, 2673 and 2674.

North Hampshire (v.c. 12): 306, 322, 357, 358, 578, 613, 891, 948, 1022, 1138, 1344, 1355, 1592, 1732, 1927, 2037, 2049, 2053, 2159, 2220, 2330, 2360, 2391, 2417, 2448, 2499, 2506, 2515, 2526, 2575, **2634**, 2668, 2669, 2670, 2671, 2672, 2673 and 2674.

West Sussex (v.c. 13): 116, **117**, 129, 161, 225, 255, 306, 357, 358, 409, **427**, 433, 460, 461, 469, 471, 505, 547, 550, 563, 574, 578, 582, 591, 613, 622, 720, 860, 861, 863, 865, 920, 923, 931, 948, 949, 960, 963, 970, 991, 1021, 1022, 1032, 1035, 1059, 1060, 1063, 1068, 1110, 1111, 1129, 1138, 1148, 1195, 1226, 1239, 1249, 1253, 1271, 1272, 1279, 1299, 1304, 1309, 1318, 1343, 1344, 1345, 1352, 1354, 1430, 1448, 1458, 1460, 1485, 1516, 1593, 1610, 1635, 1643, 1684, 1728, 1729, 1732, 1733, 1777, 1927, 1928, 1934, 1935, 1976, 2011, 2037, 2038, 2049, 2053, 2077, 2098, 2102, 2159, 2167, 2220, 2221, 2307, 2312, 2321, 2322, 2323, 2327, 2330, 2360, 2361, 2389, 2390, 2391, 2397, 2434, 2435, 2445, 2448, 2480, 2500, 2502, 2515, 2523, 2526, 2668, 2669, 2670, 2671, 2672, 2673 and 2674.

East Sussex (v.c. 14): 43, **117**, 128, 129, 225, 243, 245, 246, 248, 249, 255, 296, 306, 311, 314, 315, 351, 352, 357, 358, 371, **427**, 433, 434, 460, 461, 462, 466, 469, 471, 480, 501, 505, 508, 510, 547, 563, 574, 578, 582, 591, 622, 703, 720, 807, 821, 845, 860, 861, 863, 865, 923, 931, 940, 948, 949, 960, 965, 970, 1022, 1032, 1035, 1052, 1059, 1060, 1068, 1078, 1084, 1124, 1138, 1149, 1158, 1195, 1226, 1229, 1270, 1274, 1279, 1299, 1304, 1309, 1343, 1345, 1352, 1354, 1356, 1378, 1415, 1448, 1458, 1485, 1510, 1516, 1578, 1593, 1635, 1642, 1643, 1683, 1684, 1716, 1728, 1729, 1732, 1733, 1760, 1761, 1762, 1834, 1927, 1932, 1934, 1935, 1976, 1995, 1997, 1998, 1999, 2000, 2001, 2002, 2009, 2011, 2037, 2049, 2053, 2077, 2098, 2102, 2159, 2167, 2183, 2220, 2221, 2228, 2253, 2307, 2322, 2330, 2360, 2389, 2390, 2391, 2434, 2435, 2445, 2448, 2490, 2496, 2502, 2515, 2523, 2526, 2600, 2668, 2669, 2670, 2671, 2672, 2673, 2674 and 2695.

East Kent (v.c. 15): 13, 95, 96, 127, 128, 129, 161, 287, 289, 291, 293, 296, 297, 299, 301, 303, 304, 306, 307, 309, 310, 311, 312, 313, 314, 315, 317, 319, 320, 357, 358, 371, 372, 408, 433, 461, 482, 491, 546, 578, 703, 720, 806, 923, 931, 948, 949, 963, 970, **1037**, **1051**, 1059, 1060, 1084, 1086, 1097, 1137, 1149, 1161, 1163, 1229, 1234, 1299, 1304, 1308, 1318, 1319, 1352, 1355, 1514, 1532, 1592, 1635, 1643, 1671, 1679, 1683, 1728, 1732, 1758, 1829, 1837, 1838, 1927, 1935, 2005, 2006, 2011, 2014,

2015, 2019, 2037, 2049, 2104, 2105, 2159, 2180, 2181, 2307, 2330, 2360, 2363, 2397, 2404, 2432, 2434, 2442, 2445, 2448, 2457, 2526, 2601 and 2628.

West Kent (v.c. 16): 96, 127, 128, 129, 160, 229, 287, 289, 290, 291, 292, 293, 295, 296, 301, 302, 306, 307, 308, 309, 311, 312, 314, 318, 357, 358, 371, 372, 426, 433, 461, 466, 482, 578, 582, 591, 613, 703, 720, 721, 722, 842, 910, 920, 923, 931, 970, 1022, **1037**, **1051**, 1059, 1060, 1086, 1092, 1137, 1138, 1150, 1158, 1163, 1234, 1304, **1309**, **1311**, 1343, 1345, 1351, 1352, 1355, 1356, 1544, 1635, 1643, 1683, 1732, 1758, 1765, 1812, 1837, 1838, 1927, 1935, 1952, 1976, 2004, 2011, 2014, 2015, 2019, 2037, 2049, 2053, 2097, 2159, 2182, 2183, 2184, 2220, 2221, 2330, 2335, 2360, 2397, 2399, 2434, 2448, 2526, 2668, 2669, 2670, 2671, 2672, 2673 and 2674.

Surrey (v.c. 17): 46, 138, 139, 158, 161, 242, 255, 284, 306, 357, 358, 426, 477, 480, 492, 509, 510, 555, 574, 578, 585, 591, 613, 720, 730, 735, 737, 766, 772, 924, 969, 970, 1022, **1045**, 1059, 1060, 1061, 1086, 1087, 1088, 1092, 1138, 1152, 1160, 1195, 1196, 1199, 1229, 1234, 1239, 1253, 1270, 1271, 1295, 1296, 1299, 1300, 1304, **1309**, **1311**, 1315, 1345, 1356, 1422, 1443, 1635, 1668, 1671, 1706, 1729, 1730, 1755, 1765, 1812, 1856, 1926, 1930, 1935, 1976, 2011, 2019, 2037, 2097, 2098, 2158, 2159, 2203, 2209, 2220, 2221, 2300, 2301, 2321, 2330, 2335, 2360, 2386, 2390, 2391, 2405, 2434, 2442, 2448, 2453, 2456, 2498, 2500, 2509, 2515, 2519, 2526, 2658, 2668, 2669, 2670, 2671, 2672, 2673, 2674 and 2695.

South Essex (v.c. 18): 6, 138, 139, 174, 255, 306, 357, 358, 426, 435, 457, 461, 463, 466, 511, 578, 582, 645, 649, 720, 923, 931, 960, 970, 1021, 1059, 1060, 1079, 1097, 1147, 1195, 1272, 1304, **1309**, **1311**, 1315, 1345, 1448, 1535, 1635, 1758, 1763, 1765, 1774, 1775, 1776, 1925, 1932, 1935, 2037, 2049, 2053, 2097, 2098, 2159, **2168**, **2192**, 2193, 2209, 2220, 2221, 2330, 2335, 2360, 2361, 2390, 2391, 2500, 2526 and 2695.

North Essex (v.c. 19): 24, 26, 174, 306, 357, 358, 436, 461, 511, 578, 720, 915, 931, 949, 960, 970, 1021, 1044, 1059, 1060, 1097, 1205, 1304, 1309, 1345, 1535, 1643, 1763, 1765, 1925, 1934, 2033, 2049, 2053, 2159, **2168**, 2220, 2330, 2360, 2391, 2399, 2442, 2448, 2500, 2515 and 2526.

Hertfordshire (v.c. 20): 53, 54, 55, 56, 201, 306, 357, 358, 426, 499, 578, 582, 591, 613, 767, **829**, 923, 931, 960, 970, 1042, 1043, 1056, 1065, 1086, 1129, 1138, 1309, 1343, 1438, 1441, 1441a, 1764, **1767**, 1769, 1829, 1834, 1868, 1976, 2065, 2096, 2159, 2220, 2307, 2309, 2312, 2321, 2330, 2360, 2391, 2432, 2442, 2448, 2502, 2526, 2669, 2670, 2671, 2672, 2673, 2674 and 2695.

Middlesex (v.c. 21): 1, 137, 152, 206, 238, 357, 358, 411, 426, 432, **456**, 462, 463, 476, 482, 487, 499, 510, 544, 585, 591, 604, 605, 613, 778, 780, 788, 806, 927, 940, 950, 969, 1021, 1022, 1059, 1060, 1086, 1092, 1097, 1162, 1165, 1231, **1309**, **1311**, 1313, 1315, 1319, 1352, 1441a, 1448, 1472, 1592, 1598, 1635, 1643, 1683, 1723, 1744, 1758, 1812, 1845, 1925, 1926, 1927, 1949, 1952, 1970, 1975, 1976, 2037, 2097, 2098, 2159, 2220, 2330, 2360, 2390, 2391, 2422, 2432, 2434, 2448, 2502, 2526, 2668, 2669, 2670, 2671, 2672, 2673 and 2674.

Berkshire (v.c. 22): 138, 139, 140, 215, 255, **265**, 268, 461, 574, 611, 617, 622, 686, 720, 970, 1017, 1019, 1097, 1138, 1147, 1208, 1242, 1310, 1320, 1356, 1729, 1816, 1951,

1954, 1957, 2037, 2038, 2054, 2098, 2159, 2213, 2214, 2220, 2330, 2345, 2390, 2391, 2422, 2463, 2505, 2509, 2510, 2515, 2526, 2672, 2673 and 2674.

Oxfordshire (v.c. 23): 123, **213**, 267, 270, 306, 357, 358, 578, 591, 609, 687, 827, 848, 949, 970, 1016, 1018, 1019, 1022, 1059, 1060, 1086, 1087, 1088, 1208, 1300, 1345, 1518, 1558, 1635, 1771, 1879, 1927, 1932, 1956, 2049, 2159, 2179, 2220, 2221, 2309, 2330, 2337a, 2391, 2422, 2434, 2463, 2498, 2509, 2515, 2517, 2526, 2600, 2668, 2669, 2670, 2671, 2672, 2673 and 2674.

Buckinghamshire (v.c. 24): 498, 578, **610**, 720, 1022, 1064, 1271, 1304, 2016, 2037, 2159, 2220, 2221, 2309, 2310, 2312, 2321, 2330, 2391, 2422, 2434, 2442, 2480, 2502, 2505, 2506, 2515, 2526 and 2577.

East Suffolk (v.c. 25): 106, 244, 247, 255, 285, 288, 296, 299, 306, 346, 434, 436, 461, 467, 547, 569, 574, 578, 614, 622, 630, 673, 720, 915, 920, 923, 931, 949, 950, 963, 965, 970, 981, 1022, 1059, 1060, 1086, 1129, 1195, 1283, 1304, 1318, 1319, 1343, 1345, 1419, 1510, 1531, 1533, 1545, 1583, **1584**, 1585, 1741, 1809, 1927, 1946, 2007, 2008, 2022, 2023, 2033, 2038, 2049, 2053, 2098, 2159, 2220, 2221, 2330, 2360, 2386, 2388, 2390, 2442, 2445, 2448, 2515, 2526, 2669, 2670, 2671, 2672, 2673 and 2674.

West Suffolk (v.c. 26): 244, 247, 255, 262, 285, 288, 296, 306, 408, 436, 461, 505, 547, 574, 578, 585, 622, 673, 688, 689, 690, 691, 720, 770, 807, 920, 923, 931, 949, 970, 981, 1022, 1059, 1060, 1229, 1270, 1281, 1292, 1304, 1345, 1419, 1434, 1518, 1531, 1533, **1584**, 1768, 1829, 1847, 1927, 1934, 1935, 2008, 2022, 2040, 2049, 2098, 2159, 2220, 2221, 2304, 2309, 2312, 2330, 2335, 2336, 2360, 2389, 2390, 2434, 2448, 2500, 2502, 2504, 2509, 2515, 2526, 2532, 2533, 2534, 2535, 2536, 2537, 2538, 2669, 2670, 2671, 2672, 2673 and 2674.

East Norfolk (v.c. 27): 102, 155, 157, **244**, 255, 306, 332, 333, 383, 436, 437, 461, **467**, **516**, 574, 578, 585, 593, 614, 630, 669, 670, 673, 720, 737, 923, 931, 949, 950, 963, 965, 970, 999, 1022, 1053, 1054, 1059, 1060, 1092, 1147, 1154, 1234, 1254, 1270, 1273, 1282, 1283, 1284, 1304, 1310, 1319, 1321, 1345, 1356, 1431, 1434, 1482, 1526, 1527, 1528, 1529, 1533, 1545, 1643, 1661, 1667, 1726, 1727, 1741, 1811, 1829, 1837, 1838, 1929, 1950, 2008, 2022, 2026, 2040, 2049, 2066, 2067, 2098, 2159, 2219, 2220, 2221, 2257, 2330, 2334, 2335, 2360, 2386, 2388, 2389, 2390, 2391, 2434, 2448, 2490, 2492, 2494, 2498, 2500, 2502, 2509, 2515, 2526, 2568, 2621, 2622, 2669, 2670, 2671, 2672, 2673 and 2674.

West Norfolk (v.c. 28): 155, **244**, 255, 306, 332, 333, 383, 436, 439, 503, 505, **516**, 553, 574, 578, 580, 584, 630, 671, 672, 673, 688, 689, 690, 691, 720, 737, 762, 915, 923, 931, 949, 950, 963, 970, 1053, 1054, 1059, 1060, 1094, 1154, 1234, 1281, 1283, 1284, 1302, 1319, 1332, 1356, 1415, 1419, 1482, 1526, 1528, 1529, 1533, 1617, 1644, 1661, 1667, 1671, 1710, 1726, 1727, 1729, 1754, 1768, 1808, 1825, 1827, 1837, 1838, 1847, 1950, 2008, 2022, 2026, 2066, 2067, 2159, 2219, 2220, 2221, 2299, 2334, 2335, 2336, 2360, 2390, 2391, 2397, 2424, 2432, 2434, 2442, 2448, 2490, 2492, 2494, 2502, 2509, 2515, 2532, 2533, 2534, 2621, 2622, 2669, 2670, 2671, 2672, 2673 and 2674.

Cambridgeshire (v.c. 29): 305, 357, 358, 461, 466, 469, 501, 505, 578, 585, 591, 677, 684, 720, 791, 915, 923, 931, 950, 970, 1020, 1022, 1032, 1059, 1060, 1086, 1226, 1292, 1299, 1304, 1316, 1319, 1343, 1419, 1623, 1624, 1643, 1708, 1729, 1733, 1895,

1919, 1925, 1926, 1927, 1931, 1935, 1938, 1939, 1940, 1941, 1942, 1943, 2022, 2049, 2098, 2159, 2220, 2221, 2330, 2391, 2434, 2445, 2448, 2517, 2526, 2528, 2584, 2600, 2668, 2669, 2670, 2671, 2672, 2673 and 2674.

Bedfordshire (v.c. 30): 2, 3, **4**, 466, 469, 471, 775, 776, 843, 923, 927, 931, 970, **1046**, 1367, 1673, 1723, 1834, 1932, 2019, 2097, 2098, 2159, 2220, 2330, 2391, 2434 and 2528.

Huntingdonshire (v.c. 31): 650, **771**, 950, 970, 1129, 1299, 1300, **1301**, 1304, 1314, 1317, 1850, 2159, 2330, 2434, 2442, 2445, 2448 and 2528.

Northamptonshire (v.c. 32): 238, 306, 357, 358, **597**, 612, 791, 795, 950, 970, 1021, 1022, 1078, 1223, 1297, 1298, 1299, 1300, 1302, 1304, 1305, 1630, 1671, 1931, 1932, 1946, 2159, 2203, 2304, 2330, 2391, 2434, 2442, 2448, 2500, 2501, 2526, 2528 and 2631.

East Gloucestershire (v.c. 33): 140, 141, 142, 143, 144, 145, 146, 147, 148, 150, 153, 161, 162, 163, 306, 357, 358, 480, 482, 488, 491, 505, 578, 613, 622, 923, 931, 1021, 1079, 1102, 1103, 1216, **1255**, 1302, 1331, 1382, 1558, 1626, 1701, 1729, 1733, 1766, 1834, 1879, 1955, 1976, 2068, 2159, 2203, 2315, 2330, 2360, 2365, 2402, 2432, 2434, 2435, 2442, 2448, 2488, 2490, 2492, 2494, 2496, 2498, 2500, 2502, 2506, 2509, 2515, 2523, 2526, 2528, 2600 and 2687.

West Gloucestershire (v.c. 34): 101, 140, 141, 143, 144, 146, 147, 148, 149, 151, 153, 156, 161, 164, 165, 166, 168, 221, 224, 225, 306, 345, 357, 358, 403, 461, 491, 578, 591, 622, 720, 761, 923, 931, 970, 1021, 1022, 1032, 1059, 1060, 1086, 1087, 1088, 1092, 1216, 1229, 1245, **1255**, 1318, 1331, 1343, 1346, 1354, 1382, 1459, 1490, 1642, 1643, 1684, 1728, 1729, 1733, 1772, 1879, 1927, 1931, 1976, 2010, 2051, 2089, 2100, 2159, 2215, 2220, 2221, 2309, 2330, 2360, 2390, 2391, 2392, 2402, 2432, 2434, 2435, 2480, 2487, 2490, 2494, 2496, 2498, 2500, 2502, 2505, 2506, 2509, 2510, 2515, 2526, 2528, 2600, 2668, 2669, 2670, 2671, 2672, 2673, 2674 and 2686.

Monmouthshire (v.c. 35): 154, 578, 591, 923, 931, 1229, 1230, 1245, 1728, 1878, 2010, 2034, 2159, 2330, 2432, 2434, 2442, 2448, 2502, 2509, 2515, 2526 and 2528.

Herefordshire (v.c. 36): 19, 101, **105**, 125, 148, 154, 256, 357, 358, 375, 467, 578, 585, 587, 591, 632, 633, 970, 974, 1032, 1059, 1060, 1086, 1087, 1088, 1138, 1252, 1255, 1257, 1304, 1326, 1329, 1330, 1331, 1344, 1349, 1355, 1367, 1415, 1434, 1521, 1635, 1643, 1728, 1729, 1770, 1820, 1837, 1838, 1879, 1934, 1935, 1976, 2012, 2024, 2025, 2027, 2031, 2039, 2047, 2049, 2059, 2098, 2215, 2220, 2221, 2330, 2391, 2434, 2480, 2492, 2500, 2515, 2526, 2528, 2600, 2668, 2669, 2670, 2671, 2672, 2673 and 2674.

Worcestershire (v.c. 37): 19, **28**, 45, 132, **134**, 306, 461, 466, 507, 546, 578, 920, 923, 931, 950, 967, 970, 1032, 1229, 1230, 1240, 1279, 1304, 1324, 1325, 1326, 1328, 1329, 1330, 1331, 1344, 1345, 1346, 1415, 1610, 1629, 1635, 1643, 1728, 1729, 1810, 1837, 1838, 1878, 1879, 1947, 1969, 2013, 2018, 2039, 2059, 2098, 2159, 2220, 2221, 2237, 2330, 2360, 2390, 2391, 2434, 2435, 2445, 2448, 2494, 2500, 2523, 2526, 2528, 2600, 2668, 2669, 2670, 2671, 2672, 2673 and 2674.

Warwickshire (v.c. 38): 118, 119, 120, 127, 130, 131, 132, **133**, **135**, 400, 408, 505, 507, 546, 594, 595, 596, 923, 927, 931, 970, 1021, 1208, 1304, 1345, **1406**, 1407, 1628,

1629, 1854, 1878, 1879, 1920, 1921, 1925, 2012, 2056, 2075, 2159, 2330, 2360, 2434, 2442, 2500, 2515, 2528, 2530, 2636, 2637, 2670, 2671, 2672, 2673 and 2674.

Staffordshire (v.c. 39): 72, 125, **136**, 336, 357, 358, 507, 578, 658, **758**, 906, 923, 931, 944, 950, 953, 954, **958**, 967, 970, 1208, 1229, 1307, 1355, 1812, 1833, 1836, 1839, 1854, 1924, 1926, 1933, 2013, 2019, 2047, 2159, 2209, 2330, 2360, 2391, 2434, 2480, 2482, 2499, 2500, 2515, 2526, 2528, 2670, 2671, 2672, 2673 and 2674.

Shropshire (v.c. 40): 16, 52, 94a, 225, 372, 461, 462, 480, 501, 533, 546, 574, 578, 590, 591, 613, 740, 906, 923, 924, 931, 949, 1019, 1022, 1059, 1060, 1086, 1092, 1138, 1208, 1229, 1230, 1252, 1271, 1277, 1279, 1302, 1304, 1318, 1343, 1344, 1345, 1346, 1347, 1349, 1350, 1355, 1360, 1363, 1364, 1367, 1371, 1375, 1378, 1382, 1394, 1418, 1419, 1431, 1433, 1434, 1485, 1513, 1514, 1557, 1610, 1627, 1643, 1728, 1729, 1733, 1735, 1819, 1821, **1826**, 1828, 1841, 1878, 1879, 1927, 1934, 1935, 1936, 1946, 1954, 1976, 2039, 2047, 2059, 2069, 2070, 2071, 2078, **2080**, 2083, 2086, 2098, 2159, 2189, 2198, 2220, 2309, 2330, 2335, 2360, 2391, 2424, 2434, 2442, 2448, 2496, 2500, 2502, 2509, 2523, 2526, 2528, 2588, 2668, 2669, 2670, 2671, 2672, 2673, 2674, 2694 and 2695.

Glamorganshire (v.c. 41): 152, 306, 578, 593, 618, 712, 720, 834, 850, 851, 923, 927, 931, 948, 1121, 1229, 1230, 1302, 1804, 1881, 1882, 1883, 1884, 1885, 1887, 1888, 1890, 1893, 2159, 2220, 2313, 2314, 2330, 2390, 2427, 2428, 2430, 2432, 2433, 2434, 2438, 2441, 2442, 2444, 2445, 2446, 2447, 2448, 2449, **2450**, 2496, 2500, 2502, 2509, 2510, 2515, 2523, 2526 and 2528.

Breconshire (Brecknockshire) (v.c. 42): 121, 538, 620, 720, 923, 931, 1229, 1230, 1271, 1273, 1279, 1302, 1440, 2034, 2077, 2159, 2220, 2309, 2330, 2432, 2434, 2442, 2445, 2448, 2490, 2492, 2494, 2498, 2499, 2500, 2502, 2515, 2523, 2526, 2528 and 2635.

Radnorshire (v.c. 43): 224, 225, 357, 358, 510, 538, 591, 923, 931, 1086, 1087, 1088, 1229, 1230, 1271, 1367, 1509, 1510, 1558, 1618, 1729, 1934, 2034, 2053, 2059, 2079, 2159, 2330, 2390, 2391, 2432, 2434, 2442, 2445, 2448, 2492, 2500, 2509, 2515, 2526, 2528, **2635**, 2668, 2669, 2670, 2671, 2672, 2673 and 2674.

Carmarthenshire (v.c. 44): 343, 720, 920, 923, 931, 1229, 1230, 1269, 1302, 1635, 1885, 1886, 2034, 2159, 2309, 2330, 2432, 2434, **2437**, 2442, 2445, 2448, 2502, 2509, 2515, 2526 and 2528.

Pembrokeshire (v.c. 45): 196, 207, 306, 353, 461, 578, 582, 585, 587, 622, 683, **701**, **704**, 711, 720, 802, 803, 804, 805, 834, 923, 931, 1059, 1060, 1129, 1130, 1143, 1219, 1229, 1230, 1232, 1270, 1274, 1302, 1304, 1318, 1324, 1343, 1345, 1355, 1379, 1382, 1510, 1610, 1639, 1640, 1642, 1643, 1651, 1729, 1905, 1935, 1955, 2034, 2042, 2043, 2049, 2053, 2077, 2091, 2098, 2159, 2160, 2162, **2165**, 2201, 2203, 2220, 2221, 2307, 2314, 2321, 2330, 2391, **2431**, 2432, 2434, 2435, 2440, 2442, 2443, 2445, 2448, 2452, 2498, 2505, 2523, 2526, 2528, 2673 and 2674.

Cardiganshire (v.c. 46): 255, 267, 460, 466, 467, 562, 565, 591, 682, 720, 738, 818, 865, 923, 924, 931, 1059, 1060, 1087, 1088, 1143, 1229, 1230, 1273, 1324, 1610, 1620, 1635, 1643, 1729, 1739, 1818, 2032, 2034, 2049, 2059, 2086, 2159, 2220, 2221, 2330,

2335, 2390, 2391, 2432, 2434, 2442, 2445, 2494, 2498, 2500, 2502, 2509, 2515, 2523, 2526, 2528, 2600, 2670, 2672, 2673 and 2674.

Montgomeryshire (v.c. 47): 36, 111a, 906, 923, 924, 931, 994, 1229, 1230, 1302, 1318, 1343, 1344, 1350, 1367, 1394, 1948, 2086, 2159, 2220, 2221, 2318, 2330, 2333, 2391, **2423**, 2434, 2442, 2448, 2509, 2515, 2517, 2526, 2528, 2670, 2671, 2672, 2673, 2674 and 2686.

Merionethshire (v.c. 48): 36, 107, 108, 109, 110, 111, 236, 237, 306, 321, 354, 357, 358, 377, 393, 402, 404, 408, 460, 461, 462, 469, 476, 487, 505, 538, 546, 577, 578, 582, 590, 591, 613, 654, 675, 679, 680, 681, 682, 738, 740, 865, 903, 923, 924, 931, 932, 949, 994, 997, 1005, 1047, 1086, 1088, 1091, 1127, 1130, **1131**, 1138, 1206, 1229, 1230, 1267, 1269, 1273, 1277, 1302, 1304, 1310, 1319, 1324, 1343, 1344, 1345, 1346, 1348, 1350, 1355, 1356, 1358, 1363, 1373, 1382, 1391, 1393, 1418, 1431, 1440, 1485, 1510, 1513, 1518, 1610, 1625, 1626, 1635, 1639, 1643, 1671, 1685, 1686, 1728, 1729, 1732, 1739, 1759, 1827, 1845, 1880, 1935, 1955, 1976, 1994, 2038, 2049, 2059, 2069, 2074, 2077, 2082, 2085, **2087**, 2091, 2102, 2159, 2161, 2162, 2198, 2201, 2203, 2220, 2221, 2294, 2304, 2307, 2313, 2322, 2330, 2333, 2335, 2389a, 2390, 2391, 2424, 2432, 2434, 2435, 2440, 2442, 2445, 2448, 2480, 2488, 2490, 2492, 2494, 2496, 2498, 2500, 2502, 2509, 2510, 2514, 2515, 2517, 2523, 2526, 2528, 2586, 2600, 2601, **2633**, 2668, 2669, 2670, 2671, 2672, 2673, 2674, 2691 and 2695.

Caernarvonshire (v.c. 49): 9, 11a, 13, 36, 222, 223, 224, 225, 236, 255, 282, 306, 354, 357, 358, 379, 393, 394, 404, 461, 462, 476, 487, 491, 501, 505, 508, 510, 538, 548, 565, 574, 578, 590, 591, 593, 613, 680, 711, 720, 736, 738, 739, 791, **836**, 856, 923, 924, 931, 932, 949, 1018, 1022, 1059, 1060, 1086, 1087, 1088, 1091, 1092, 1114, 1127, 1138, 1141, 1156, 1157, 1170, 1171, 1196, 1211a, 1211b, 1226, 1229, 1230, 1232, 1244, 1269, 1272, 1277, 1302, 1304, 1319, 1324, 1344, 1345, 1346, 1350, 1355, 1363, 1367, 1368, 1369, 1373, 1375, 1380, 1382, 1390, 1400, 1419, 1431, 1434, 1435, 1558, 1610, 1635, 1639, 1643, 1666, 1670, 1671, 1716, 1728, 1729, 1734, 1735, 1818, 1822, 1864, 1865, 1894, 1907, 1909, 1925, 1927, 1931, 1934, 1935, 1936, 1946, 1954, 1960, 1976, 2049, 2059, 2070, 2071, 2077, 2098, 2133, 2159, **2164**, 2198, 2220, 2294, 2307, 2322, 2327, 2328, 2330, 2360, 2389a, 2390, 2391, 2417, 2422, 2432, 2434, 2435, 2436, 2442, 2448, 2480, 2484, 2488, 2490, 2492, 2494, 2496, 2498, 2499, 2500, 2502, 2515, 2517, 2523, 2526, 2528, 2530, 2550, 2586, 2600, 2607, 2648a, 2649, 2649a, 2649b, 2651, 2666, 2668, 2669, 2670, 2671, 2672, 2673, 2674, 2675 and 2692.

Denbighshire (v.c. 50): 236, 306, 357, 358, 578, 740, 920, 923, 931, 1059, 1060, 1092, 1128, 1229, 1230, 1244, 1318, 1343, 1344, 1369, 1433, 1434, 1514, 1635, 1643, 1728, 1794, 1934, 1935, 1946, 2049, 2053, 2220, 2221, 2299, 2309, 2327, 2330, 2333, 2391, 2402, 2434, 2440, 2448, 2480, 2482, 2492, 2494, 2498, 2499, 2500, 2502, 2509, 2515, 2523, 2526, 2528, 2601, 2670, 2671, 2672, 2673, 2674 and 2686.

Flintshire (v.c. 51): 73, 236, 255, 378, **551**, **552**, 564, 923, 1092, 1273, 1489, 1799, 2071, 2220, 2221, 2309, 2391, 2420a, 2432, 2434, 2480, 2494, 2502, 2509, 2515, 2528, 2600, 2671, 2672, 2673 and 2674.

Anglesey (v.c. 52): 255, 282, 357, 358, 380, 401, 534, 535, 537, 564, 565, 566, 578, 579, 591, 630, 686, 700, 709, 710, 711, 736, 738, 814, 815, **836**, 838, 854, 863, 920,

923, 931, 932, 999, 1020, 1059, 1060, 1092, 1114, 1129, 1130, 1132, 1140, 1143, 1149, 1196, 1273, 1344, 1345, 1355, 1367, 1397, 1399, 1419, 1635, 1639, 1643, 1865, 1901, 1902, 1930, 2053, 2082, 2091, 2098, 2104, 2133, 2159, 2160, 2220, 2221, 2307, 2313, 2327, 2328, 2330, **2373**, 2386, 2391, 2412, 2416, 2420, 2432, 2434, 2442, 2445, 2448, 2500, 2505, 2509, 2515, 2523, 2526, 2528, 2548, 2668, 2669, 2670, 2671, 2672, 2673 and 2674.

South Lincolnshire (v.c. 53): 381, 578, 630, 665, 791, 950, 960, 1053, 1338, 1658, 1859, 1952, 2108, 2109, 2111, 2113, 2116, 2118, 2119, 2120, 2122, 2123, 2124, **2127**, 2128, 2131, 2138, 2139, 2141, 2434, 2528, 2678 and 2680.

North Lincolnshire (v.c. 54): 155, 429, 578, 630, 923, 931, 950, 960, 969, 970, 1053, 1054, 1139, 1304, 1332, 1337, 1338, 1557, 1723, 1750, 1754, 1793, 1796, 1952, 2108, 2109, 2110, 2111, 2112, 2113, 2114, 2115, 2116, 2117, 2118, 2120, 2121, 2122, 2124, **2127**, 2128, 2131, 2134, 2135, 2138, 2139, 2140, 2141, 2169, 2196, 2297, 2448, 2528, 2676, 2677, 2679, 2681, 2682 and 2683.

Leicestershire (including Rutland) (v.c. 55): 102, 125, 200, 251, 306, 336, 357, 358, 415, 451, 461, 466, 505, 508, 510, 574, 578, 591, 665, 920, 923, 924, 927, 929, 931, 933, 939, 949, 950, 963, 970, **972**, 973, 1021, 1022, 1059, 1060, 1070, 1072, 1073, 1074, 1075, 1079, 1086, 1226, 1230, 1273, 1300, 1302, 1304, 1306, 1310, 1315, 1343, 1344, 1345, 1346, 1349, 1350, 1355, 1367, 1408, 1419, 1434, 1557, 1634, 1635, 1643, 1658, 1716, 1729, 1733, 1853, 1870, 1871, 1872, 1873, 1874, 1875, 1926, 1927, 1931, 1932, 1946, 1976, 2059, 2159, 2239, **2243**, **2245**, **2246**, 2330, 2360, 2390, 2391, 2426, 2434, 2442, 2515, 2526, 2528, 2530, 2669, 2670, 2671, 2672, 2673 and 2674.

Nottinghamshire (v.c. 56): 255, 357, 358, 367, **368**, 375, 574, 576, 923, 931, 970, **1083**, 1812, 1926, 2052, 2139, 2330, 2391 and 2528.

Derbyshire (v.c. 57): 152, 197, 215, 255, 306, 336, 341, 357, 358, 449, 460, 461, 462, 510, 568, 574, 578, 585, 586, 591, 601, 646, 658, 780, 798, 807, 816, 817, 839, 916, 918, **920**, 923, 924, 928, 931, 933, 943, **944**, **946**, 950, 954, 960, 963, 970, 989, 1085, 1092, 1196, 1229, 1230, 1243, 1274, 1302, 1318, 1343, 1345, 1413, 1415, 1485, 1489, 1635, 1729, 1745, 1831, 1832, 1833, 1835, 1839, 1879, 1920, 1921, 1946, 1954, 1966, 1967, 2019, 2098, 2170, 2172, 2206, 2218, 2220, 2221, 2330, 2334, 2339, 2360, 2391, 2421, 2432, 2434, 2448, 2480, 2482, 2488, 2490, 2498, 2500, 2502, 2509, 2526, 2528, 2530, 2658, 2668, 2669, 2670, 2671, 2672, 2673 and 2674.

Cheshire (v.c. 58): 61, 65, 69, 70, 71, 73, 74, 77, 78, 157, 167, 236, 449, **556**, 658, 839, 840, 923, 931, 940, 944, 954, 958, 970, 1229, 1230, 1346, 1356, 1542, 1543, 1671, 1778, 1833, 1931, 1946, 1953, 1961, 2069, 2360, 2368, 2372, **2374**, 2375, 2390, 2391, 2420a, 2434, 2445, 2448, 2470, 2480, 2492, 2496, 2499, 2500, 2502, 2509, 2515, 2526, 2528, 2546, 2547, 2600, 2611, 2669, 2670, 2672, 2673 and 2674.

South Lancashire (v.c. 59): 53, 65, 174, 225, 226, 236, 255, 357, 358, 449, 578, 591, 622, 630, 707, 712, 839, 841, 923, 931, 940, 950, 960, 970, 1028, 1059, 1060, 1079, 1086, 1087, 1088, 1092, 1230, 1389, 1435, 1440, 1542, 1543, 1592, 1635, 1643, 1921, 1922, 1931, 1932, 1946, 1968, 1974, 1976, 2069, 2070, 2071, 2098, 2099, 2159, 2203, 2209, 2220, 2221, 2254, 2255, 2256, 2306, 2314, 2322, 2330, 2332, 2334, 2368, 2370,

2376, 2391, 2420a, 2432, 2434, 2481, 2500, 2526, 2528, 2591, 2595, 2597, 2598, 2600, 2610, **2611**, 2612, 2613, 2669, 2670, 2671, 2672, 2673 and 2674.

West Lancashire (v.c. 60): 62, 66, 225, 294, 357, 433, 574, 578, 622, 707, 923, 924, 931, 1059, 1060, 1079, 1092, 1225, 1226, 1229, 1230, 1273, 1304, 1435, 1440, 1554, 1592, 1635, 1666, 1739, 1780, 1930, 1932, 1946, 1976, 2159, 2209, 2220, 2221, 2313, 2330, 2334, 2432, 2434, 2436, 2442, 2498, 2500, 2502, 2503, 2509, 2515, 2526, 2528, 2592, 2594, 2596, 2599, 2600, 2602, 2614, **2615**, 2619, 2644, 2645, 2673 and 2674.

South-east Yorkshire (v.c. 61): **176**, 218, 233, 255, 306, 357, 358, 387, 438, 466, 574, 578, 585, 613, 724, 960, 1022, 1086, 1090, 1092, 1139, 1229, 1346, 1355, 1364, 1367, 1389, 1401, 1402, 1433, 1485, 1567, 1572, 1576, 1592, 1611, 1749, 1752, 1753, 1754, 1807, 1837, 1838, 1920, 1921, 1927, 1934, 1935, 1978, 1981, 2098, 2102, 2125, 2151, 2171, 2220, 2221, 2318, 2330, 2360, 2390, 2424, 2435, 2436, 2445, 2448, 2480, 2499, 2500, 2517, **2521**, 2528, 2544, 2545, 2549, 2576, 2601, 2658, 2669, 2670, 2671, 2672, 2673 and 2674.

North-east Yorkshire (v.c. 62): 9, 175, **176**, 255, 306, 357, 358, 371, 372, 388, 389, 393, 414, 437, 461, 466, 482, 487, 491, 532, 533, 574, 577, 578, 582, 585, 613, 651, 661, 696, 780, 860, 923, 924, 931, 960, 970, 990, 1022, 1031, 1086, 1092, 1130, 1138, 1139, 1196, 1229, 1230, 1250, 1272, 1302, 1304, 1344, 1345, 1346, 1347, 1348, 1349, 1350, 1355, 1364, 1367, 1375, 1389, 1401, 1402, 1419, 1429, 1430, 1431, 1433, 1434, 1485, 1486, 1509, 1510, 1513, 1514, 1516, 1518, 1568, 1576, 1592, 1611, 1635, 1636, **1641**, 1642, 1643, 1644, 1665, 1728, 1729, 1732, 1786, 1787, 1789, 1807, 1828, 1837, 1838, 1920, 1921, 1927, 1934, 1935, 1936, 1955, 1964, 1976, 2030, 2052, 2059, 2098, 2102, 2125, 2133, 2135, 2140, 2159, 2161, 2162, 2198, 2201, 2203, 2220, 2221, 2294, 2295, 2318, 2330, 2335, 2340, 2341, 2345, 2360, 2367, 2390, 2397, 2399, 2424, 2434, 2435, 2436, 2448, 2478, 2480, 2490, 2492, 2496, 2498, 2500, 2502, 2505, 2512, 2514, 2515, 2517, **2521**, 2523, 2526, 2528, 2543, 2549, 2561, 2578, 2600, 2601, 2658, 2669, 2670, 2671, 2672, 2673, 2674, 2694 and 2695.

South-west Yorkshire (v.c. 63): 41a, 41b, 122, **176**, 227, 255, 257, 279, 306, 324, 357, 358, 370, 389, 429, 437, 461, 466, 505, 517, 518, 520, 570, 574, 578, 585, 591, 613, 640, 642, 643, 646, 647, 652, 658, 839, 916, 923, 931, 944, 950, 960, 969, 970, 977, 1022, 1027, 1086, 1087, 1088, 1092, 1230, **1337**, 1343, 1346, 1355, 1364, 1367, 1389, 1401, 1402, 1433, 1440, 1485, 1510, 1534, 1569, 1576, 1579, 1592, 1595, 1597, 1608, 1609, 1611, 1619, 1631, 1632, 1633, 1643, 1748, 1751, 1754, 1788, 1791, 1807, 1817, 1833, 1837, 1838, 1920, 1921, 1927, 1930, 1932, 1934, 1935, 1952, 1976, 1979, 1980, 1983, 1987, 1988, 1989, 1990, 2019, 2055, 2098, 2102, 2125, 2129, 2130, 2132, 2134, 2135, **2144**, 2146, 2150, 2154, 2156, 2157, 2172, 2220, 2221, 2235, 2255, 2256, 2318, 2330, 2341, 2360, 2390, 2424, 2434, 2435, 2448, 2472, 2477, 2480, 2490, 2496, 2500, 2502, 2505, 2512, 2521, 2526, 2528, 2540, 2549, 2556, 2557, 2562, 2564, 2565, 2566, 2576, 2578, 2583, 2600, 2601, 2620, 2658, 2669, 2670, 2671, 2672, 2673, 2674 and 2695.

Mid-west Yorkshire (v.c. 64): 10, 31, 34, 175, **176**, 177, 214, 227, 255, 306, 324, 325, 342, 349, 355, 357, 358, 370, 384, 385, 386, 389, 431, 437, 461, 466, 570, 574, 578, 582, 585, 591, 606, 613, 647, 664, 692, 723, 739, 769, 824, 923, 924, 931, 948, 960, 969, 979, 992, 1022, 1059, 1060, 1078, 1086, 1087, 1088, 1090, 1092, 1126, 1130,

1138, 1195, 1196, 1207, 1229, 1230, 1269, 1299, 1302, 1307, 1318, 1333, 1334, **1337**, 1339, 1340, 1346, 1355, 1364, 1367, 1389, 1401, 1402, 1419, 1431, 1433, 1434, 1440, 1476, 1485, 1570, 1576, 1579, 1592, 1595, 1596, 1597, 1611, 1619, 1643, 1671, 1728, 1729, 1751, 1754, 1795, 1800, 1802, 1807, 1830, 1832, 1837, 1838, 1897, 1920, 1921, 1927, 1931, 1932, 1934, 1935, 1946, 1952, 1976, 1981, 1982, 1984, 1985, 2017, 2055, 2098, 2102, 2125, 2126, 2129, 2130, 2132, 2133, 2134, 2135, 2140, **2144**, 2146, 2150, 2151, 2156, 2157, 2159, 2172, 2188, 2220, 2221, 2233, 2235, 2236, 2248, 2251, 2309, 2315, 2318, 2324, 2325, 2327, 2328, 2330, 2341, 2360, 2371, 2390, 2391, 2424, 2432, 2434, 2435, 2442, 2445, 2448, 2480, 2490, 2492, 2494, 2496, 2498, 2500, 2502, 2503, 2505, 2509, 2510, 2512, 2515, 2517, 2521, 2526, 2528, 2542, 2549, 2552, 2555, 2560, 2567, 2576, 2578, 2580, 2582, 2583, 2587, 2600, 2601, 2620, 2646, 2658, 2664, 2665, 2668, 2669, 2670, 2671, 2672, 2673, 2674 and 2695.

North-west Yorkshire (v.c. 65): 9, 115, 173, **176**, 177, 217, 255, 277, 278, 306, 357, 358, 389, 431, 437, 461, 466, 487, 547, 570, 574, 578, 582, 585, 613, 622, 641, 662, 663, 664, 692, 768, 791, 792, 822, 865, 923, 924, 931, 960, 970, 1022, 1059, 1060, 1086, 1087, 1088, 1092, 1113, 1138, 1195, 1196, 1229, 1230, 1277, 1302, 1333, 1335, 1336, **1337**, 1343, 1345, 1346, 1355, 1364, 1367, 1389, 1401, 1402, 1433, 1485, 1554, 1556, 1566, 1571, 1576, 1579, 1592, 1597, 1611, 1639, 1642, 1643, 1754, 1785, 1790, 1807, 1830, 1837, 1838, 1864, 1920, 1921, 1927, 1930, 1934, 1935, 1976, 2017, 2029, 2030, 2098, 2102, 2125, 2133, 2155, 2159, 2220, 2221, 2230a, 2248, 2318, 2330, 2341, 2360, 2390, 2391, 2424, 2435, 2448, 2464, 2480, 2498, 2500, 2512, 2514, 2515, **2521**, 2528, 2541, 2549, 2551, 2553, 2554, 2558, 2559, 2563, 2576, 2578, 2587, 2600, 2646, 2647, 2648, 2658, 2659, 2660, 2669, 2670, 2671, 2672, 2673 and 2674.

Durham (v.c. 66): 6, 115, 217, 261, 277, 347, 431, 437, 461, 467, 469, 476, 482, 498, 505, 546, 574, 578, 602, 622, 646, 662, 740, 775, 780, 782, 788, 792, 826, 837, 854, 865, 887, 920, 923, 924, 931, 970, 994, 1030, 1059, 1060, 1112, 1113, 1138, 1151, 1155, 1166, 1175, 1180, 1195, 1196, 1197, 1198, 1200, 1215, 1226, 1229, 1230, 1277, 1304, 1343, 1346, 1363, 1364, 1510, 1512, 1513, 1635, 1642, 1643, 1644, 1671, 1728, 1729, 1790, 1830, **1852**, 1864, 1945, 1975a, 2030, 2049, 2052, 2098, 2102, 2133, 2159, 2191, 2201, 2220, 2221, 2312, 2330, 2340, 2360, 2386, 2390, 2391, 2397, 2399, 2402, 2424, 2434, 2442, 2448, 2502, 2510, 2517, 2526, 2528, 2563, 2589, 2659, 2660, **2661**, 2663, 2672, 2673, 2674, 2687 and 2695.

South Northumberland (v.c. 67): 59, 127, 129, 186, 206, 261, 357, 358, 461, 476, 503, 505, 510, 574, 577, 578, 646, 773, 774, 775, 776, 778, 780, 781, 782, 783, 784, 788, 798, 834, **865**, 884, 920, 923, 924, 927, 931, 939, 948, 969, 970, 1018, 1059, 1060, 1117, 1166, 1167, 1170, 1171, 1172, 1174, 1180, 1181, 1182, 1183, 1185, 1192, 1193, 1195, 1196, 1230, 1309, 1315, 1343, 1485, 1510, 1636, 1643, 1674, 1712, 1729, 1887, 1888, 1934, 1976, 2017, 2019, 2052, 2097, 2133, 2220, 2221, 2330, 2391, 2399, 2432, 2434, 2442, 2448, 2460, 2517, 2526, 2528, 2659, **2661**, 2663, 2672, 2673 and 2674.

North Northumberland (Cheviotland) (v.c. 68): 129, 224, 225, 264, 357, 358, 443, 461, 574, 578, 582, 780, **786**, 834, **865**, 866, 867, 872, 877, 880, 881, 882, 884, 920, 923, 924, 931, 939, 1055, 1059, 1060, 1180, 1182, 1185, 1196, **1202**, 1203, **1204**, 1230, 1313, 1343, 1346, 1415, 1485, 1577, 1832, 1837, 1838, 1971, 1976, 2059, 2133, 2187,

2220, 2221, 2330, 2339, 2351, 2352, 2391, 2434, 2448, 2460, 2517, 2526, 2528, 2659, **2661**, 2663, 2673, 2674 and 2695.

Westmorland (including the Furness district of north Lancashire) (v.c. 69): 6, 68, 216, 224, 225, 241, 255, 261, 294, 306, 323, 357, 358, 421, 431, 437, 461, 469, 476, 480, 482, 487, 505, 508, 510, 545, 546, 547, 548, 578, 621, 622, 630, 644, 653, 662, 663, 678, 707, 779, 920, 923, 931, 948, 949, 963, 970, 1047, 1059, 1060, 1064, 1076, 1078, 1086, 1087, 1088, 1137, 1138, 1151, 1152, 1153, 1155, 1167, 1195, 1196, 1229, 1230, 1234a, 1269, 1273, 1277, 1343, 1346, 1355, 1364, 1415, 1434, 1435, 1449, 1508, 1510, 1511, 1513, 1550, 1551, 1552, 1553, **1554**, 1555, 1556, 1557, 1558, 1559, 1560, 1561, 1562, 1563, 1565, 1575, 1610, 1635, 1643, 1650, 1662, 1666, 1698, 1713, 1715, 1716, 1717, 1718, 1719, 1729, 1733, 1739, 1780, 1781, 1792, 1797, 1805, 1830, 1879, 1920, 1921, 1930, 1931, 1932, 1935, 1936, 1946, 1976, 2021, 2030, 2036, 2041, 2044, 2049, 2052, 2059, 2098, 2133, 2159, 2220, 2221, 2240, 2241, 2242, 2244, 2302, 2307, 2309, 2330, 2360, 2391, 2432, 2434, 2435, 2442, 2445, 2448, 2490, 2494, 2498, 2500, 2502, 2504, 2515, 2523, 2526, 2528, 2541, 2563, 2570, 2571, 2573, 2574, 2586, 2605, 2616, 2644, 2645, **2650**, 2653, 2654, 2658, 2669, 2670, 2671, 2672, 2673, 2674 and 2695.

Cumberland (v.c. 70): 5, 57, 76, 94a, 112, 125, 261, 306, 357, 358, 395, 401, 461, 477, 503, 505, 508, 510, 524, 546, 547, 574, 578, 582, 585, 587, 602, 620, 622, 662, 675, 678, 711, 718, 731, 740, 779, 852, 854, 855, 923, 924, 928, 931, 970, 1041, 1059, 1060, 1086, 1130, 1137, 1138, 1151, 1167, **1168**, 1170, 1171, 1173, 1177, 1180, 1181, 1182, 1183, 1184, 1185, 1191, 1192, 1195, 1196, 1201, 1226, 1229, 1230, 1241, 1269, 1273, 1277, 1304, 1324, 1343, 1344, 1345, 1346, 1355, 1356, 1393, 1415, 1419, 1449, 1455, 1502, 1510, 1516, 1551, 1553, 1554, 1557, 1558, 1560, 1564, 1610, 1635, 1639, 1643, 1650, 1662, 1663, 1712, 1715, 1729, 1732, 1733, 1735, 1744a, 1780, 1792, 1805, 1830, 1930, 1931, 1935, 1936, 1946, 2021, 2036, 2038, 2041, 2044, 2048, 2049, 2052, 2059, 2088, 2133, 2159, 2172, 2209, 2220, 2221, 2241, 2307, 2321, 2330, 2339, 2361, 2390, 2391, 2399, 2402, 2417, 2432, 2434, 2435, 2442, 2445, 2448, 2488, 2498, 2500, 2502, 2505, 2507, 2509, 2515, 2523, 2526, 2528, 2571, 2574, 2586, 2600, 2601, 2658, 2660, **2662**, 2669, 2670, 2671, 2672, 2673 and 2674.

Isle of Man (v.c. 71): 57, 255, 433, **648**, 760, 830, 904, **905**, 931, 1007, 1129, 1130, 1137, 1182, 1228, 1229, 1273, 1277, 1302, 1304, 1413, 1415, 1522, 1554, 1558, 1635, 1650, 1666, 1728, 1729, 1739, 1805, 1878, 1930, 1946, 2159, 2307, 2321, 2330, 2335, 2432, 2434, 2442, 2445, 2448, 2498, 2500, 2502, 2515, 2523, 2526, 2528, 2603, 2604, 2606, 2608, 2632, 2669, 2670, 2672, 2673 and 2674.

Dumfriesshire (v.c. 72): 225, 255, 274, 357, 358, 393, 461, 715, 825, 863, 923, 931, 980, 1025, 1059, 1060, 1088, 1137, 1196, 1229, 1304, 1310, 1345, 1401, 1402, 1415, 1418, 1419, 1431, 1434, **1461**, 1462, 1463, **1469**, 1502, 1610, 1635, 1643, 1657, 1910, 2059, 2061, 2098, 2102, 2106, 2133, **2142**, 2198, 2211, 2220, 2221, 2259, 2275, 2289, 2328, 2330, 2360, 2434, 2442, 2448, 2505, 2509, 2510, 2517, 2526, 2528, 2669, 2670, 2671, 2672, 2673 and 2674.

Kirkcudbrightshire (v.c. 73): 13, 16, 273, 510, 546, 578, 807, 861, 863, 923, 924, 931, 932, 1130, 1135, 1196, 1226, 1229, 1230, 1270, 1274, 1277, 1460, **1461**, 1463, 1464, 1466, **1469**, 1470, 1473, 1487, 1488, 1610, 1671, 1728, 1914, 1935, 2059, 2133,

2153, 2159, 2234, 2281, 2282, 2285, 2286, 2287, 2289, 2291, 2330, 2434, 2435, 2442, 2445, 2448, 2502, 2505, 2517, 2526, 2528, 2581 and 2600.

Wigtownshire (v.c. 74): 198, 371, 461, 620, 809, 819, 923, 931, 1395, **1461**, 1463, 1464, 1465, 1468, **1469**, 1639, 1954, 2434, 2440, 2442, 2445, 2448, 2505 and 2528.

Ayrshire (v.c. 75): 181, 271, 272, 276, 340, 578, 666, 705, 711, 808, 923, 931, 949, 1229, 1287, 1344, 1345, 1367, 1401, 1402, 1415, 1419, 1610, 1723, 1724, 1728, 1837, 1838, 2059, 2159, 2275, 2326, 2330, 2400, 2434, 2435, 2445, 2505, 2519, 2526 and 2528.

Renfrewshire (v.c. 76): 306, 348, 607, 666, 667, 931, 1066, 1138, 1613, 1721, 1723, 1724, 2224, 2275 and 2528.

Lanarkshire (v.c. 77): 180, 187, 276, 306, 357, 358, 461, 578, 666, 694, 695, 743, 832, 923, 931, 1066, 1229, 1231, 1280, 1384, 1402, 1415, 1419, 1431, 1444, 1445, 1467, 1610, 1650, 1721, 1723, 1724, 1739, 1837, 1838, 2098, 2133, 2220, 2275, 2330, 2515, 2517, 2519, 2526, 2528, 2669, 2670, 2671, 2672, 2673 and 2674.

Peeblesshire (v.c. 78): 187, 357, 358, 461, 694, 695, 713, 832, 869, 876, 923, 931, 1269, 1401, 1402, 1415, 1419, 1431, 1434, 1444, 1445, 1467, 1502, 1573, 1574, 1837, 1838, 2062, 2098, 2159, 2220, 2442, 2494, 2498, 2505, 2517, 2526, 2528, 2669, 2670, 2671, 2672, 2673 and 2674.

Selkirkshire (v.c. 79): 923, 931, 1431, 2339 and 2445.

Roxburghshire (v.c. 80): 188, 619, 780, **865**, 871, 877, 878, 879, 885, 1159, **1204**, 1230, 1235, 1299, 1415, 1419, 2187, 2288, 2434, 2505 and 2528.

Berwickshire (v.c. 81): 35, 170, 339, 365, 366, 401, 711, 864, **865**, 870, 873, 875, 880, 923, 931, 985, 1137, 1196, **1202**, 1203, **1204**, 1229, 1345, 1419, 1431, 1434, 1575, 1577, 1635, 1896, 2059, 2159, 2187, 2330, 2352, 2434, 2448, 2496, 2498, 2502, 2509, 2517, 2526, 2528 and 2695.

East Lothian (Haddingtonshire) (v.c. 82): 180, 182, 184, 187, 712, 786, 832, 833, 834, 868, 874, 883, 907, 924, 1059, 1060, 1230, 1313, 1343, 1367, 1401, 1402, 1415, 1419, 1433, 1434, 1643, 1669, 1896, 2159, 2220, 2221, 2330, 2424, 2434, 2442, 2445, 2505, 2509, 2510, 2517, 2526, 2528, 2669, 2670 and 2672.

Midlothian (Edinburghshire) (v.c. 83): 178, 179, 180, **187**, 255, 306, 357, 358, 461, 567, 574, 578, 593, 602, 623, 694, 695, 832, 833, 834, 907, 923, 931, 949, 950, 1059, 1060, 1090, 1229, 1230, 1231, 1345, 1356, 1401, 1402, 1411, 1415, 1419, 1431, 1434, 1444, 1445, 1459, 1467, 1471, 1490, 1581, 1610, 1635, 1639, 1643, 1650, 1656, 1728, 1837, 1838, 1952, 2059, 2062, 2098, 2159, 2178, 2218, 2220, 2313, 2318, 2330, 2390, 2434, 2442, 2492, 2502, 2505, 2510, 2517, 2526, 2528, 2669, 2670, 2671, 2672, 2673 and 2674.

West Lothian (Linlithgowshire) (v.c. 84): 187, 832, 834, 907, 923, 931, 1229, 1401, 1402, 1419, 1723, 2330, 2517, 2526, 2528, 2672, 2673 and 2674.

Fifeshire (including Kinross-shire) (v.c. 85): 187, 444, 445, 446, 482, 578, 634, 659, 832, 834, 907, 917, 918, 923, 931, 939, 960, 1024, 1025, 1026, 1118, 1229, 1279, 1322,

1350, 1401, 1402, 1415, 1419, 1431, 1433, 1434, 1590, 1603, 1656, 1720, 1723, 1916, 2060, 2063, 2159, **2166**, 2445, 2448, 2476, 2505, 2509, 2510, 2515, 2517, 2519, 2528, 2656, 2669, 2670, 2672, 2673 and 2674.

Stirlingshire (v.c. 86): 49, 255, 258, 357, 358, 457, 461, 463, 578, 666, 759, 923, 931, 1059, 1060, 1066, 1067, 1115, 1195, 1226, 1229, 1230, 1277, 1359, 1401, 1402, 1415, 1418, 1419, 1429, 1610, 1687, 1698, 1723, 1724, 1733, 1814, 1837, 1838, 1944, 2098, 2220, 2221, 2223, 2275, 2330, 2349, 2362, 2442, 2445, 2505, 2517, 2526 and 2528.

West Perthshire (including Clackmannanshire) (v.c. 87): 1, 66, 81, 187, 371, 372, 461, 463, 477, 487, 548, 578, 582, 615, 666, 675, 793, 923, 931, 1059, 1060, 1090, 1138, 1229, 1230, 1269, 1273, 1277, 1279, 1351, 1387, 1415, 1417, 1508, 1518, 1558, 1589, 1590, 1610, 1639, 1671, 1682, 1698, 1723, 1725, 1728, 1729, 1732, 1837, 1838, 1863, 1931, 1946, 1991, 2035, 2049, 2159, 2220, 2275, 2330, 2355, 2356, 2397, 2409, 2434, 2435, 2442, 2473, 2480, 2498, 2500, 2502, 2505, 2509, 2510, 2517, 2519, 2523, 2526, 2528, **2618** and 2623.

Mid Perthshire (v.c. 88): 1, 13, 20, 38, 51, 60, 64, 66, 180, 190, 191, 192, 193, 275, 337, 357, 358, 371, 372, 373, 393, 394, 401, 402, 406, 457, 458, 459, 460, 461, 462, 463, 464, 467, 470, 474, 476, 477, 480, 481, 482, 484, 485, 486, 487, 488, 491, 501, 503, 505, 510, 538, 547, 574, 582, 585, 588, 602, 620, 626, 630, 663, 685, 714, 742, 748, 791, 793, 794, 857, 865, 909, 923, 924, 931, 948, 949, 960, 978, 980, 982, 983, 994, 997, 1001, 1025, 1026, 1032, 1053, 1054, 1059, 1060, 1092, 1112, 1127, **1134**, 1137, 1138, 1144, 1151, 1195, 1221, 1226, 1229, 1230, 1262, 1270, 1272, 1273, 1277, 1279, 1302, 1303, 1304, 1318, 1343, 1344, 1345, 1346, 1351, 1352, 1355, 1357, 1362, 1387, 1401, 1402, 1411, 1413, 1415, 1417, 1418, 1419, 1420, 1423, 1430, 1431, 1433, 1434, 1440, 1484, 1485, 1486, 1489, 1502, 1503, 1507, 1508, 1510, 1512, 1518, 1520, 1521, 1590, 1591, 1600, 1603, 1610, 1626, 1635, 1642, 1643, 1650, 1656, 1671, 1676, 1677, 1679, 1682, 1683, 1685, 1686, 1689, 1691, 1695, 1697, 1698, 1699, 1702, 1708, 1711, 1728, 1729, 1732, 1733, 1734, 1735, 1798, 1829, 1837, 1838, 1850, 1851, 1863, 1930, 1932, 1934, 1935, 1946, 1985, 1992, 2059, 2062, 2098, 2140, 2159, 2200, 2201, 2209, 2220, 2221, 2225, 2229, 2230, 2233, 2262, 2266, 2267, 2268, 2269, 2271, 2273, 2275, 2276, 2279, 2283, 2284, 2286, 2290, 2291, 2306, 2307, 2309, 2311, 2312, 2322, 2328, 2330, 2335, 2336, 2361, 2390, 2403, 2405, 2409, 2424, 2429, 2432, 2434, 2435, 2442, 2448, 2465, 2473, 2480, 2490, 2496, 2498, 2499, 2500, 2502, 2505, 2509, 2510, 2514, 2515, 2517, 2519, 2523, 2526, 2528, 2579, 2590, 2592, 2600, **2618**, 2669, 2670, 2671, 2672, 2673, 2674 and 2695.

East Perthshire (v.c. 89): 193, 198, 221, 275, 306, 337, 357, 358, 392, 407, 408, 412, 413, 461, 462, 464, 472, 475, 476, 477, 480, 482, 486, 487, 494, 501, 503, 505, 510, 578, 579, 620, 664, 714, 747, 751, 752, 754, 755, 793, 794, 917, 923, 924, 931, 960, 998, 1024, 1025, 1026, 1087, 1088, 1090, 1116, 1118, 1130, 1138, 1151, 1215, 1229, 1230, 1270, 1273, 1277, 1279, 1304, 1318, 1344, 1355, 1356, 1367, 1387, 1415, 1417, 1418, 1419, 1420, 1431, 1433, 1434, 1440, 1502, 1507, 1508, 1510, 1512, 1514, 1518, 1558, 1590, 1600, 1603, 1643, 1666, 1671, 1694, 1696, 1698, 1700, 1708, 1711, 1715, 1728, 1729, 1732, 1733, 1768, 1863, 1864, 2059, 2098, 2159, 2162, 2203, 2220, 2275, 2276, 2309, 2321, 2330, 2334, 2335, 2419, 2424, 2434, 2435, 2442, 2445, 2448,

2465, 2473, 2500, 2509, 2510, 2515, 2517, 2519, 2523, 2526, 2528, 2579, 2600, 2617, **2618**, 2625, 2669, 2670, 2671, 2672, 2673, 2674 and 2687.

Angus (Forfarshire) (v.c. 90): 30, 255, **280**, 281, 306, 337, 393, 416, 461, 463, 480, 491, 578, 582, 602, 603, 620, 621, 622, 631, 664, **697**, 744, 745, 746, 747, 749, 750, 751, 752, 753, 754, 755, 756, 756a, 756b, 827, 923, 924, 931, 949, 960, 980, 984, 1025, 1026, 1059, 1060, 1118, 1137, 1138, 1226, 1229, 1230, 1270, 1274, 1277, 1279, 1304, 1323, 1344, 1346, 1356, 1367, 1390, 1396, 1415, 1419, 1433, 1434, 1440, 1502, 1503, 1507, 1590, 1603, 1604, 1610, 1635, 1639, 1642, 1643, 1644, 1656, 1671, 1673, 1716, 1954, 1955, 1993, 2059, 2140, 2159, 2220, 2221, 2226, 2275, 2284, 2312, 2314, 2330, 2334, 2348, 2360, 2434, 2440, 2442, 2445, 2448, 2498, 2505, 2509, 2510, 2515, 2517, 2519, 2523, 2526, 2528, 2671, 2672, 2673 and 2674.

Kincardineshire (v.c. 91): 13, 84, 202, 216, 458, 459, 460, 461, 462, 463, 466, 470, 476, 491, 493, 498, 510, 533, 546, 578, 582, 583, 620, 621, 622, 623, 624, 629, 711, 807, 808, 920, 923, 931, 939, 948, 965, 1118, 1137, 1229, 1230, 1273, 1279, 1304, 1367, 1440, 1510, 1558, 1599, 1635, 1639, 1660, 1671, 1695, 1728, 1729, 1733, 1962, 1966, 1967, 1972, 2053, 2059, 2159, 2160, 2162, 2309, 2330, 2434, 2440, 2442, 2445, 2448, 2498, 2502, 2510, 2517, 2523, 2526, 2528 and 2569.

South Aberdeenshire (v.c. 92): 1, 15, 32, 183, 198, 205, 216, 235, 237, 344, 358, 371, 372, 393, 412, 424, 450, 454, 455, 457, 458, 459, 460, 461, 462, 463, 467, 468, 469, 470, 471, 475, 476, 477, 480, 487, 491, 498, 505, 506, 508, 509, 510, 536, 546, 578, 582, 583, 585, 602, 620, 755, 756b, 807, 823, 920, 923, 924, 931, 949, 997, 1000, 1004, 1059, 1060, 1087, 1088, 1112, 1119, 1127, 1138, 1194, 1226, 1229, 1230, 1270, 1272, 1273, 1277, 1279, 1304, 1306, 1346, 1351, 1355, 1356, 1362, 1367, 1371, 1376, 1416, 1418, 1419, 1420, 1429, 1430, 1431, 1433, 1434, 1440, **1480**, 1486, 1489, 1501, 1503, 1506, 1507, 1510, 1558, 1594, 1599, 1626, 1635, 1643, 1644, 1652, 1671, 1682, 1689, 1690, 1693, 1695, 1696, 1698, 1728, 1729, 1733, 1739, 1851, 1863, 1908, 1935, 1946, 1955, 1966, 1967, 1996, 2059, 2098, 2159, 2163, 2185, 2220, 2221, 2233, 2262, 2266, 2275, 2298, 2314, 2322, 2330, 2397, 2399, 2408, 2424, 2434, 2442, 2445, 2448, 2451, 2465, 2466, 2494, 2496, 2509, 2510, 2517, 2519, 2523, 2526, 2528, 2569, 2572, 2579, 2638, 2669, 2670, 2671, 2672, 2673 and 2674.

North Aberdeenshire (v.c. 93): 216, 578, 583, 807, 808, 923, 931, 993, 1087, 1088, 1138, 1229, 1230, 1289, 1419, 1431, 1436, 1682, 1852, 1860, 1861, 1863, 2098, 2220, 2434, 2442, 2445, 2448, 2528, 2569 and 2627.

Banffshire (v.c. 94): 344, 450, 487, 510, 547, 582, 583, 602, 608, 622, 700, 755, 808, 809, 823, 923, 931, 1053, 1054, 1059, 1060, 1078, 1119, 1226, 1229, 1270, 1274, 1299, 1300, 1304, 1356, 1419, 1489, 1490, 1502, 1503, 1506, 1507, 1594, 1635, 1643, 1644, 1652, 1728, 1729, 1863, 1912, 1996, 2059, 2220, 2298, 2330, 2434, 2436, 2473, 2515, 2517, 2519, 2526, 2528, 2671, 2672, 2673 and 2674.

Morayshire (Elgin, including part of Inverness-shire) (v.c. 95): 15, 306, 371, 372, 461, 593, 620, 711, 923, 931, 987, 1019, 1024, 1226, 1229, 1230, 1246, 1248, 1272, 1300, 1302, 1419, 1507, 1603, 1635, 1671, 1733, 1757, 1825, 1863, 1934, 1935, 1959, 2035, 2059, 2098, 2159, 2220, 2262, 2265, 2269, 2284, 2330, 2390, 2418, 2434, 2445, 2517, 2526, 2528, 2652, 2655, 2672, 2673 and 2674.

192

East Inverness-shire (Easterness, including Nairnshire) (v.c. 96): 94a, 154, 172, 198, 344, 363, 401, 413, 435, 461, 467, 469, 547, 574, 578, 582, 608, 620, 622, 675, 711, 756b, 790, 798, 807, 809, 835, 918, 923, 924, 931, 939, 960, 962, 980, 984, 1025, 1053, 1059, 1060, 1076, 1078, 1112, 1119, 1211, 1229, 1230, 1246, 1248, 1272, 1273, 1274, 1277, 1299, 1300, 1304, 1306, 1346, 1356, 1405, 1413, 1415, 1419, 1420, 1425, 1434, 1440, 1486, 1489, 1490, 1493, 1497, 1501, 1502, 1503, 1506, 1507, 1594, 1610, 1635, 1643, 1644, 1652, 1671, 1728, 1729, 1739, 1744a, 1850, 1851, 1863, 1864, 1908, 1930, 1932, 1946, 1959, 1966, 1967, 2059, 2098, 2159, 2221, 2262, 2269, 2275, 2298, 2307, 2330, 2334, 2335, 2354, 2360, 2390, 2405, 2432, 2434, 2435, 2440, 2442, 2445, 2448, 2465, 2494, 2500, 2505, 2509, 2510, 2515, 2517, 2519, 2523, 2526, 2528, 2579, 2600, 2624, 2652 and 2674.

West Inverness-shire (Westerness) (v.c. 97): 1, 13, 169, 267, 357, 358, 371, 413, 433, 461, 469, 485, 521, 527, 534, 535, 538, 546, 547, 578, 582, 602, 620, 622, 664, 675, 711, 785, 787, 788, 793, 831, 835, 860, 863, 918, 923, 924, 928, 931, 939, 948, 960, 1025, 1053, 1054, 1076, 1078, 1129, 1132, 1135, 1138, 1226, 1229, 1230, 1260, 1273, 1277, 1299, 1304, 1310, 1318, 1346, 1355, 1356, 1397, 1401, 1402, 1415, 1418, 1419, 1420, 1425, 1434, 1440, 1502, 1503, 1507, 1603, 1610, 1626, 1635, 1643, 1647, 1671, 1721, 1723, 1728, 1729, 1737, 1837, 1838, 1946, 1962, 2028, 2038, 2049, 2059, 2064, 2073, 2098, 2159, 2167, 2220, 2234, 2247, 2275, 2307, 2309, 2312, 2315, 2316, 2318, 2322, 2330, 2335, 2390, 2397, 2403, 2405, 2407, 2410, 2432, 2434, 2435, 2436, 2442, 2448, 2465, 2502, 2505, 2509, 2510, 2517, 2519, 2526, 2528, 2579 and 2600.

Main Argyllshire (v.c. 98): 15, 51, 198, 199, 255, 267, 364, 371, 373, 402, 460, 461, 462, 469, 476, 477, 480, 482, 487, 491, 492, 493, 494, 500, 501, 505, 510, 533, 547, 574, 577, 578, 582, 630, 666, 675, 833, 923, 924, 931, 932, 984, 994, 1019, 1025, 1059, 1060, 1090, 1137, 1138, 1141, 1212, 1226, 1229, 1230, 1270, 1274, **1275**, 1277, 1304, 1306, 1318, 1346, 1356, 1374, 1397, 1398, 1404, 1415, 1418, 1419, 1429, 1431, 1434, 1440, 1481, 1486, 1491, 1492, 1507, 1577, 1610, 1613, 1629, 1635, 1643, 1650, 1698, 1705, 1706, 1708, 1728, 1729, 1732, 1733, 1808, 1837, 1838, 1935, 1945, 1954, 1958, 1994, 2028, 2035, 2049, 2053, 2059, 2098, 2159, 2194, 2212, 2220, 2221, 2262, 2263, 2269, 2270, 2272, 2274, 2275, 2278, 2283, 2284, 2285, 2330, 2331, 2335, 2355, 2356, 2390, 2397, 2419, 2424, 2432, 2434, 2435, 2442, 2448, 2465, 2473, 2480, 2494, 2496, 2498, 2500, 2502, 2505, 2510, 2515, 2517, 2519, 2423, 2526, 2528, 2579, 2586, 2600, 2629, 2686 and 2695.

Dunbartonshire (v.c. 99): 198, 258, 276, 369, 461, 462, 480, 578, 591, 666, 759, 923, 931, 1059, 1060, 1066, 1115, 1229, 1230, 1310, 1343, 1415, 1419, 1429, 1575, 1577, 1601, 1610, 1723, 1724, 1946, 2059, 2098, 2159, 2220, 2221, 2271, 2274, 2275, 2362, 2434, 2496, 2498, 2500, 2502, 2515, 2517, 2519, 2528, 2539, 2600, 2669, 2672, 2673, 2674 and 2695.

Clyde Isles (Buteshire, islands of Bute, Arran, etc.) (v.c. 100): 15, 185, 189, 357, 358, 445, 446, 466, 574, 666, 1059, 1060, 1229, 1230, 1273, 1285, 1287, 1288, 1346, 1401, 1402, 1419, 1494, 1577, 1650, 1837, 1838, 1946, 1973, 2159, 2275, 2330, **2369**, 2434, 2442, 2445, 2448, 2500, 2505, 2517, 2523, 2526, 2528, 2586, 2600, **2609**, 2630 and 2695.

Kintyre (part of Argyllshire) (v.c. 101): 541, 577, 666, 711, 923, 931, 1025, 1140, 1144, 1152, 1229, 1230, 1264, 1398, 1418, 1610, 1635, 1946, 2035, 2049, 2159, 2275, 2285, 2292, 2330, 2445, 2448, 2490, 2505 and 2526.

South Ebudes (southern Western Isles of Argyllshire) (v.c. 102): 561, 860, 923, 931, 1069, 1261, 1398, 1401, 1402, 1442, 1495, 1932, 2275, 2280, 2528, 2669, 2670, 2672, 2673 and 2674.

Mid Ebudes (northern Western Isles of Argyllshire) (v.c. 103): 394, 447, 547, 622, 711, 717, 827, 854, 923, 931, 943, 948, 994, 1005, 1019, 1059, 1060, 1138, 1140, 1144, 1220, 1229, 1230, 1310, 1312, 1401, 1402, 1610, 1639, 1671, 1733, 1954, 2098, 2221, 2321, 2330, 2406, 2407, 2413, 2414, 2432, 2434, 2448, 2509, 2517, 2519, 2523, 2526, 2528, 2626, 2669, 2670, 2672 and 2673.

North Ebudes (Skye, Rona, Rhum, etc.) (v.c. 104): 50, 125, 204, 231, 232, 357, 358, 360, 361, 401, 461, 546, 578, 581, 620, 622, 660, 791, 923, 931, 949, 1059, 1060, 1138, 1229, 1330, 1277, 1304, 1368, 1398, 1401, 1402, 1415, 1418, 1419, 1429, 1431, 1433, 1434, 1490, 1507, 1546, 1610, 1643, 1671, 1739, 2059, 2082, 2098, 2159, 2216, 2221, 2330, 2390, 2434, 2435, 2445, 2448, 2502, 2505, 2509, 2510, 2517, 2519, 2526, 2528, 2626, 2669, 2670, 2671, 2672, 2673 and 2674.

West Ross-shire (Wester Ross, west Ross and Cromarty) (v.c. 105): 161, 198, 199, 306, 357, 363, 408, 409, 523, 547, 561, 578, 582, 598, 620, 622, 630, 761, 807, 813, 918, 923, 924, 931, 939, 948, 960, 980, 1059, 1060, 1129, 1133, 1134, 1138, 1211, 1229, 1230, 1277, 1279, 1302, 1304, 1397, 1401, 1402, 1415, 1496, 1497, 1498, 1499, 1502, 1507, 1605, 1610, 1626, 1728, 1729, 1739, 1829, 1851, 1863, 1955, 2053, 2059, 2098, 2159, 2220, 2221, 2262, 2275, 2305, 2319, 2330, 2335, 2434, 2445, 2448, 2519, 2528, 2669, 2670, 2671, 2672 and 2674.

East Ross-shire (Easter Ross, east Ross and Cromarty) (v.c. 106): 216, 357, 363, 547, 578, 620, 787, 918, 923, 924, 931, 939, 960, 1025, 1059, 1060, 1211, 1229, 1277, 1279, 1302, 1304, 1401, 1402, 1415, 1507, 1602, 1605, 1671, 1733, 1851, 1863, 1908, 1930, 2059, 2163, 2220, 2319, 2330, 2434, 2435, 2440, 2442, 2445, 2448, 2519, 2523, 2528, 2669, 2670, 2671, 2672, 2673 and 2674.

East Sutherland (v.c. 107): 442, 578, 582, 620, 675, 711, 923, 924, 931, 963, 1227, 1229, 1424, 1431, 1507, 1863, 2059, 2275, 2319, 2330, 2424, 2432, 2434, 2442, 2445, 2517, 2526 and 2528.

West Sutherland (v.c. 108): 253, 286, 401, 467, 533, 578, 622, 664, 698, 699, 711, 793, 807, 808, 809, 811, 923, 924, 931, 963, 1025, 1129, 1130, 1132, 1141, 1142, 1227, 1229, 1230, 1397, 1424, 1440, 1502, 1507, 1602, 1626, 1728, 1783, 1829, 1851, 1931, 1935, 2059, 2159, 2220, 2247, 2275, 2276, 2304, 2307, 2309, 2314, 2318, 2319, 2330, 2335, 2405, 2434, 2442, 2445, 2448, 2492, 2515, 2517, 2519 and 2528.

Caithness (v.c. 109): 1, 263, 338, 393, 620, 711, 808, 865, 918, 923, 931, 970, 1226, 1229, 1230, 1277, 1397, **1403**, 1424, 1440, 1507, 1510, 1635, 1729, 1734, 1930, 1931, 1932, 1934, 1965, 2059, 2159, 2247, 2275, 2276, 2330, 2335, 2350, 2424, 2432, 2434, 2442, 2448, 2496, 2502, 2505, 2517, 2519, 2526, 2528, 2600 and 2624.

Outer Hebrides (v.c. 110): 194, 195, 362, 482, 711, 797, 800, 801, 808, 812, 923, 931, 1006, 1130, 1229, 1230, 1269, 1273, 1276, 1277, 1304, 1423, 1424, 1425, 1427, 1478, 1479, 1500, **1530**, 1739, 1806, 1855, 2159, 2200, 2442, 2445, 2448, 2474, 2475, 2502, 2505, 2509, 2510, 2517, 2519 and 2528.

Orkney Islands (v.c. 111): 12, 29, 33, 198, 199, 209, 216, 405, 466, 574, 582, 616, 622, 834, 923, 931, 963, 1059, 1060, 1230, 1268, 1304, 1343, 1398, 1415, 1423, 1427, 1440, 1648, 1649, 1653, 1671, 2059, 2102, 2159, 2220, 2221, 2358, 2432, 2434, 2445, 2448, 2453, 2502, 2505, 2517, 2528, 2668, 2669, 2670, 2671, 2672, 2673 and 2674.

Shetland Islands (Zetland) (v.c. 112): 21, 203, 234, 256, 371, 461, 544, 578, 622, 625, 628, 630, 655, 656, 657, 711, 807, 820, 911, 912, 913, **914**, 917, 920, 921, 922, 923, 931, 939, 959, 963, 1120, 1125, 1132, 1229, 1230, 1385, 1413, 1415, 1423, 1424, 1440, 1446, 1447, 1522, 1643, 1666, 2159, 2249, 2358, 2432, 2434, 2442, 2448, 2502, 2505, 2509, 2510, 2528 and **2585**.

Ireland. The following titles have either appeared since the publication of Mitchell's (1971) *Bibliography* or were omitted by him. No special search for Irish records has been made by us and this listing is certainly far from complete: 16, 22, 47, 155, 159, 206, 211, 306, 357, 402, 433, 522, 539, 574, 579, 582, 622, 644, 681, 693, 702, 716, 834, 838, 850, 918, 923, 927, 931, 949, 964, 965, 988, 994, 998, 1001, 1003, 1004, 1097, 1108, 1140, 1144, 1213, 1214, 1231, 1265, 1430, 1433, 1457, 1477, 1518, 1519, 1614, 1615, 1629, 1647, 1666, 1674, 1723, 1730, 1731, 1732, 1733, 1735, 1738, 1768, 1829, 1857, 1926, 1954, 2088a, 2091, 2098, 2102, 2145, 2147, 2148, 2149, 2161, 2190, 2220, 2221, 2247, 2294, 2295, 2344, 2357, 2360, 2397, 2409, 2422, 2424, 2434, 2462, 2490, 2515, 2523, 2526, 2672, 2673, 2674 and 2686.

Distribution Maps. The following titles include maps showing either detailed distributions of particular species within the British Isles (and Ireland), or those of the European or world ranges of particular species indicating their presence in the British Isles: 12, 15, 17, 18, 157, 306, 321, 391, 392, 393, 394, 395, 396, 397, 398, 433, 434, 435, 531, 534, 535, 538, 548, 559, 577, 578, 599, 646, 676, 740, 780, 788, 791, 792, 793, 794, 860, 861, 862, 888, 923, 931, 932, 939, 940, 941, 942, 943, 945, 952, 955, 960, 963, 965, 968, 969, 970, 993, 994, 996, 997, 1002, 1124, 1138, 1140, 1141, 1143, 1144, 1145, 1146, 1147, 1148, 1149, 1150, 1151, 1152, 1153, 1212, 1231, 1266, 1277, 1278, 1391, 1393, 1450, 1459, 1497, 1510, 1523, 1586, 1610, 1722, 1723, 1724, 1737, 1844, 1846, 1952, 2019, 2053, 2059, 2088a, 2095, 2130, 2136, 2143, 2145, 2146, 2161, 2162, 2353, 2361 and 2690.

General. The following titles include references to lichens in the British Isles which are not sufficiently well localized to ascribe to any particular county or vice-county: 5, 6, 7, 8, 9, 12, 14, 15, 16, 17, 22, 23, 27, 33, 40, 48, 58, 97, 100, 113, 114, 124, 125, 152, 155, 171, 205, 206, 208, 210, 215, 221, 224, 225, 228, 230, 237, 250, 255, 255a, 256, 260, 266, 283, 295, 296, 298, 300, 314, 316, 324, 357, 359, 360, 361, 390, 402, 410, 417, 418, 422, 423, 425, 433, 453, 464, 465, 478, 479, 483, 490, 495, 497, 498, 502, 504, 506, 507, 509, 512, 513, 514, 515, 524, 525, 526, 528, 529, 530, 531, 535, 538, 540, 545, 554, 557, 558, 560, 561, 574, 577, 578, 582, 586, 587, 589, 591, 592, 593, 598, 600, 605, 608, 613, 622, 627, 630, 637, 638, 639, 646, 675,

676, 711, 712, 719, 732, 733, 734, 741, 757, 764, 765, 766, 773, 776, 777, 780, 781, 782, 788, 789, 796, 799, 808, 809, 810, 826, 833, 846, 847, 849, 853, 854, 855, 858, 859, 860, 861, 886, 887, 907, 908, 909, 918, 919, 923, 925, 926, 927, 931, 941, 942, 945, 955, 960, 963, 968, 969, 970, 974, 975, 987, 988, 992, 994, 995, 997, 999, 1002, 1004, 1021, 1022, 1023, 1028, 1029, 1033, 1038, 1040, 1052a, 1053, 1054, 1057, 1058, 1059, 1060, 1062, 1071, 1077, 1080, 1081, 1082, 1086, 1089, 1092, 1093, 1095, 1102, 1103, 1108, 1116, 1117, 1124, 1133, 1135, 1137, 1138, 1141, 1164, 1169, 1176, 1178, 1179, 1186, 1187, 1188, 1189, 1217, 1218, 1222, 1224, 1225, 1226, 1229, 1230, 1231, 1233, 1236, 1258, 1260, 1262, 1263, 1266, 1286, 1291, 1304, 1306, 1310, 1319, 1324, 1341, 1342, 1353, 1364, 1365, 1366, 1370, 1375, 1377, 1381, 1382, 1383, 1387, 1390, 1397, 1398, 1401, 1402, 1409, 1410, 1412, 1414, 1415, 1418, 1420, 1421, 1423, 1425, 1426, 1428, 1430, 1431, 1432, 1433, 1435, 1440, 1444, 1445, 1447, 1448, 1450, 1452, 1453, 1454, 1457, 1458, 1459, 1475, 1490, 1502, 1503, 1504, 1505, 1506, 1508, 1510, 1511, 1512, 1513, 1515, 1516, 1517, 1518, 1522, 1549, 1550, 1578, 1582, 1586, 1587, 1588, 1590, 1598, 1610, 1626, 1627, 1628, 1629, 1638, 1643, 1645, 1646, 1659, 1666, 1672, 1673, 1674, 1675, 1680, 1681, 1685, 1686, 1689, 1690, 1691, 1693, 1694, 1699, 1700, 1708, 1721, 1722, 1723, 1724, 1728, 1729, 1730, 1731, 1732, 1733, 1734, 1735, 1736, 1737, 1740, 1745, 1758, 1768, 1773, 1779, 1782, 1784, 1788, 1805, 1812, 1813, 1815, 1823, 1824, 1829, 1837, 1838, 1839, 1840, 1841, 1842, 1843, 1847, 1858, 1869, 1873, 1898, 1911, 1912, 1913, 1915, 1922, 1923, 1926, 1927, 1929, 1934, 1935, 1936, 1937, 1952, 1976, 2003, 2019, 2038, 2049, 2053, 2058, 2068, 2072, 2085, 2088a, 2090, 2092, 2093, 2094, 2095, 2098, 2101, 2102, 2103, 2104, 2107, 2129, 2130, 2136, 2137, 2143, 2145, 2152, 2161, 2162, 2173, 2174, 2175, 2176, 2177, 2184, 2190, 2199, 2202, 2204, 2205, 2207, 2208, 2210, 2217, 2220, 2221, 2223, 2227, 2252, 2258, 2262, 2264, 2277, 2293, 2294, 2295, 2296, 2308, 2311, 2313, 2318, 2320, 2321, 2322, 2329, 2330, 2332, 2334, 2335, 2337, 2338, 2353, 2358, 2359, 2360, 2366, 2383, 2384, 2385, 2387, 2390, 2393, 2394, 2397, 2398, 2401, 2402, 2422, 2424, 2435, 2442, 2445, 2448, 2465, 2467, 2468, 2483, 2485, 2489, 2504, 2512, 2423, 2527, 2528, 2530, 2531, 2642, 2643, 2657, 2668, 2669, 2670, 2671, 2672, 2673, 2674, 2687, 2689, 2690, 2691, 2692, 2693 and 2695.

Titles not abstracted. The following titles have not been indexed by county or vice-county either because they have not been seen by us, or because of the time that would be required to abstract them (see also p. 44): 171, 250, 260, 453, 483, 515, 550, 1033, 1057, 1291, 1342, 1366, 1370, 1383, 1409, 1582, 1876, 2076, 2076a, 2197, 2204, 2205, 2210, 2217, 2384, 2385, 2425 and 2657.

HERBARIA

In order to verify literature records, it is necessary to re-examine the specimens on which they were based. Although various compilations indicating the locations of the herbaria of lichenologists who described new taxa have been published (e.g. Grummann, 1974; Hawksworth, 1974), no comprehensive list dealing with all those whose material has been used in regional listings within the British Isles has been prepared. Work on this listing has continued throughout the preparation of the bibliographic section of this book and we have visited numerous British herbaria personally over this period. In addition, P. M. Earland-Bennett kindly solicited data from museums in 1975 through the Museums' Association *Bulletin*, B. J. Coppins carefully analyzed the lichen herbarium in E, and many museums answered letters written to enquire if they had lichen material of particular collectors. Where we are uncertain as to whether a particular museum or other institution still has particular material a ' ?' precedes the herbarium abbreviation; in cases where the main collection is in one herbarium with important parts or duplicates elsewhere, or the main set is lost, collections apart from the main one appear in parenthesis. Small numbers of packets of major collectors inevitably find their way into a large number of institutions; these have not been included here as such packets almost invariably duplicate those in the collector's main herbarium. It should be noted that some of the information in Desmond (1977) on the locations of herbaria is outdated.

For deceased collectors, dates of birth and death have been included where known. In the case of early workers in particular, discrepancies often exist between different listings of their dates. In such instances we have generally followed Barnart (1965), Henrey (1975) or Stafleu and Cowan (1976).

Collectors

Abbot, C. 1761–1817 (LINN)
Adams, J. 1872–1948 DBN
Alvin, K. L. Private
Atwood, M. M. *fl*. 1850–1870 BM (NOT)

Babington, C. 1821–1889 BM
Bagnall, J. E. 1830–1918 BIRA, BM
Bailey, R. H. Private
Baker, J. G. 1834–1920 Destroyed 1864
Balfour, J. H. 1808–1884 E
Barnard, A. M. 1825–1911 NWH
Bloxam, A. 1801–1878 BM, LSR
Bohler, J. 1797–1872 (BIRA)
Borrer, W. 1781–1862 K-Borr. (BM, E, LINN)
Bowen, H. J. M. Private (DOR, E, Read., Ayles.)
Bradbury, J. 1768–1823 (LINN)
Brightman, F. H. BLS
Brinklow, R. DEE (BEL)

Brodie, J. 1744–1824 E (LINN)
Broome, C. E. 1812–1886 BTH (K)
Brown, D. H. Private
Brown, R. 1773–1858 BM
Brunton, W. 1775–1806 LINN
Buchanan, F. 1762–1829 LINN
Buddle, A. ?1660–1715 BM-Sloane (OXF)
Burn, R. NWH (CGE)

Calder, M. *fl*. 1880–1910 ?GRK
Campbell, I. S. C. 194?5–1970 BLS
Carr, J. W. 1862–1939 NOT
Carrington, B. 1827–1893 E
Carroll, I. 1828–1880 BM (CRK, DBN, E)
Chandler, J. H. Private
Coppins, B. J. E (BM)
Croall, A. 1809–1885 STI (E)
Crombie, J. M. 1831–1906 BM (E, H-Nyl.)

197

Crouch, J. F. 1801?–1889 HFD
Curnow, W. ?1809–1887 BM

Dahl, E. O (BM, CGE, LTR)
Darbishire, O. V. 1870–1934 BRIST
Davey, S. R. WCR
Davies, H. 1739–1821 BM (LINN)
Deakin, R. 1809–1873 BM
Degelius, G. Private
Dickie, G. 1812–1882 BM, E
Dickson, J. 1738–1822 BM (E, LINN)
Dillenius, J. J. 1684–1747 OXF (BM-Sloane)
Dillwyn, L. W. 1778–1855 LINN
Dobson, F. S. Private
Don, G., sr. 1764–1814 E (DBN, LINN)
Doody, S. 1656–1705 BM-Sloane
Duncan, U. K. Private (BM, E)

Earland-Bennett, P. M. HFX
Ellis, E. A. NWH
Evans, W. E. 1882–1963 E

Fenton, A. F.-G. Private
Fisher, H. 1860–1935 NOT
Forster, E. 1765–1849 BM (LINN)

Gage, T. 1781–1820 ?CGE
Gardiner, W. 1808–1852 DEE, E (St. Albans Mus.)
Gilbert, J. L. Private
Gilbert, O. L. Private (BM, E, IMI)
Gilchrist, J. 1813–1885 DFS, E
Gliddon, W. A. Destroyed
Graham, G. G. Private
Gray, S. F. 1766–1828 None kept
Greville, R. K. 1794–1866 E (?GL)
Griffith, J. E. 1843–1933 NMW
Griffith, J. W. 1763–1834 (LINN)

Hardy, J. 1815–1898 Part destroyed (E)
Harriman, J. 1760–1831 (BM, H-Ach., LINN)
Harrold, P. E
Hartley, J. W. 1866–1939 (BON)
Hawksworth, D. L. IMI (BM, LSR, LTR)
Hebden, T. 1849–1931 KGY
Henderson, A. Private
Henderson, D. M. E
Hiern, W. P. 1839–1925 RAMM
Hitch, C. J. B. Private
Holl, H. B. 1820–1886 BM (NOT)
Holmes, E. M. 1843–1930 NOT (BM, DOR, E)
Hooker, J. D. 1817–1911 BM
Hooker, W. J. 1785–1865 BM (E, LINN)
Horwood, A. R. 1879–1937 NMW (BM, LSR)
Hudson, W. 1734–1793 Destroyed 1783 (few in LINN)

James, P. W. BM
Johnson, W. 1844–1919 LDS (HAMU)
Johnston, G. 1797–1855 E
Jones, T. 1790–1868 DBN (BM)
Joshua, W. 1828–1898 BM (NOT)

Knight, H. H. 1862–1944 NMW (LIV)
Knowles, M. C. 1864–1933 DBN

Lamb, I. M. BM (E)
Lambley, P. W. NWH
Larbalestier, C. du B. 1838–1911 BM (CRK, K, FI)
Laundon, J. R. BM
Lees, F. A. 1847–1921 CMM
Leighton, W. A. 1805–1889 BM
Lett, H. W. 1838–1920 DBN
Leyland, R. 1784–1847 Destroyed
Lightfoot, J. 1735–1788 Main lichen herbarium lost (few in LINN)
Lillie, D. ?–? GL (E)
Lindsay, D. C. Private (LSR)
Lindsay, W. L. 1829–1880 E (BM)
Livens, H. M. 1860–19?48 BON (LIV)
Lyell, C. 1767–1849 BM (E, LINN)

M'Andrew, J. 1836–1917 E
McIvor, W. G. ?–1876 DBN
Macmillan, H. 1833–1903 E
Maingay, A. C. 1836–1869 E
Manning, S. A. Private (E, NWH)
Marquand, E. D. 1848–1918 (E)
Martindale, J. A. 1837–1914 KDL (BM)
Mason, P. B. 1842–1903 BON
Maughan, R. 1769–1844 E
Mayfield, A. 186?8–1956 NWH
Menzies, A. 1754–1842 E
Merrett, C. 1614–1695 BM-Sloane
Mitchell, A. H. 1794–1882? MSE
Mitchell, M. E. Private
Moore, D. 1807–1879 E (DBN)
Morgan-Jones, G. IMI
Morison, R. 1620–1683 OXF
Mudd, W. 1830–1879 BM (E)

Parfitt, E. 1820–1893 BM
Parsons, H. F. 1846–1913 ?CYN
Paulson, R. 1857–1935 BM (Epp., Ess.)
Peterken, J. H. G. 1893–1973 BLS
Petiver, J. 1664?–1718 BM-Sloane
Phillips, R. A. 1866–1945 DBN
Piggot, H. 1821–1913 BM
Pigott, C. D. LANC
Plukenet, L. 1642–1706 BM-Sloane
Porter, L. E. DBN
Praeger, R. Ll. 1865–1953 DBN
Pulteney, R. 1730–1801 (BM, LINN)

Ralfs, J. 1807–1890 BM
Ranwell, D. S. Furz., Coast.
Reader, H. P. 1850–1929 BRIST (LSR, STO)
Rhodes, P. G. M. 1885–1934 BIRA
Richardson, D. H. S. Private (DBN, IMI)
Richardson, R. 1663–1741 BM-Sloane
Rimington, F. C. Destroyed by fire
Robertson, W. ?–1846 HAMU
Rogers, R. fl1880 ?KRG
Rose, F. BM (Private, E)

Sadler, J. 1837–1882 E
Salisbury, G. Private (BM, LIV)
Salt, J. 1759–1810 SCM
Salwey, T. 1791–1877 ?BM (perhaps lost)

Scannell, M. J. P. DBN
Seaward, M. R. D. Private (DBN, E, LCN)
Shackleton, A. 1830–1916 KGY
Skinner, J. F. Private (STD)
Smith, A. L. 1854–1937 BM (DBN)
Smith, J. E. 1759–1828 LINN (BM)
Sowerby, J. 1757–1822 BM (E, LINN)
Sowter, F. A. 1899–1972 LSR (LTR)
Stelfox, A. W. 1883–1972 DBN
Stewart, P. R. HFX
Stirton, J. 1833–1917 GLAM (BM, E)
Stone, T. S. BIRA
Stuart, J. 1743–1821 LINN
Swinscow, T. D. V. Private (BM)

Taylor, T. c. 1787–1848 FH (BM, DBN)
Teesdale, R. ?–1804 YRK (LINN)
Tellam, R. V. 1826–1908 TRU destroyed 1953 (BM, E, NOT)
Topham, P. B. Private
Travis, W. G. 1877–1958 LIV
Trevelyan, W. C. 1797–1879 HAMU

Trimmer, K. 1804–1887 NWH
Turner, D. 1775–1858 BM (K-Borr., E, LINN)

Varenne, E. G. 1811–1887 ?Ess.

Wade, A. E. NWM (LSR)
Wallace, E. C. Private
Wallace, N. 1907–1972 BLS
Waterfall, W. B. 1850–1915 CMM (E)
Watson, W. 1872–1960 BM (TTN)
Wattam, W. E. L. 1872–1953 HDD
West, W. 1848–1914 E (DBN, LIV)
Wheldon, J. A. 1862–1924 NMW (LIV)
Wilkinson, W. H. ?–1918 BM
Williams, E. 1762–1833 (LINN)
Willisel, T. ?–1675 OXF
Wilson, A. 1862–1949 YRK (LIV)
Wilson, H. E. 1899–1968 ABD
Winch, N. J. 1768–1838 (HAMU, LINN)
Windsor, J. 1787–1868 MANCH
Withering, W. 1741–1799 BM
Wood, W. 1745–1808 (LINN)

Herbarium abbreviations

Abbreviations entirely in capital letters follow Kent (1957) and Holmgren and Keuken (1974) with the exception of BLS; those in upper and lower case lack any internationally recognized abbreviation.

ABD	Department of Botany, University of Aberdeen, Aberdeen AB9 24D.
Ayles.	Buckinghamshire County Museum, Church Street, Aylesbury HP20 2QP.
BEL	Ulster Museum, Botanic Garden, Belfast BT9 5AB.
BM	Department of Botany, British Museum (Natural History), Cromwell Road, London SW7 5BD.
BM-Sloane	Sloane herbarium at BM where it is kept separately (see Dandy, 1958).
BIRA	City Museum and Art Gallery, Congreve Street, Birmingham B3 3DH.
BLS	British Lichen Society, c/o Dr J. D. Guiterman, The Nook, Rosehill, Lostwithiel, Cornwall PL22 0DQ. [Library: c/o Dr D. H. Brown at BRIST.].
BON	Museum and Art Gallery, Civic Centre, Bolton, Lancashire BL1 1SE.
BRIST	Department of Botany, University of Bristol, Woodlands Road, Bristol BS8 1UG.
BTH	Museum of the Bath Royal Literary and Scientific Institution, Bath.
CGE	Botany School, University of Cambridge, Downing Street, Cambridge CB2 3EA.
CMM	Cartwright Hall Art Gallery and Museum, Lister Park, Bradford BD9 4NS.
Coast.	Colney Research Station, Institute of Terrestrial Ecology, Colney Lane, Norwich.
CRK	Botany Department, University College, Cork, Eire.
CYN	Croydon Natural History and Scientific Society, c/o 96a Brighton Road, South Croydon, Surrey CR2 6AD.
DEE	Dundee Museum and Art Galleries, Albert Square, Dundee DD1 1DA.
DBN	National Botanic Gardens, Glasnevin, Dublin 9, Eire.
DFS	Dumfries Museum, The Observatory, Corberry Hill, Dumfries DG2 7SW.
DOR	Dorset Natural History and Archaeological Society, Dorset County Museum, High West Street, Dorchester, Dorset DT1 1XA.
E	Royal Botanic Garden, Inverleith Row, Edinburgh EH3 5LR.
Epp.	Epping Forest Museum, Rangers Road, Chingford, Essex E4.
Ess.	Essex Field Club, now incorporated into the Passmore Edwards Museum, Romford Road, Stratford, London E15 4LZ.
FH	Farlow Herbarium and Reference Library, Harvard University, 20 Divinity Avenue, Cambridge, Massachusetts 02138, USA.

FI	Herbarium Universitatis Florentinae, Istituto Botanico, Via Lamarmora n. 4, 50121 Firenze, Italy.
Furz.	Furzebrook Research Station, Institute of Terrestrial Ecology, Furzebrook, Wareham, Dorset BH20 5AS.
GL	Department of Botany, The University, Glasgow G12 8QQ.
GLAM	Department of Natural History, Art Gallery and Museum, Kelvingrove, Glasgow G3 8AG.
GRK	Greenock Museum & Art Galleries, 9 Union Street, West End, Greenock, Renfrewshire.
H-Ach.	Herbarium of E. Acharius kept separately in H, Botanical Museum, University of Helsinki, Unioninkatu 44, SF-00170 Helsinki 17, Finland.
H-Nyl.	Herbarium of W. Nylander kept separately in H (see previous entry).
HAMU	Hancock Museum, Barras Bridge, Newcastle upon Tyne NE2 4PT.
HDD	Tolson Memorial Museum, Ravensknowle Park, Wakefield Road, Huddersfield HD5 8DJ.
HFD	City Library, Museum, Art Gallery and Old House, Broad Street, Hereford HR4 9AU.
HFX	Bankfield Museum, Boothtown Road, Halifax HX3 6HG.
IMI	Commonwealth Mycological Institute, Ferry Lane, Kew, Surrey TW9 3AF.
K	Royal Botanic Gardens, Kew, Surrey TW9 3AE; most lichens formerly here are now in BM.
K-Borr.	Herbarium of W. Borrer kept as a separate collection in K.
KDL	Borough of Kendal Museum, Station Road, Kendal, Westmorland.
KGY	City of Bradford Metropolitan Council, Cliffe Castle Art Gallery and Museum, Keighley, Yorkshire BD20 6LH.
KRG	Kettering Borough Public Library Art Gallery & Museum, Sheep Street, Kettering, Northamptonshire.
LANC	Department of Biological Sciences, University of Lancaster, Bailrigg, Lancaster LA1 4YQ.
LCN	Lincoln City and County Museum, Broadgate, Lincoln.
LDS	Department of Plant Sciences, University of Leeds, Leeds LS2 9JT.
LINN	Linnean Society of London, Burlington House, Piccadilly, London W1V 0LQ; mentions in the list of collectors above refer to the Smithian herbarium in LINN.
LIV	Merseyside County Museum, William Brown Street, Liverpool L3 8EN.
LSR	Leicestershire Museums, Museum and Art Gallery, New Walk, Leicester LE1 6TD.
LTR	Botanical Laboratories, University of Leicester, University Road, Leicester LE1 7RH.
MANCH	The Manchester Museum, The University, Manchester M13 9PL.
MSE	Montrose Natural History & Antiquarian Society Museum, Panmure Place, Montrose, Angus.
NMW	Department of Botany, National Museum of Wales, Cardiff CF1 3NP.
NOT	The Natural History Museum, Wollaton Hall, Wollaton Park, Nottingham NG8 2AE.
NWH	Norfolk Museums Service, Castle Museum, Norwich NR1 3JU.
O	Botanisk Museum, Trondheimsvn 23B, Oslo 5, Norway.
OXF	Fielding-Druce Herbarium, Department of Botany, University of Oxford, Oxford OX1 3RA.
RAMM	Royal Albert Memorial Museum, Queen Street, Exeter EX4 3RX.
Read.	Reading Museum, Biagrave Square, Reading RG1 1QL.
St.Alb.	St Albans City Museum, Hatfield Road, St Albans AL1 3RR.
SCM	Sheffield City Museum, Weston Park, Sheffield 10.
STD	Museums Service, Southend-on-Sea Borough Council, Central Museum, Victoria Avenue, Southend-on-Sea SS2 6EX.
STI	Smith Art Institute, Albert Place, Stirling FK8 2RQ.
STO	City Museum and Art Gallery, Unity House, Hanley, Stoke-on-Trent ST1 4HY.
TRU	Royal Institution of Cornwall, County Museum and Art Gallery, River Street, Truro, Cornwall.
TTN	Somerset Archaeological and Natural History Society, Somerset County Museum, Taunton Castle, Taunton TA1 4AD.
WCR	Hampshire County Museum Service, Chilcomb House, Chilcomb Lane, Bar End, Winchester SO23 8RD.
YRK	The Yorkshire Museum, Museum Gardens, York YO1 2DR.

EXSICCATAE

A published exsiccata (pl. exsiccatae) is a collection of dried specimens with printed or duplicated labels, serially numbered, identical sets of which were distributed by sale, gift or exchange to individuals or institutions. Several exsiccatae either devoted exclusively to British lichens or including them have been published and these have been listed in the *Bibliography* section of this book.

Some herbaria retain exsiccatae as intact sets separate from their main collections whilst others distribute them amongst their general collections. As a result of the latter procedure, it is not uncommon to find specimens in herbaria with printed labels and numbers which evidently belong to an exsiccata but with little means of determining to which they belonged. In order to facilitate the identification of such isolated exsiccata packets, sample labels of those including British lichens are reproduced below with the exception of those of Don's *Herbarium Britannicum*[602] and Dickson's *A Collection of Dried Plants*[586] figured on p. 14 above, Joshua's *Microscopical Slides of British Lichens*[1217], and the exisiccatae of Arnold[112] and Salwey[2076, 2085]; sets of the latter four have not been located by us. Two items by W. Gardiner[756a, 756b] may perhaps qualify as exsiccatae but as only one copy of each has so far been seen, and they have not previously been recognized as such, they are omitted here.

Lynge (1915–22) lists the numbers and names of all taxa in each lichen exsiccata he studied under the name of the exsiccata and also provides an index arranged by genera and species to all such exsiccatae. For the dates of publication of different parts of lichen exsiccatae and further bibliographical information on them, the comprehensive work of Sayre (1969) should be consulted.

In the following illustrations, the currently accepted names of the taxa figured are included in square brackets.

No. 70. RAMALINA fraxinea, *Ach. Syn. Lich.* p. 296. *Grev. Fl. Ed.* p. 348. Lichen fraxineus, *Sibth.* n. 907. E. B. t. 1781. *Purt. Midl. Fl.* v. 2. n. 838. *Dill.* t. 22. f. 59. HAB. On the trunks and branches of trees. Common.

[*Ramalina fraxinea* (L.)Ach.]

W. Baxter (1787–1871), *Stirpes cryptogamae Oxoniensis* (1825–28[213]).

LECIDEA.

Specific Character.

L. aurantiaca, Ach. *(saffron-coloured Lecidea),* crust granulated whitish-lemon-coloured, apothecia sessile rather convex orange-coloured with a yellow waved border. Sm.—Ach. *Syn.* p. 50.—*Br. Fl.* vol. ii. p. 186.—*Lichen aurantiacus,* Lightf.— *Lichen salicinus,* Schrad.—*E. Bot.* t. 1305.— *Lecanora salicina,* Ach. *Syn.* p. 175.

On the trunks of trees, especially of Willow and Poplar.

The *crust* is thin, uneven, granulated, and cracked, of a whitish-lemon colour, which seems very delicate and perishable. *Apothecia* scattered or clustered, the disk very smooth and polished; while young they are much immersed in the thallus, with a light powdery edge; as they advance in age, they rise a little above the thallus, become rather convex, of a much darker colour, the border remaining of a lemon hue, giving it much the appearance of a *Lecanora.*

[*Caloplaca flavorubescens* (Huds.)Laundon]

J. Bohler (1797–1872), *Lichenes Britannici* (1835–37[255]).

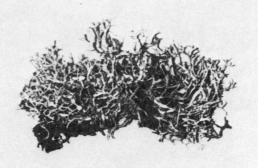

[*Heterodermia leucomelos* (L.)Poelt]

[*Teloschistes chrysophthalmus* (L.)Th.Fr.]

I. Carroll (1828–1880), *Lichenes Hibernici exsiccati* (1859).

205

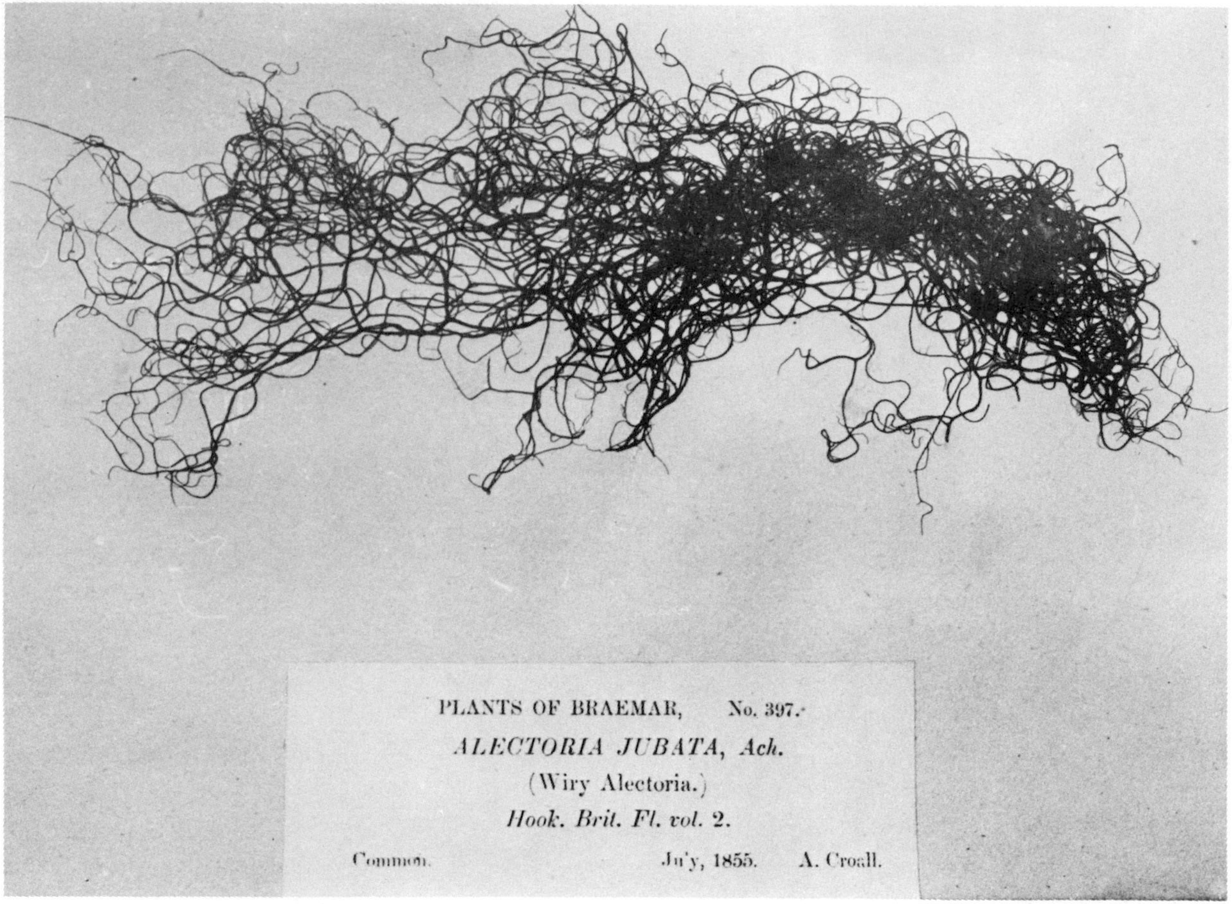
PLANTS OF BRAEMAR, No. 397.
ALECTORIA JUBATA, Ach.
(Wiry Alectoria.)
Hook. Brit. Fl. vol. 2.
Common. July, 1855. A. Croall.

A. Croall (1809–1885), *Plants of Braemar* (1857[450]). [*Bryoria fuscescens* (Gyeln.)Brodo & D. Hawksw.]

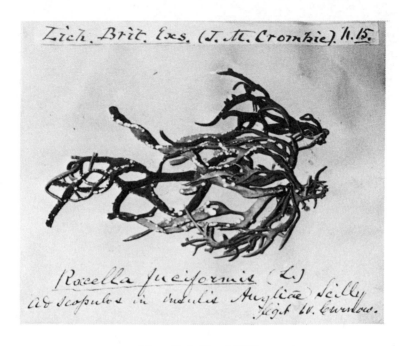

[*Roccella fuciformis* DC.]

J. M. Crombie (1831–1906), *Lichenes Britannici exsiccati* (1874–77[483]).

[*Pseudocyphellaria crocata* (L.)Vain.]

J. Dickson (1738–1822), *Hortus siccus Britannicus* (1793–1802[587]).

253. **Pannaria plumbea**, LIGHTF., Leight. L. Fl. ed., p. 154.
On trees and bushes; near Glencorbot, Connemara.

[*Parmeliella plumbea* (Lightf.)Vain.]

C. du Bois Larbalestier (1838–1911), *Lichen-Herbarium* (1879–81[1291]).

211

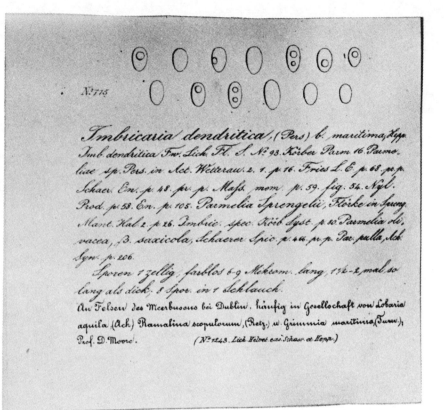

[*Parmelia pulla* Ach.]

P. Hepp (1797–1867), *Die Flechten Europas* (1853–67[988]).

82. Ricasolia lætevirens *Leight.*, Form laciniola.

Thallus more rugose and more fertile than the type; apothecia crowded; the surface of thallus and margin of apothecia more or less laciniolate.

On old tree trunks, Long Wood, Keswick.

[*Lobaria laetevirens* (Lightf.)Zahlbr.]

W. Johnson (1844–1919), *The North of England Lichen-Herbarium* (1894–1918[1196]).

1.

Cladonia alcicornis, Flk.
On dry sandy heaths. Thetford Warren, Suffolk.

[*Cladonia foliacea* (Huds.)Willd.]

C. du Bois Larbalestier (1838–1911), *Lichenes exsiccati circa Cantabrigiam collecti a C. du Bois Larbalestier* (1896[1292]).

215

94. Endocarpon miniatum. Ach., Syn., p. 101; Fries, Lich. Eur. Ref.
p. 408; Schœr., Enum., p. 231; Nyl., Lich. Gall. et Alg., p. 174,
Enum. Gen., p. 135, Pyrenoc, p. 11; Lich. Scand., p. 264; Mudd,
Man. Brit. Lich., p. 265; *Lichen miniatus*, Smith, E. Bot., t. 593 (up. fig.)

On rocks near the sea; L'Etacq, Beauport, Rozel Tower, Jersey.

[*Dermatocarpon miniatum* (L.)Mann]

C. du Bois Larbalestier (1838–1911), *Lichenes Caesarienses
et Sargienses exsiccati* (1867–72[1290]).

217

67. Opegrapha Chevallieri, LEIGHT.
Leight. Brit. Graphid. inedit.
O. lithyrga, Chev. Graphid. 54. t. 11. f. 4. 5. (excl. Ach. syn.)
Great Orme's Head, Caernarvonshire.

[*Opegrapha chevallieri* Leight.]

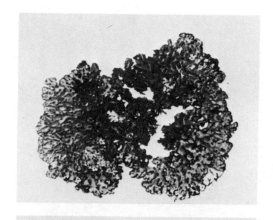

264. Parmelia diatrypa, ACH.
E. Bot. t. 1248.
Barmouth, North Wales.

[*Menegazzia terebrata* (Hoffm.)Massal.]

W. A. Leighton (1805–1889), *Lichenes Britannici exsiccati* (1851–67[1344]).

219

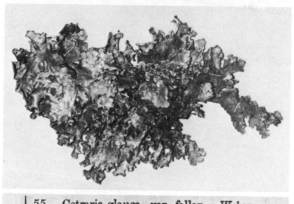

55. Cetraria glauca. var. fallax. *Weber.*
Pine trees, Ingleby Park, Cleveland.

[*Platismatia glauca* (L.)Culb. & C. Culb.]

W. Mudd (1830–1879), *Lichenum Britannicorum* (1861[1642]).

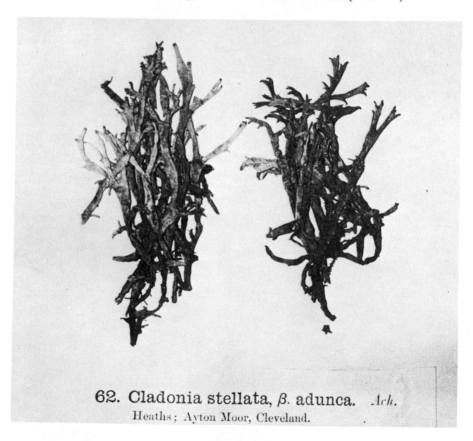

62. Cladonia stellata, β. adunca. *Ach.*
Heaths; Ayton Moor, Cleveland.

[*Cladonia uncialis* subsp. *dicraea* (Ach.)D. Hawksw.]

W. Mudd (1830–1879), *A Monograph of British* Cladoniae (1865[1644]).

221

A. VĚZDA: LICHENES SELECTI EXSICCATI

1164. **Parmelia britannica** D. Hawksworth et P. James sp. n.

Planta saxicola, subsimilis *Parmeliae revolutae* F l k., a qua differt **sorediis marginalibus** constanter caesio-nigris [haud albidis vel fuscescentibus], thallo plusminusve nitido, et lobis 1,5—2,0 mm latis, validius sinuatis, substrato densius adpressis [subsimilibus eorum *Parmeliae conspersae* [E h r h. ex A c h.] A c h.], subconvexis, apicibus soredia gerentibus vulgo ascendentibus. Apothecia et pycnidia ignota. Thallus K+ luteus; medulla K—, C+ vel KC+ rosea, atranorinum et acidum gyrophoricum continens.

MAGNA BRITANNIA. Cambria, Anglesey, Holy Island, Holyhead, inter Plas Meilw et Bod-Warren, alt. 15 m s. m. In rupibus siliceis umbrosis maritimis. [Holotypus in Herb. Musei Britannici, Londinium (BM)]. — 6. IV. 1973.

Leg. P. W. James

[*Parmelia britannica* D. Hawksw. & P. James]

A. Vězda (b. 1920), *Lichenes selecti exsiccati* (1960 onwards[2406, etc.]).

223

SOURCES

In addition to references cited in the preceding sections not included in the main *Bibliography*, selected source-books used in the preparation of the historical, bibliographical and herbarium data, or in the tracing of particular publications are mentioned here as they may also prove of value to users of the present book as sources of supplementary biographical or bibliographical data. Introductory material in floras dealing primarily with vascular plants is often a rich source of biographical information on local collectors.

AINSWORTH, G. C. (1976) *Introduction to the History of Mycology*. Cambridge: Cambridge University Press.

ALLEN, D. E. (1976) *The Naturalist in Britain*. London: Allen Lane.

BARNHART, J. H. (1965) *Biographical Notes upon Botanists*. 3 vols. Boston: Hall.

BARTHOLOMEW, J. W. & SON LTD (1966) *Gazetteer of the British Isles*. Ed. 9 (with additions). Edinburgh: Bartholomew & Son.

BRITISH MUSEUM (NATURAL HISTORY) (1904) *The History of the Collections contained in the Natural History Department of the British Museum*. Vol. 1. London: British Museum (Natural History).

——— (1903–40) *Catalogue of the Books, Manuscripts, Maps and Drawings in the British Museum (Natural History)*. 8 vols. [incl. *Supplement*]. London: British Museum (Natural History).

——— (1975) *List of Serial Publications in the British Museum (Natural History) Library*. 3 vols. [Publication No. 778.] London: British Museum (Natural History).

BRITTEN, J. and BOULGER, G. S. (1931) *A Biographical Index of deceased British and Irish Botanists*. Ed. 2 [revised and completed by A. B. Rendle.]. London: Taylor & Francis.

CHALMERS-HUNT, J. M. (1976) *Natural History Auctions 1700–1972*. London: Sotheby Parke Bernet.

CIFERRI, R. (1957–60) *Thesaurus literaturae mycologicae et lichenologicae, supplementum 1911–1930*. 4 vols. Pavia: Cortina.

CLOKIE, H. N. (1964) *An account of the Herbaria of the Department of Botany in the University of Oxford*. London: Oxford University Press.

CULBERSON, W. L. (1951→) Recent literature on lichens 1→. *Bryologist* **54**→. [A continuing series.]

DANDY, J. E. (1958) *The Sloane Herbarium*. London: British Museum (Natural History).

——— (1969) *Watsonian Vice-Counties of Great Britain*. London: Ray Society.

DESMOND, R. (1977) *Dictionary of British and Irish Botanists and Horticulturalists*. London: Taylor & Francis.

DICTIONARY OF NATIONAL BIOGRAPHY (1885–1900). 63 vols. Oxford: Oxford University Press.

GREEN, J. R. (1914) *A History of Botany in the United Kingdom from the earliest times to the end of the 19th century*. London and Toronto: Dent.

GRUMMANN, V. J. (1974) *Biographisch-bibliographisches Handbuch der Lichenologie*. Lehre: J. Cramer.

HAWKSWORTH, D. L. (1974) *Mycologist's Handbook. An introduction to the principles of taxonomy and nomenclature in the fungi and lichens*. Kew: Commonwealth Mycological Institute.

—————— (1977) A bibliographical guide to the lichen floras of the world. In *Lichen Ecology* (M. R. D. Seaward, ed.): 437–502. London, New York and San Francisco: Academic Press.

HEDGE, I. C. and LAMOND, J. M. (1970) *Index of Collectors in the Edinburgh Herbarium*. Edinburgh: H.M.S.O.

HENREY, B. (1975) *British Botanical and Horticultural Literature before 1800*. 3 vols. London: Oxford University Press.

HOLMGREN, P. K. and KEUKEN, W. (1974) *Index Herbariorum. Part I. The Herbaria of the World*. Ed. 6 [*Regnum Vegetabile* **92**.] Utrecht: Oosthoek, Scheltema & Hoekema. [Also addenda: *Taxon* **24**: 543–551 (1975); **25**: 517–524 (1976).]

KENT, D. H. (1957) *British Herbaria*. London: Botanical Society of the British Isles.

KREMPELHUBER, A. von (1867–72) *Geschichte und Litteratur der Lichenologie*. 3 vols. Munich: Wolf.

LINDAU, G. and SYDOW, P. (1908–18) *Thesaurus litteraturae mycologicae et lichenologicae*. 5 vols. Berlin: Borntraeger.

LYNGE, B. (1915–22) Index specierum et varietatum 'Lichenum exsiccatorum'. *Nyt Mag. Naturvid.* **53–60**. [Issued with separate pagination from the journal; Pars I 1916–1919; Pars II 1920–22.]

McCLINTOCK, D. (1966) *Companion to Flowers*. London: G. Bell & Sons.

MITCHELL, M. E. (1971) *A Bibliography of Books, Pamphlets and Articles relating to Irish Lichenology, 1727–1970*. Galway: privately printed.

NISSEN, C. (1966) *Die botanische Buchillustration*. Stuttgart: Hiersman.

ORDNANCE SURVEY (1953) *Gazetteer of Great Britain*. Chessington: Ordnance Survey.

PRAEGER, R. Ll. (1934) *The Botanist in Ireland*. Dublin: Hodges Figgis. [Reprinted (1974) Wakefield: EP Publishing.]

—————— (1949) *Some Irish Naturalists*. Dundalk: Dungalgan Press.

PULTENEY, R. (1790) *Historical and Biographical Sketches of the Progress of Botany in England*. 2 vols. London: Cadell.

RACKHAM, O. (1976) *Trees and Woodland in the British Landscape*. London: J. M. Dent & Sons.

ROYAL SOCIETY OF LONDON (1867–1925) *Catalogue of Scientific Papers*. 19 vols. London: Royal Society.

SAWYER, F. C. (1971) A short history of the libraries and list of MSS. and original drawings in the British Museum (Natural History). *Bull. Br. Mus. nat. Hist., hist. ser.* **4**: 77–204.

SAYRE, G. (1969) Cryptogamae exsiccatae—An annotated bibliography of published exsiccatae of Algae, Lichenes, Hepaticae and Musci. *Mem. N. Y. Bot. Gdn* **19**: 1–174.

SIMPSON, N. D. (1960) *A Bibliographical Index to the British Flora*. Bournemouth: privately published.

SMITH, A. L. (1921) *Lichens*. Cambridge: Cambridge University Press. [Reprinted (1975) Richmond: Richmond Publishing.]

—— (1922) History of lichens in the British Isles. *SEast. Nat.* **1922**: 19–35.

STAFLEU, F. A. (1967) *Taxonomic Literature. A selective guide to botanical publications with dates, commentaries and types*. [*Regnum Vegetabile* **52**.] Utrecht: International Bureau for Plant Taxonomy and Nomenclature.

—— and COWAN, R. S. (1976) *Taxonomic Literature. A selective guide to botanical publications and collections with dates, commentaries and types*. Ed. 2. Vol. 1: A-G. [*Regnum Vegetabile* **94**.] Utrecht: Bohn, Scheltema & Holkema.

WATSON, W. (1950) *Census Catalogue of British Lichens*. London: Cambridge University Press.

WORLD LIST OF SCIENTIFIC PERIODICALS PUBLISHED IN THE YEARS 1900–1960 (1963–65). Ed. 4. 3 vols. London: Butterworths.

INDEX

Biographical index to Historical Section

NOTES

NOTES

NOTES

NOTES